The Six Pillars of Calculus

Business Edition

Pure and Applied
UNDERGRADUATE TEXTS · 56

The Six Pillars of Calculus

Business Edition

Lorenzo Sadun

Providence, Rhode Island USA

EDITORIAL COMMITTEE

Giuliana Davidoff   Tara S. Holm
Steven J. Miller   Maria Cristina Pereyra

Gerald B. Folland (Chair)

2020 *Mathematics Subject Classification*. Primary 00-01, 26A06, 91-01, 00A71, 26-01.

For additional information and updates on this book, visit
www.ams.org/bookpages/amstext-56

Library of Congress Cataloging-in-Publication Data

Names: Sadun, Lorenzo Adlai, author.
Title: The six pillars of calculus : business edition / Lorenzo Sadun.
Description: Providence, Rhode Island : American Mathematical Society, [2022] | Series: Pure and applied undergraduate texts, 1943-9334 ; volume 56 | Includes index.
Identifiers: LCCN 2022028500 | ISBN 9781470469955 (paperback) | ISBN 978-1-4704-7185-9 (ebook)
Subjects: LCSH: Calculus–Textbooks. | AMS: General – Instructional exposition (textbooks, tutorial papers, etc.). | Real functions – Functions of one variable – One-variable calculus. | Game theory, economics, social and behavioral sciences – Instructional exposition (textbooks, tutorial papers, etc.). | General – General and miscellaneous specific topics – Theory of mathematical modeling. | Real functions – Instructional exposition (textbooks, tutorial papers, etc.).
Classification: LCC QA303.2 .S23 2022 | DDC 515–dc23
LC record available at https://lccn.loc.gov/2022028500

Copying and reprinting. Individual readers of this publication, and nonprofit libraries acting for them, are permitted to make fair use of the material, such as to copy select pages for use in teaching or research. Permission is granted to quote brief passages from this publication in reviews, provided the customary acknowledgment of the source is given.

Republication, systematic copying, or multiple reproduction of any material in this publication is permitted only under license from the American Mathematical Society. Requests for permission to reuse portions of AMS publication content are handled by the Copyright Clearance Center. For more information, please visit **www.ams.org/publications/pubpermissions**.

Send requests for translation rights and licensed reprints to **reprint-permission@ams.org**.

© 2023 by the author. All rights reserved.
The American Mathematical Society retains all rights
except those granted to the United States Government.
Printed in the United States of America.

∞ The paper used in this book is acid-free and falls within the guidelines
established to ensure permanence and durability.
Visit the AMS home page at **https://www.ams.org/**

10 9 8 7 6 5 4 3 2 1   28 27 26 25 24 23

To my students, and to everybody who has helped me understand calculus better. (But I repeat myself.)

Contents

Instructors' Guide and Background	xi
Chapter 1. What is Calculus? The Six Pillars	1
Chapter 2. Predicting the Future: The SIR Model	7
2.1. A Problem of Market Penetration	7
2.2. Building the SIR Model	9
2.3. Analyzing the Model Numerically	13
2.4. Theoretical Analysis: What Goes Up Has to Stop Before it Comes Down	16
2.5. Epidemics	18
2.6. Covid-19 and the SIR model	19
2.7. Chapter Summary	22
2.8. Exercises	23
Chapter 3. Close is Good Enough!	31
3.1. The Idea of Approximation	31
3.2. Functions	34
3.3. Linear Functions and Their Graphs	37
3.4. Linear Approximations and Microscopes	41
3.5. Euler's Method and Compound Interest	43
3.6. The SIR Model by Computer	46
3.7. Solving Algebraic Equations	49
3.8. Chapter Summary	51
3.9. Exercises	52

Chapter 4.	Track the Changes!	59
4.1.	The National Debt	59
4.2.	Marginal Cost, Revenue, and Profit	62
4.3.	Local Linearity and Microscopes	64
4.4.	The Derivative	67
4.5.	A Global View	72
4.6.	Chapter Summary	76
4.7.	Exercises	77
Chapter 5.	Computing and Using Derivatives (What Goes Up has to Stop Before it Comes Down)	85
5.1.	Building Blocks	86
5.2.	Adding, Subtracting, Multiplying, and Dividing Functions	92
5.3.	The Chain Rule	97
5.4.	Optimization	102
5.5.	The Shape of a Graph	105
5.6.	Newton's Method	114
5.7.	Chapter Summary	120
5.8.	Supplemental Material: Small Angle Approximations	122
5.9.	Exercises	124
Chapter 6.	Models of Growth and Oscillation	135
6.1.	Modeling with Differential Equations	137
6.2.	Exponential Functions and Logarithms	144
6.3.	Simple Models of Growth and Decay	150
6.4.	Two Models of Oscillation	153
6.5.	More Sophisticated Models	157
6.6.	Chapter Summary	161
6.7.	Supplemental Material: A Crash Course in Trigonometry	163
6.8.	Exercises	174
Chapter 7.	The Whole Is the Sum of the Parts	185
7.1.	Slicing and Dicing	186
7.2.	Riemann Sums	195
7.3.	The Definite Integral	201
7.4.	The Accumulation Function	205
7.5.	Chapter Summary	208
7.6.	Exercises	210

Chapter 8.	The Fundamental Theorem of Calculus (One Step at a Time)	221
8.1.	Three Different Quantities	222
8.2.	FTC2: The Integral of the Derivative	225
8.3.	FTC1: The Derivative of the Accumulation	228
8.4.	Anti-Derivatives and Ballistics	233
8.5.	Computing Anti-Derivatives	237
8.6.	Chapter Summary	238
8.7.	Exercises	240
Chapter 9.	Methods of Integration	249
9.1.	Integration by Substitution	250
9.2.	Integration by Parts	254
9.3.	Numerical Integration	260
9.4.	Chapter Summary	266
9.5.	Exercises	268
Chapter 10.	One Variable at a Time!	281
10.1.	Partial Derivatives	282
10.2.	Linear Approximations	288
10.3.	Double Integrals and Iterated Integrals	291
10.4.	Chapter Summary	298
10.5.	Exercises	299
Chapter 11.	Taylor Series	313
11.1.	What Does $\pi = 3.14159265\cdots$ Mean?	314
11.2.	Power Series	317
11.3.	Taylor Polynomials and Taylor Series	319
11.4.	Sines, Cosines, Exponentials, and Logs	323
11.5.	Tests for Convergence	330
11.6.	Intervals of Convergence	336
11.7.	Chapter Summary	340
11.8.	Exercises	341
Index		351

Instructors' Guide and Background

The Purpose of This Book. In 1952, George B. Thomas was asked to revise a publisher's existing calculus textbook. Instead, he wrote his own. The many editions of Thomas's *Calculus and Analytic Geometry*, and the numerous subsequent adaptations of his work, have dominated university classrooms ever since. The names on the textbooks have changed over time, with each new author bringing his or her own distinctive style. (I am especially a fan of James Stewart's precision and attention to detail.) However, the overall blueprint is the same. As a result, today's calculus classrooms largely resemble those of the 1990s, of the 1970s, and even of the 1950s.

For some students (including this author in his youth) and especially for those who wish to pursue careers in math, physics, and engineering, this approach has worked well. But for increasingly many other students, it has failed. To those students, calculus is a weed-out course, where success depends mostly on memorizing a large number of unmotivated definitions, formulas, and problem-solving techniques, and applying them without any regard for what the variables and functions actually represent. Lacking any unifying principle, these elements are quickly forgotten when the semester is over. Even successful alumni often look back at their calculus classes as painful and useless ordeals.

It doesn't have to be that way! Calculus is built on a handful of very simple ideas that everybody can and should understand. Calculus is connected to every aspect of day-to-day life—from making business decisions to addressing environmental problems to building bridges that don't fall down—in ways that should interest and inspire everybody. Calculus can also be enjoyable if students are brought into the process of discovery instead of just being handed results. This text and the course at the University of Texas that it was written for are an attempt to reclaim calculus and make it comprehensible, memorable, practical, and even fun.

So what's so different about this text?

- Instead of presenting theorems first and applications later, concepts are developed by tackling realistic problems. For instance, we begin with the SIR system of coupled differential equations that model the market penetration of a new phone app (the same equations that model the spread of a disease and, in particular, the Covid pandemic) long before we ever define what a derivative is! Instead of starting with formal definitions of limits and derivatives, we generate *rate equations* from an informal understanding of what *rate of change* means. This leads naturally to linear approximations and to solving difficult problems one step at a time. Students discover that getting an answer through many small steps requires more work than taking a few big steps, but gives more accurate results. Exact answers emerge as the limit of this process, motivating the concept of a derivative.

- Everything in the book is connected explicitly to six central ideas that I call the *Six Pillars of Calculus*:
    (1) Close is good enough (approximation and limits)
    (2) Track the changes (derivatives)
    (3) What goes up has to stop before it comes down (optimization and turning points)
    (4) The whole is the sum of the parts (integration)
    (5) One step at a time (the Fundamental Theorem of Calculus)
    (6) One variable at a time (the key to multi-variable calculus)
  Making sense of these six ideas is both a goal in itself and an organizing principle for every other concept. Having such an overall conceptual framework to build on makes learning easier and forgetting harder.

- We stress multiple ways of looking at each concept. Is a derivative a limit of a difference quotient, a rate of change, the slope of a tangent line, the best linear approximation to a function, or the gain on a tuning knob? To pass a quiz, a student might memorize one of these things, but the real power of a derivative comes from it being all of these at the same time.

- We emphasise "when" and "why" over "how". Recognizing that a real-world phenomenon involves a derivative or an integral is much more important than being able to compute that derivative or integral from a formula.

- We also put a heavy emphasis on numerical approximations and programming. Difference quotients approximate derivatives, but we can just as well use derivatives to approximate difference quotients. Riemann sums approximate integrals, and integrals approximate Riemann sums. By studying approximations on their own terms, and not just as stepping stones toward the definitions of derivatives and integrals, students begin to grasp what derivatives, integrals, and infinite series are trying to accomplish.

    Programming also serves a pedagogical role. The best way to learn anything is to explain it to somebody else. That somebody else can be a computer, and the required explanation can be a MATLAB program. Students are not necessarily expected to write their own MATLAB programs from scratch, but they are expected to modify examples in the book to fit different circumstances.

- We use differential equations and modeling as recurring themes. Many, if not most, quantities in the world—from the number of customers for our new phone app to the size of our bank balance to the rabbit population of Australia—are governed by differential equations. The goal is not just to teach students calculus, but to help them recognize calculus in the world around us.

- We embrace heuristic reasoning. While no lies are told and everything in this book *can* be justified rigorously, most of the time we just present an intuitive explanation and stop there. In particular, we skip over the formal machinery of limits almost completely, relying instead on a basic "what we are getting closer and closer to" notion that Newton, Leibniz, and Euler would have immediately recognized.

- We place integrals before anti-derivatives. In most calculus classes, anti-derivatives and the Fundamental Theorem of Calculus are presented to students before they fully understand what integrals actually are, so they come away with the impression that integrals are just anti-derivatives. Later on, double integrals make no sense to them, because you cannot take a double anti-derivative with respect to x and y. Here, integrals are developed as generalized sums, period. This allows us to apply integration to a host of problems that bear little superficial resemblance to the area under a curve or a distance traveled. Only after we learn the fundamental theorem of calculus do we turn to the anti-derivative, both as the simplest way to compute most integrals, and as an important and useful concept in its own right.

This book was designed for a one-semester course at the University of Texas. Naturally, when fitting what has been traditionally been a two-semester sequence into one semester, something has to give. As noted above, we skip much of the formal structure of limits. We give short shrift to continuity and differentiability, merely presenting a few examples of discontinuous or nondifferentiable functions to build intuition. We also omit many of the calculational techniques that appear in traditional calculus texts. You will find no mention of l'Hôpital's rule, of implicit differentiation, of logarithmic differentiation, of integration by trig substitution, or of integration by partial fractions. In an age when you can pick up your phone to look up derivatives and integrals online, knowing how to compute a complicated derivative or integral by hand is much less important than it used to be.

Origins. This book did not emerge from a vacuum. Much of its structure and terminology (especially Chapters 2 and 3) is based on the 1994 text, *Calculus in Context*, by James Callahan, David Cox, Kenneth Hoffman, Donal O'Shea, Harriet Pollatsek, and Lester Senechal, who compose the Five Colleges Calculus Project. That in turn was based on a course developed by Michael Sutherland and Kenneth Hoffman at Hampshire College.

Calculus in Context stresses the role of calculus as "a language and a tool for exploring the whole fabric of science," with differential equations as a unifying theme. Fortunately for us, the Five Colleges Calculus Project had the support of the National Science Foundation, with the condition that the results be made available for others to adopt and adapt. I am grateful for the opportunity (and the explicit permission!) to do so.

At the University of Colorado, Eric Stade developed his own course on Calculus, Stochastics, and Modeling (Math 1310). This course was aimed at biology students, using *Calculus in Context* as the primary text, but it also used a large number of homegrown worksheets and miniprojects that walk students through the difficulties. When Bill Wolesensky and I learned about Math 1310, we immediately wanted to teach a similar course at the University of Texas.

As a result, in 2014 Bill and I developed a one-semester calculus course for biologists (M408R) at the University of Texas at Austin that was explicitly modeled on Math 1310. Eric Stade generously shared all of his course materials, which we adapted for use by Texas students. Over time, we (mostly Bill) developed our own materials, and our course developed its own identity, with less emphasis on stochastics and differential equations and more emphasis on the Six Pillars, while maintaining *Calculus in Context*'s commitment to learning through tackling hard, realistic problems.

M408R was so successful that the McCombs School of Business asked us to develop a similar course for business students. At this point, it was clear that we would need our own textbook, since the content of *Calculus in Context* is almost entirely drawn from the natural sciences, and since none of the existing business calculus books followed our approach. I took on the task of writing this text, the first writing an edition for business students and then an edition for biology students. Eventually, we intend to develop versions of the course and editions of this text for students in physics, engineering, and pure math.

Chapter-by-Chapter Summary. **Chapter 1** addresses the question "What is calculus?" The short answer is "the study of things that change". A longer answer involves the Six Pillars. This is the shortest chapter in the book, but it is also the most important, as it frames everything that comes after.

In **Chapter 2** we start predicting the future with mathematical models. We develop the SIR equations in two settings. The first setting is taken from marketing, where $S(t)$ is the number of potential users of a hypothetical phone app that the reader has developed, $I(t)$ is the number of active users, and $R(t)$ is the number of users who have rejected the product. The second setting (which historically came first) is from epidemiology, where $S(t)$ is the number of individuals who are susceptible to a disease, $I(t)$ is the number of infected individuals, and $R(t)$ is the number of recovered individuals. The resulting equations are identical in both settings, because S's become I's through contact with I's, and I's become R's through the mere passage of time (attrition or recovery). In fact, most modeling of the Covid-19 pandemic has involved extensions of the SIR model, making this model extremely relevant.

In either setting, knowing where things stand today allows us to compute how fast things are changing, which allows us to make projections for the near future. Repeating the process allows us to predict the more distant future. Studying when $I'(t) = 0$ allows us to understand the conditions for peak market penetration or the peak of an epidemic.

Instructors' Guide and Background

In **Chapter 3** we take up the 1st Pillar of Calculus (*Close is good enough*), in a more systematic way. After discussing approximate methods for solving algebraic equations, we turn to functions and ask how quickly they change. We study linear functions, where the rate of change is constant. We then approximate nonlinear functions by linear functions over short stretches where the rate of change is nearly constant. Euler's method for solving differential equations then emerges as a natural consequence of linear extrapolation. This approximation, and others introduced in this chapter, get better and better as the time step is reduced, which sets up the concept of a derivative. We close with the bisection method for finding approximate roots of a function and the amazingly accurate Babylonian method for approximating square roots.

In **Chapter 4** we turn to the 2nd Pillar (*Track the changes*) and define derivatives. First we approximate rates of changes with forward, backward, and centered difference quotients. Marginal costs, revenue, and profit are introduced similarly, and we develop the approximation

$$(0.1) \qquad f(x) \approx f(a) + f'(a)(x - a),$$

which we call the *microscope equation*, because any smooth graph looks straight if you zoom in enough. We define the derivative $f'(a)$ as

(1) the rate of change of the function at a,

(2) the slope of the tangent line,

(3) the multiplier in the best linear approximation near a, and

(4) a limit of difference quotients,

and we see why these are all the same. We look at the derivative *function $f'(x)$*, and we see how the value of $f'(x)$ relates to how fast $f(x)$ is changing. Using algebra, we compute the derivatives of a few simple functions, such as x^2 and $1/x$.

Chapter 5, on computing and using derivatives, is similar to a chapter in a standard calculus text. We obtain the derivatives of polynomials, exponentials, sines, and cosines from the definitions. We then develop the product, quotient, and chain rules. By looking at where a derivative is 0, we see how to maximize or minimize a function, invoking the 3rd Pillar (*What goes up has to stop before it comes down*). We use the signs of $f'(x)$ and $f''(x)$ to understand the shape of a graph and to develop the first and second derivative tests for maxima and minima. We finish with Newton's method as an example of the power of linear approximation.

In **Chapter 6** we use differential equations to understand many common functions. Exponentials arise as solutions to $dy/dt = ry$, which describes everything from money earning interest to population growth to the cooling of a cup of coffee. Logarithms appear when we solve problems involving exponentials. Sines and cosines arise as ratios of sides of triangles, as locations of points on the unit circle, as solutions to the differential equations that describe the rotation of a wheel, and in simple harmonic motion. More complicated models of growth and oscillation include the logistic model of limited growth (as well as the zombie apocalypse) and the Lotka-Volterra model of interactions between predators and prey.

Chapter 7 is about integration as a process of computing a bulk quantity by breaking it into pieces, estimating the pieces, and adding up the pieces. This is the 4th Pillar of Calculus: *The whole is the sum of the parts*. Mostly, we work with Riemann sums and definite integrals, but we also consider the accumulation function $A(x) = \int_0^x f(s)\,ds$, and note that $\int_a^b f(x)\,dx = A(b) - A(a)$.

Chapter 8 is about the Fundamental Theorem of Calculus. First we define the three key quantities: definite integrals, accumulation functions, and anti-derivatives. We have already seen how accumulation functions and definite integrals are related. The second Fundamental Theorem of Calculus (FTC2) relates anti-derivatives and definite integrals. If $F'(x) = f(x)$, then $\int_a^b f(x)dx = F(b) - F(a)$. This is proved directly with the 5th Pillar of Calculus: *One step at a time*. We divide the change in $F(x)$ from $x = a$ to $x = b$ into pieces, estimate each piece as $f(x)\Delta x$, and add up the pieces to get $\int_a^b f(x)\,dx$. The first Fundamental Theorem of Calculus (FTC1) states that the accumulation function is an anti-derivative, and it is proved both directly and as a corollary of FTC2. Only after the meaning of integration and the importance of FTC2 are established do we take up anti-derivatives, turning many of our known facts about derivatives into facts about anti-derivatives.

Chapter 9 is about methods of integration. The first half of the chapter is about finding anti-derivatives through integration by substitution and integration by parts. (We do *not* treat trigonometric substitutions, partial fractions, or any but the simplest integrals of trigonometric functions.) The second half of the chapter is about numerical integration, in particular the trapezoidal rule and Simpson's rule. The two halves are grouped together in a single chapter to emphasize that integrals are generalized sums and that anti-derivatives are (mostly) just a tool for computing integrals.

In **Chapter 10** we consider functions of two or more variables and study them with the 6th and final pillar: *One variable at a time*. The main theme is that multi-variable calculus is not fundamentally different from single-variable calculus. The main ideas are the same, but sometimes we need to do things several times, once for each variable. For instance, standard scientific practice is to experiment with varying one control parameter at a time. This leads naturally to partial derivatives, linear approximations for functions of multiple variables, and the multi-dimensional chain rule. Double integrals are the same as single integrals, in that we break up the region of integration into small pieces, estimate the contribution of each piece, and add up the pieces. By organizing our pieces, summing first over one variable and then over the other, we convert a double integral into an iterated integral on which FTC2 can be applied.

Finally, in **Chapter 11** we return to the 1st Pillar with sequences and series. We define an infinite sum as the limit of finite sums. Saying that a certain quantity *is* an infinite sum is just a shorthand way of saying that it is *approximated* better and better by finite sums. We see that the geometric series $\sum_{n=0}^{\infty} r^n$ converges to $1/(1-r)$ when $|r| < 1$ and it diverges otherwise. We then consider power series. By manipulating the power series $\sum_{n=0}^{\infty} x^n = 1/(1-x)$, we obtain series expansions for $\ln(1-x)$ and $\tan^{-1}(x)$. We then consider Taylor polynomials, Taylor series, and Taylor's theorem with remainder, and we use Taylor polynomials to approximate functions and their integrals. We compute the power series of several common functions and derive Euler's formula $e^{ix} = \cos(x) + i\sin(x)$. Finally, we develop the main tests for convergence, in particular the root and ratio tests, and we apply them to power series to obtain radii of convergence for Taylor series.

Dependencies and possible course structures. The course M408Q that we teach at the University of Texas does *not* cover the entire book. We usually cover Chapters 1–9. The remaining chapters are provided in case we (or you!) encounter an exceptionally strong class or can devote more than one semester to the material. For instance, at a school that uses quarters rather than semesters, Chapters 1–6 would make a coherent one-quarter course on differential calculus, to be followed by a second quarter covering Chapters 7–11.

The first eight chapters are laid out in a very definite sequence. It is hard to imagine skipping any of these chapters without completely losing the flow. The first two chapters are short and set the stage for the ideas that follow. Chapter 3 develops the concept of successive approximation that drives almost all of calculus. Chapter 4 develops the concept of the derivative, and Chapter 5 shows how to use derivatives. Exponential functions and trigonometric functions are introduced early on, but they only really come into their own in Chapter 6. Chapter 7 explains what integrals are, and Chapter 8 explains the Fundamental Theorem of Calculus. There is no getting around those core subjects.

As a result, almost nothing in Chapters 1, 4, 7, or 8 can really be skipped. However, it is certainly possible to skip some topics in Chapters 2, 3, 5, and 6 without losing continuity. In Chapter 2, there is no need to cover the SIR equations twice, once for market penetration and once for epidemics. While the material on epidemics, and in particular on Covid, has considerable real-world significance, it is included mainly to whet students' appetite for modeling. Section 3.7, on the bisection and Babylonian methods of solving algebraic equations, can also be skipped. If you are downplaying the use of computers (an option that is *not* recommended!), then you may find Section 3.6, on implementing Euler's method by computer, to be superfluous. Chapter 5 contains three applications of derivatives, for optimization (Section 5.4), graph sketching (Section 5.5), and Newton's method (Section 5.6). Most instructors include the first two and skip the third, but the choice is yours. In Chapter 6, most students will not need the supplementary material Section 6.7 (A Crash Course in Trigonometry), and especially well-prepared students may already know the material in Section 6.2 (Exponential Functions and Logarithms). If you are not interested in population dynamics, then Section 6.5, on the logistic and Lotka-Volterra models, can be safely skipped.

Chapters 9, 10, and 11 are stand-alone introductions to methods of integration, to multi-variable calculus, and to series. Depending on the interests and background of your students, it is reasonable to design a calculus course that includes any subset of these topics (including the empty subset).

In Chapter 9, the analytic and numerical sections are independent of one another. On the analytic side, I strongly recommend covering at least integration by substitution; integration by parts is optional. For the numerical methods, it is reasonable to include the trapezoidal rule and skip Simpson's rule. If the rest of Chapter 9 is skipped, then Section 9.1 (Integration by Substitution) can be folded into Section 8.5 (Computing Anti-Derivatives).

Within Chapter 10, it is possible to do partial derivatives without doing double integrals, or to do double integrals without partial derivatives. Within the sections on partial derivatives, it is possible to skip the discussion of higher partial derivatives, the "mixed partials are equal" theorem, and quadratic approximations.

Chapter 11 is aimed at a more sophisticated audience than the rest of the book, covering a lot of ground in relatively few pages. The first four sections are essential to understanding what series and power series are and how to use them. Sections 11.5 and 11.6 are about tests for convergence, determining when and where the material of the first three sections can be meaningfully applied. The bulk of a traditional treatment of sequences and series is devoted to convergence; testing for convergence is a wonderful exercise in open-ended mathematical thinking. However, many successful scientists and engineers have gotten away with cavalier attitudes toward convergence, especially since the use of Taylor *polynomials* (as opposed to series) only requires differentiability, not analyticity.

Course development is a collaborative process. If you have questions on how to use this book in your course, and especially if you have feedback on what did and didn't work with your students, please contact me directly.

Acknowledgments. This book would not have been possible without the efforts of Bill Wolesensky, who has been my partner in developing the courses M408R (Calculus for Biologists) and M408Q (Calculus for Business) at the University of Texas. The structure of this book reflects the structure of those courses, which in turn reflects Bill's ideas as much as my own.

This book would also not have been possible without the pioneering work of the Five Colleges Calculus Project (James Callahan, David Cox, Kenneth Hoffman, Donal O'Shea, Harriet Pollatsek, and Lester Senechal). Not only are the early chapters of this book consciously modeled on their wonderful text, *Calculus in Context*, but the courses at the University of Colorado and the University of Texas that directly led to the book you are holding would never have been designed without *Calculus in Context*. The Five Colleges Calculus Project in turn was supported by the National Science Foundation. I am grateful both for the inspiration that *Calculus in Context* provided, and for permission to adapt material from that book. [1] In particular, many of the exercises in this book are modeled on exercises in *Calculus in Context*. Even more are inspired by *Calculus in Context*'s general approach to the subject. Several computer programs in this book, including SIREulers, Riemann, and the unnamed implementation of the Babylonian method, are MATLAB implementations of similar programs in *Calculus in Context*.

---

[1] Source: Wikimedia Commons, author: Callahan, James; Cox, David; Hoffman, Kenneth; O'Shea, Donal; Pollatsek, Harriet; and Senechal, Lester, "Calculus in Context" (2008). Open Educational Resources: Textbooks, Smith College, Northampton, MA, licensed under the Creative Commons Attribution 4.0 International (CC BY-4.0) (https://creativecommons.org/licenses/by/4.0/) license.

1. What is Calculus? The Six Pillars

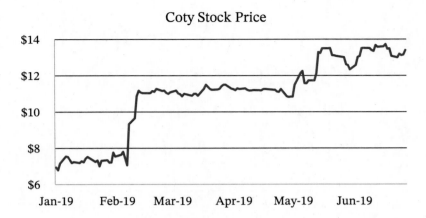

Figure 1.1. Coty's stock went up in the first half of 2019.

Figure 1.2. Nordstrom's didn't.

the rate of change of the profit curve and figuring out where that rate of change is 0, we can determine the optimal production level.

(4) **The whole is the sum of the parts.** What will the national debt be next year? That's a complicated question, but we can break things down year by year. The national debt next year is the national debt this year plus this year's budget deficit.[1] Similarly, this year's debt is last year's debt plus last year's deficit. Working backward, year by year, we see that the national debt this year is the sum of the national budget deficits every year going back to 1776.

If we plot the budget deficit as a function of time, as in Figure 1.4, the sum of all those values is the same as the area under the curve. We're going to spend a lot of time talking about area, but it isn't because we're obsessed with geometry.

---

[1] Or at least it would be if the same accounting practices were used for both debt and deficit. See Section 4.1 for the difference in how Social Security is usually counted. In this section, the national debt is adjusted to take that difference into account.

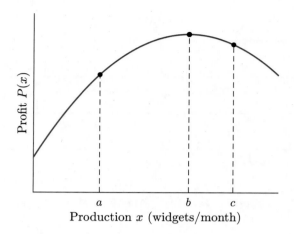

Figure 1.3. Monthly profit as a function of production level

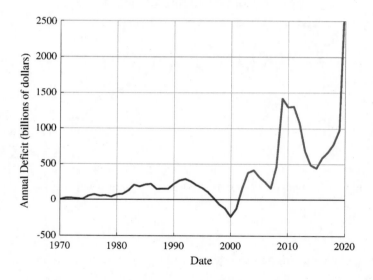

Figure 1.4. US budget deficit by year, 1970–2020

It's because most bulk quantities, like the total power used by the city of Austin in 2007 or the total amount of water carried by the Colorado River in 2011, can be treated exactly like national debt or area. Break the quantity you're studying into little pieces, estimate each piece, and add up the pieces. This process is called **integration**.

Note that this is very different from how integration is often taught. In most calculus classes, students are taught that integration is in some sense the opposite of differentiation. That isn't exactly *wrong*, as we'll see with the 5th Pillar, but it misses the point. Integration is about adding up the pieces, which is why it applies to a host of problems. Anti-derivatives are a great tool for actually *computing* integrals, but they don't explain why such-and-such quantity is represented by such-and-such integral. For that, we need the 4th Pillar.

(5) **One step at a time.** A famous Chinese proverb says that "a journey of a thousand miles begins with a single step". The journey then continues with about two million additional steps. By understanding what happens at each step, we can understand the entire journey.

Instead of asking what the national debt *is*, we can ask how much it *changed* in a short period of time, like a year. Geometrically, the change per year is the **slope** of the debt curve, shown in Figure 1.5. The steeper the curve is, the faster the debt is changing.

In the late 1990s, the deficit was negative (also known as running a surplus), and the debt came down. In the Great Recession of 2009–2012, and again in the Covid pandemic of 2020, the budget deficit was large and the debt shot up. Between 1950 and 1975 (only part of which is shown on the graph) the deficit was close to 0 and the debt was nearly constant.

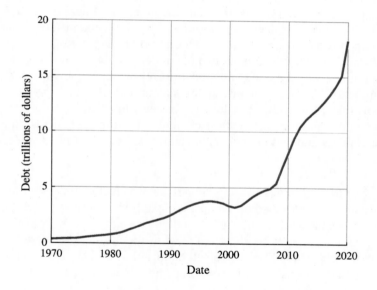

Figure 1.5. US National debt by year 1970–2020

How are Figures 1.4 and 1.5 related? The deficit each year is the same as the change in the debt from that year to the next. This means that the debt is a running total of the deficit and corresponds to the **area** under the deficit curve. The deficit is the rate at which the debt is changing, and corresponds to the **slope** of the debt curve. Mathematically, the debt is the **integral** of the deficit, and the deficit is the **derivative** of the debt. The **Fundamental Theorem of Calculus** (FTC) relates derivatives and integrals in general. With it, we can use what we already know about derivatives to understand integrals. In fact, it is the single most powerful tool we have for evaluating integrals.

The Fundamental Theorem of Calculus comes from thinking about changes one step at a time, but there is more to "one step at a time" than just the Fundamental Theorem. For instance, suppose that we invest $1000 at 6% interest, and we want to know how much money we will have in 30 years. From our initial balance and the interest rate, we can estimate the interest we receive in the first year and compute our bank balance a year from now. From that, we can estimate the interest we receive in the second year and get our bank balance two years from now. Continuing the process, step by step, we can accurately project our bank balance far into the future.

Finally, "one step at a time" is great slogan for how to approach calculus in general. Many problems are way too complicated to be done all at once. By breaking big problems into sequences of smaller problems, and by solving these smaller problems one at a time, we can accomplish wonders.

(6) **One variable at a time.** Many functions involve two or more input variables. The boiling point of water depends on our elevation and on how much salt we put in the water. In the summer, the heat index depends on temperature and humidity. In the winter, the wind chill factor depends on temperature and wind speed. The price of widgets depends on supply and demand. The value of an oil field depends on how much oil it produces and on the price of oil.

To understand functions of two (or more) variables, we always hold everything but one variable fixed and study just that variable. Asking how a change in temperature changes the heat index is a question about a function of just one variable that we already know how to answer. Similarly, we can figure out how changing the humidity affects the heat index. Putting the two answers together, we can understand how changing both temperature and humidity affects the heat index.

It's tempting to say "that's all there is!", but that isn't really true. Over the centuries, lots of really smart people have cooked up lots of really smart ways to solve lots of really hard problems. Now, with the help of computers, we can solve even more problems. In the next ten chapters, we're going to follow in the footsteps of these masters and learn some of their results. Yes, there will be formulas to memorize and techniques to practice and algorithms to implement on computers. And yes, it will take work to really absorb everything.

But hopefully this voyage of discovery won't be a mystery. Every new formula or technique or algorithm will lead straight back to the Six Pillars, so it will be connected to every other formula or technique or algorithm. If you remember the pillars, you'll have a framework for organizing all the little details. Over time, you'll forget many of those details, and that's OK. As long as you remember the pillars and are willing to look up the details as needed, you'll be able to use calculus for your whole life, not just in your classes and in your job, but in understanding the wild and complicated world we live in.

Chapter 2

Predicting the Future: The SIR Model

In this chapter, we will use four of the pillars to get a handle on a real-world problem: product adoption.

- It's impossible to keep track of the behavior of every customer in the world, so we devise a simplified model to describe consumer behavior. *Close is good enough!*
- This involves understanding the *rate* at which consumers start using a product and the rate at which they abandon that product. *Track the changes!*
- Once we understand where things stand and how fast they are changing, we can make realistic predictions about what is likely to happen in the future, as well as what we can do to improve that trajectory. Our projections are only accurate for a short time, so we combine a lot of short-term projections to get a long-term projection. *One step at a time!*
- Even without these long-term projections, we can understand how our market penetration will peak by comparing the factors that increase penetration with those that decrease penetration. *What goes up has to stop before it comes down!*

2.1. A Problem of Market Penetration

Imagine that you work for a company that has just introduced a hot new phone app. People are learning about it by word of mouth, and more and more people are using your product. Of course, that growth can't go on forever. Eventually, people will get tired of your app and will move on to the Next Great Thing. In order to make the most of your product's popularity, you need to forecast usage for the next year or two. You hire a mathematical consultant (your Friendly Author), and together we attack the problem.

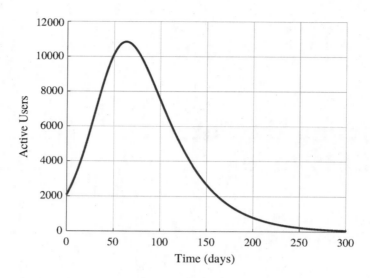

Figure 2.1. Active users as a function of time

The first thing we do is define our quantities. Let t denote time, measured in days, and let $I(t)$ be the number of active users at time t. We call t and I **variables** since they change. The time t is our **input variable**, while I is an **output variable**. The graph of I as a function of t will look something like the plot in Figure 2.1. There is an initial phase where more and more people use the product, a moment of peak usage, a decline in usage, and a long tail. To be successful, we want the rise to be as fast as possible, the peak to be as high as possible, and the decline to be as slow as possible.

Before we try to control I, we need to understand what makes I change over time. That is, we need to make a mathematical **model** for what is going on. The model needs to take into account all the *important* features of what's happening in the world, while being simple enough to be solvable. We are going to ignore a *lot* of details, because *close is good enough!*

Once we have our model, we have to **analyze** it. In this step, we don't care where our equations came from. We just want to solve them. Maybe we can find a formula for the answer. More likely, we can't find a formula, but we can run the model on a computer to generate accurate predictions. Once again, *close is good enough!*

Finally, we need to **interpret** our results. Math can tell us that such-and-such variable will have such-and-such value at such-and-such time, but we need to understand business to say what that means for the success of our company.

In other words, predicting the future is a three step process:

(1) **Model** our system mathematically. Define appropriate variables and write down some equations that describe how these variables change with time. This step requires real-world understanding as well as math.

(2) **Solve** the model to determine what each variable will be at some future time. This step is 100% math, using techniques that we are about to develop.

(3) **Interpret** the results. Take our mathematical results and make real-world sense of them.

We'll tackle these one at a time, which is a lesson in itself. *One step at a time!* We need to break hard problems into bite-sized tasks, and then do each task in turn.

2.2. Building the SIR Model

To understand the adoption of our product, we divide our population of potential customers into three groups, as in Figure 2.2.

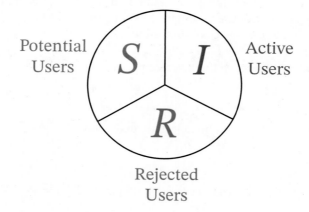

Figure 2.2. Three categories of users

- **Potential** users, or *Potentials*: These are people who might use the product in the future if they hear enough good things about it from their friends.
- **Active** users, or *Actives*: These are people who have already adopted the product. Not only are they using it, but they are spreading the word about it.
- **Rejected** users, or *Rejecteds*: These are people who won't ever use the product. Maybe they just aren't interested. Maybe they used to use the product but got tired of it. Maybe our app doesn't even work on their brand of phone. For whatever reason, they're no longer reachable.

Next, we define our variables. Let $S(t)$ be the number of Potentials at time t, let $I(t)$ be the number of Actives at time t, and let $R(t)$ be the number of Rejecteds at time t. The letters S and I, which obviously don't stand for Potential and Active, are historical.[1]

Obviously, there are differences among the users in each category, but we're going to ignore those differences. This is a model, not a complete description of reality! Instead, we're going to talk about the behavior of the *average* Potential, the *average* Active, and the *average* Rejected. We then ask two questions:

(1) At what rate do Potentials adopt the product and become Actives?

---

[1] The SIR model was originally developed to study epidemics, as we will see in Section 2.5. In that context, S stands for *Susceptible*, I stands for *Infected*, and R stands for *Recovered* or *Removed*.

(2) At what rate do Actives stop using the product and become Rejecteds?

From these rates, we can figure out the rates S', I', and R' at which the quantities S, I, and R are changing, and from those we can predict the future.

Losing customers: Attrition. Customers don't use an app forever. Sooner or later they get tired of it or switch to a competitor's app. Some apps, like navigation tools and browsers, keep their customers for months or years. Others, like games, can lose their customers after just a couple of weeks.

Suppose that we are analyzing a new game that users keep using for an average of 30 days. Among the Active users, roughly 1/30 of them will grow tired of the game today and will be among the Rejecteds tomorrow. That is,

(2.1) $$\text{today's change in the Rejected population} = \frac{I(\text{today})}{30}.$$

Likewise,

(2.2) $$\text{tomorrow's change in the Rejected population} = \frac{I(\text{tomorrow})}{30},$$

and in general

(2.3) $$\text{the change in the Rejected population on day } t = \frac{I(t)}{30}.$$

Note the units in this equation. $I(t)$ and $R(t)$ are numbers of *people*, and the number of new R's on any given day is also measured in people. But the *rate R'* at which R is changing is measured in people/day. That is,

$$R'(t) = \frac{I(t)}{30 \text{ days}}.$$

The numerator has units of people, the denominator has units of days, and the ratio has units of people/day.

This is an example of a **rate equation**. A rate equation describes the rate at which something is changing in terms of other data. In this case, it describes R' in terms of I.

More generally, if T is the average length of time that people use a product, then we expect

(2.4) $$R' = \frac{I}{T}.$$

If we define $b = 1/T$, then we can write our rate equation without fractions:

(2.5) $$R' = bI.$$

See Figure 2.3. R' has units of people/day, I has units of people, and b has units of (1/day)s. When the SIR model is used in biology, b is called the **recovery coefficient**. In our business example we'll call it the **attrition coefficient**. If you prefer, you can think of it as the **boredom coefficient**.

2.2. Building the SIR Model

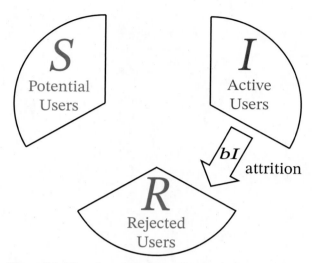

Figure 2.3. Through attrition, active users become rejected users.

The number b is called a **parameter**. It isn't a **variable**, because it doesn't change during the product run. However, it may take on different values for different products. By adjusting the value of b, we can use the same rate equation to model games, navigation aids, music players, you name it.[2]

Gaining customers: Transmission. Losing customers is grim. If our app currently has 2,100,000 users and is only used for an average of 30 days, then we're losing users at a rate of

$$(2.6) \qquad \frac{2{,}100{,}000 \text{ people}}{30 \text{ days}} = \frac{70{,}000 \text{ people}}{\text{day}}.$$

That's a lot! If we're going to stay in business, we need to get new customers to replace the old ones.

In our model, we assume that we gain customers by word of mouth. Let's look at this from the perspective of a single Potential user, who we'll call Joe. Joe will hear about our app from a certain fraction p of the Active users each day, for a total of pI contacts. The fraction p is typically very small, but the Active population I can be very large. The more Active users there are, the more times Joe will hear about our product. Every time that Joe hears about our product, there is a probability q that he will be motivated to download it and start using it. That is, there is a probability pqI per day of Joe becoming Active.

We don't actually care about p and q separately, since all that matters is their product. We define $a = pq$, and call a the **transmission coefficient**. Like b, this is a parameter, not a variable. Different products in different communities will have different values of a, but we can use the same reasoning for all of them.

---

[2]There are some exceptions, such as social media, that follow a different pattern. People don't stop using social media platforms after a certain amount of time. They stop using them when their friends stop using them. That's a much more complicated system to model!

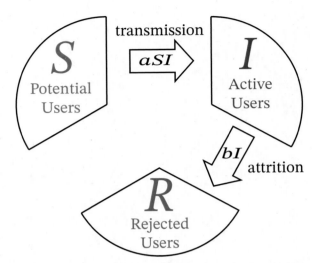

Figure 2.4. Through word of mouth, Potential users become Active users.

Finally, we look at the entire population of Potentials. If there are S Potentials, each of whom has a probability aI (per day) of becoming Active, then each day there will be approximately aSI Potentials who become Actives. That is,

$$(2.7) \qquad S' = -aSI.$$

See Figure 2.4. S and I have units of people and S' has units of people/day, so a must have units of 1/(people × days). Also note the minus sign in our equation! The *more* Potentials become Actives, the *fewer* Potentials are left.

What about I? Completing the model. So far we have figured out how the numbers of Rejecteds and Potentials change, but what about the number of Actives? We have

$$\begin{aligned} I' &= + \text{(Rate at which people adopt the product)} \\ &\quad - \text{(Rate at which they abandon the product)} \\ (2.8) \qquad &= aSI - bI. \end{aligned}$$

Another way to see this is that $S + I + R$ is the total number of people out there, and that doesn't change. Since $S' + I' + R' = 0$, we must have $I' = -S' - R'$.

Putting everything together, we have a system of three rate equations. Together, they're called the **SIR model**:

$$(2.9) \qquad \begin{aligned} S'(t) &= -aS(t)I(t), \\ I'(t) &= aS(t)I(t) - bI(t), \\ R'(t) &= bI(t), \end{aligned}$$

where

- t is the time. Depending on the setting, you may want to measure t in days, weeks, months, quarters, or years.
- $S(t)$ is the number of Potentials at time t, measured in people.
- $I(t)$ is the number of Actives at time t, also measured in people.
- $R(t)$ is the number of Rejecteds at time t, also measured in people.
- b is the attrition coefficient, which is the reciprocal of the average time T that a user keeps using the product before moving on. The units of b are the reciprocal of whatever units we are using for time. This parameter can vary from product to product.
- a is the transmission coefficient. This can depend both on the product and on the market. If you are marketing the same product in several different places, b will be more or less the same in all markets, but a will typically be bigger in the smaller markets (where each Potential knows a greater fraction of the Actives) and smaller in the bigger markets. The units of a are $1/(\text{time} \times \text{people})$.

2.3. Analyzing the Model Numerically

Now that we have our model, let's use it to predict the future. Suppose that we measure t in days, that $a = 0.000002$/person-per-day and $b = \frac{1}{30}$/day, and that we start with $S(0) = 40,000$, $I(0) = 2100$, and $R(0) = 7900$. What will S, I, and R be in two days? In five days? In ten days? For that matter, what were the values yesterday? A week ago? A month ago?

A naive approach is to use the SIR equations to compute S', I', and R' once and for all:

$$
\begin{aligned}
S' &= -0.000002(40,000)(2100) &= -168 \text{ people/day}, \\
I' &= 0.000002(40,000)(2100) - 2100/30 &= 98 \text{ people/day}, \\
R' &= 2100/30 &= 70 \text{ people/day}.
\end{aligned}
$$
(2.10)

If we have 2100 Actives on day 0, and if I is growing at a rate of $I' = 98$ people/day, then we should expect $2100 + 98 = 2198$ Actives tomorrow. We should expect $2198 + 98 = 2100 + 2(98) = 2296$ Actives the day after tomorrow. After a week, we should expect $2100 + 7(98) = 2784$ Actives, after a month we should have $2100 + 30(98) = 5040$ Actives, and after a year we should have $2100 + 365(98) = 37,870$ Actives. In general, after t days we should have

(2.11) $$I(t) \approx I(0) + I'(0)t = 2100 + 98t.$$

This is called a **linear approximation**. Instead of finding the exact equation for $I(t)$, we found the equation of the line that has the right value and the right slope at $t = 0$. We then approximate the value of the true $I(t)$ function by the value of the linear function $2100 + 98t$.

> Before we move on, let's recall some basic facts about things that change at a constant rate. If we are driving at a constant speed of 57 miles per hour and pass milepost 253 at 3:00, where will we be at 4:00? At 5:00? At time t?
>
> From 3:00 to time t is $t-3$ hours. Since rate × time = distance, and since we are going at 57 MPH, we will travel $57(t-3)$ miles in that time. Adding that to our starting point at milepost 253, we will find ourselves at milepost
>
> $$(2.12) \qquad x(t) = 253 + 57(t-3)$$
>
> at time t. (At least if we're traveling in the direction where the mile markers are increasing. If we're heading in the opposite direction, we will find ourselves at milepost $253 - 57(t-3)$.)
>
> The same idea works for any quantity that is changing at a constant rate, not just for position. If a quantity Q is changing at rate Q', and if Q starts at a value Q_0 at time t_0, then what will Q be at time t? Since it grows at rate Q' for time $(t-t_0)$, it will increase by $Q' \times (t-t_0)$. Adding that to our starting value of Q gives
>
> $$(2.13) \qquad Q(t) = Q_0 + Q' \times (t-t_0).$$
>
> This is the equation of a straight line in **point-slope form**. The linear approximation (2.11) is a special case of this, with I instead of Q, and with starting time $t_0 = 0$. We'll have a lot more to say about equations of lines and linear approximations in Chapter 3.

Returning to the SIR model, we can use the linear approximation (2.11) to study the past as well as the future. According to this approximation, yesterday we had around $2100 - 98 = 2002$ Actives, a week ago we had $2100 - 7(98) = 1414$ Actives, and a month ago we had $2100 - 30(98) = -840$ Actives.

How much do you trust those numbers? You should take them with a grain (or more) of salt. In particular, the estimate $I(-30) \approx -840$ is absurd. You can't have a negative number of Actives!

In making that estimate, something went *seriously* wrong. You should get into the habit of asking whether answers make sense. Reality check! Before reading on, take a minute to think about what went wrong with the method we used to get our negative answer.

The problem with our linear approximation is that it assumed that the rate of change of I is 98 people/day, it always was 98 people/day, and it always will be 98 people/day. In reality, we know that $I' = 98$ people/day *today*, and it's realistic to assume that I' won't change much in the next few days, but in a few weeks, or a few months, it could change by a lot. This means that we can trust our linear approximation when $t = 1$ or $t = 2$ or $t = -1$ or $t = -2$, but we shouldn't trust it when $t = 30$ or 365 or -30. The actual situation is shown in Figure 2.5. The linear approximation tracks $I(t)$ very closely for $-5 < t < 5$, but it does not account for the curvature in the graph of $I(t)$. As t gets more and more negative, the graph of $I(t)$ flattens out and stays positive, while

2.3. Analyzing the Model Numerically

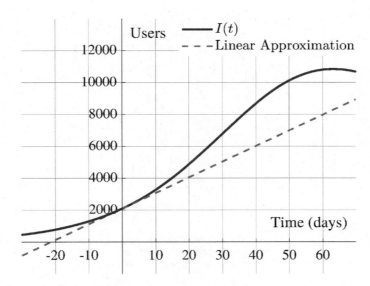

Figure 2.5. Active users over time. The linear approximation is negative when $t < -21.5$, but the actual $I(t)$ isn't.

the linear approximation goes negative. Meanwhile, as t gets more and more positive, $I(t)$ goes up and then down, while the linear approximation just keeps on growing.

To do better than our naive approximation, we need a way of estimating S' and I', and R', not just right now, but in the future and the past. Here's a way to estimate $S(10)$, $I(10)$, and $R(10)$.

First, we need to decide how far we can trust our linear approximation. Since I seems to be changing at about 5% each day, trusting it for a couple of days is reasonable. We're going to pick a time interval of two days, which we'll call Δt days, and predict the future $\Delta t = 2$ days at a time.

(1) Use the initial values $S(0)$, $I(0)$, and $R(0)$ together with the SIR equations to estimate $S'(0)$, $I'(0)$, and $R'(0)$.

(2) Use a linear approximation to estimate $S(2) \approx S(0) + 2S'(0)$, as well as $I(2) \approx I(0) + 2I'(0)$ and $R(2) \approx R(0) + 2R'(0)$.

(3) Use the values of $S(2)$, $I(2)$, and $R(2)$ together with the SIR equations to estimate $S'(2)$, $I'(2)$, and $R'(2)$.

(4) Use a linear approximation to estimate $S(4) \approx S(2) + 2S'(2)$, etc.

(5) Use *those* values and the SIR equations to approximate $S'(4)$, etc.

(6) Lather, rinse, repeat. At each time t, plug the estimated values of $S(t)$, $I(t)$, and $R(t)$ into the SIR equations to get estimated values of $S'(t)$, $I'(t)$, and $R'(t)$. Then use a linear approximation to compute $S(t + \Delta t) \approx S(t) + \Delta t\, S'(t)$, etc. Table 2.1 shows the results for the first ten days.

Likewise, we can go backward in time, using a small time step. We can take $\Delta t = -2$ days and go from $t = 0$ to $t = -2$ to $t = -4$, etc., all the way back to to

Table 2.1. Projecting forward with $\Delta t = 2$

t	S(t)	I(t)	R(t)	S'(t)	I'(t)	R'(t)
0	40,000	2,100	7,900	-168	98	70
2	39,664	2,296	8,040	-182	105.5	76.5
4	39,300	2,507	8,193	-197	113.5	83.5
6	38,906	2,734	8,360	-213	121.5	91.5
8	38,480	2,977	8,543	-229	130	99
10	38,022	3,237	8,741			

Table 2.2. Projecting backward with $\Delta t = -2$

t	S(t)	I(t)	R(t)
0	40,000	2,100	7,900
-6	40,923	1,559	7,518
-12	41,621	1,143	7,236
-18	42,139	832	7,029
-24	42,519	602	6,879
-30	42,796	433	6,771

$t = -30$. The results are shown in Table 2.2, with some times skipped and the values of S', I', and R' omitted to save space. As you can see, $I(-30)$ isn't negative at all.

This method still doesn't give exact answers, but *close is good enough!* If we want to compute $I(10)$ with greater accuracy, we can use the same algorithm with $\Delta t = 1$ instead of $\Delta t = 2$, for an answer of $I(10) = 3261$ instead of 3237. Of course, that requires ten iterations instead of five. If we want even more accuracy, we can take $\Delta t = 0.1$ and do 100 iterations, getting $I(10) = 3282$, or take $\Delta t = 0.01$ and do 1000 iterations, getting $I(10) = 3284$. If we're willing to do the extra work, or if we program a computer to do the extra work for us, we can have as much accuracy as we want. (However, our model is only an approximation of the real world, so even if we can solve our model to great accuracy, that doesn't necessarily mean that we can predict the future with that much accuracy.)

This is the 5th Pillar of Calculus: *One step at a time.* Every big change is made up of many little changes. If we can understand each little change, we can put the pieces together to understand the big change.

2.4. Theoretical Analysis: What Goes Up Has to Stop Before it Comes Down

So far we have used the 1st, 2nd, and 5th Pillars of Calculus. By making approximations and tracking the changes in our variables S, I, and R, and by putting a lot of short-term linear approximations together, we figured out how to obtain good projections of the

2.4. Theoretical Analysis: What Goes Up Has to Stop Before it Comes Down

future, and we were able to use those same projections to understand the past. Now we're going to tackle the question:

What is happening when $I(t)$ reaches its peak?

The key fact is that *the sign of I' tells you whether I is increasing or decreasing*. Whenever $I'(t) > 0$, $I(t)$ must be increasing. Whenever $I'(t) < 0$, I must be decreasing. At the very top of the curve, when $I(t)$ has stopped increasing and hasn't yet started decreasing, $I'(t)$ transitions from positive to negative. At that instant of time, we must have $I'(t) = 0$.

Let's figure out what is happening at that time. The SIR equations tell us that

$$I'(t) = aS(t)I(t) - bI(t) = I(t)(aS(t) - b). \tag{2.14}$$

Since $I(t)$ is always positive, the sign of $I'(t)$ is the same as the sign of $aS(t) - b$. As long as $S(t) > b/a$, $I'(t)$ will be positive and $I(t)$ will increase. But eventually we will run out of Potential customers. When $S(t)$ drops below b/a, $I'(t)$ will become negative and $I(t)$ will start to decrease.

The number b/a is called a **threshold**. In our example, it was 16,667. This is **not** the value of $I(t)$ at the peak. Rather, it is the value of $S(t)$ when $I(t)$ hits its peak. In Figure 2.6, $I(t)$ hits its peak value of 10,870 when $t = 63.2$, which is when $S(t)$ passes through 16,667.

If the threshold b/a is large and there aren't many Potentials to begin with, our product will fizzle from the start. If b/a is small and $S(0)$ is large, we will have a long period of growth before we saturate the market. As a marketer, our considerations for launching a product are the following.

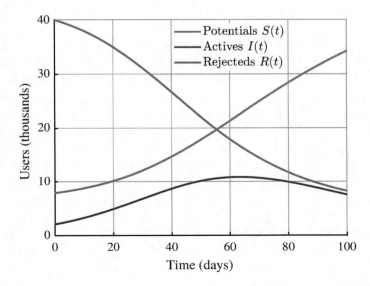

Figure 2.6. Market penetration via the SIR model. Over time, the number of Potentials drops, the number of Rejecteds rises, and the number of Actives rises, reaches a peak, and then falls,

(1) Make b as small as possible. Equivalently, make the usage time $T = 1/b$ as big as possible. This has to do with the quality of our product. The more exciting our product is and the more we provide updates and other continuing benefits to the users, the lower the attrition coefficient will be.

(2) Make a as big as possible. Do everything possible to encourage word-of-mouth communication between Active and Potential users.

(3) Pick a market where $S(0)$ is as big as possible. Even the best products can fizzle if they're directed at the wrong market. A lemonade stand will do a lot better in Texas in August than in Minnesota in February.

2.5. Epidemics

The SIR model originally came from epidemiology, not from business, and was developed to model the spread of disease. In that setting S stands for **Susceptible**, not Potential; I stands for **Infected**, not Active; and R usually stands for **Recovered**, not Rejected.[3] Infected people turn into Recovered people (with immunity against reinfection) when the disease has run its course. If we're talking about the flu, which lasts about one week, then the **recovery coefficient** b is $1/(7 \text{ days})$ and $R' = bI$. The rate at which Susceptible people become Infected is proportional to the number of times that an Infected person sneezes on a Susceptible person, which is proportional to the product $S(t)I(t)$. That is, $S'(t) = -aS(t)I(t)$, where a is still called the **transmission coefficient**.

In other words, the equations for market penetration and for the spread of disease are *exactly the same*! To a mathematician, they're the same problem. The difference is in interpreting the results and deciding what to do. With an epidemic we're trying to keep people from becoming Infected, while in marketing we're trying to get people to become Active.

In particular, if we are trying to contain an outbreak of disease, then we want to the threshold b/a to be as **big** as possible, and we want the number S of Susceptibles to be as **small** as possible. The main tools for doing this are the following.

(1) Vaccination. This transforms Susceptibles (who might get sick) directly into Recovereds. These are people who aren't sick, won't get sick, and can't spread the disease. If we can get the starting value of S to be below the threshold, then we can prevent the epidemic from even starting. This is called **herd immunity**. People often think that it's important to get a shot to protect themselves, and it is. But it's even more important for them to get that shot to protect the rest of us!

(2) Reduce the transmission coefficient a. Usually this is done through voluntary measures like encouraging sick people to stay home, wearing masks, keeping a safe distance from others, and avoiding handshakes. If a lot of people are sick, public health officials may need to take more drastic measures, such as closing schools, canceling large public events, quarantining infected individuals, or issuing shelter-in-place orders.

---

[3] When dealing with fatal illnesses or with quarantines, R sometimes stands for **Removed**. Either way, we're talking about individuals who can no longer spread the disease.

2.6. Covid-19 and the SIR model

(3) It would also be good to increase b, which is the same as decreasing the time that people are sick and infectious. Unfortunately, this is difficult. For instance, antiviral drugs can sometimes reduce the length of the flu by a small amount, but only by a small amount. For a disease like Covid-19 that can be spread by people who don't have symptoms, it takes extensive testing and quarantining to get infectious people out of circulation. Even if somebody hasn't *recovered*, we can try to *remove* them from the general population until they stop being infectious.

The bottom line is that our best defense against epidemics is vaccination, which usually has to be done in advance. Once an epidemic hits, the focus usually turns to improving sanitation and limiting person-to-person contact. Vaccines and face masks provide a lot more bang for the buck than anti-viral drugs.

2.6. Covid-19 and the SIR model

In December 2019, a strange form of pneumonia started affecting residents of Wuhan, China. The virus causing this disease was quickly identified, and the disease was named COVID-19 for "**CO**rona**VI**rus **D**isease 20**19**".[4] By April 2020 Covid-19 had spread around the world and had mutated into a more transmissible form, leading to drastic lockdowns to control the spread. Schools closed or went online, most people were told to stay home except for essential errands like buying food, and everybody was told to practice "social distancing". Despite these measures, Covid-19 continued to spread, and by mid-2022 had killed at least 6 million people worldwide (with some estimates being twice that big) including over one million Americans.

Everybody wanted to know what would happen next, which kept mathematical modelers (including the author) very busy. The details of their models could be complicated, but the main ideas were exactly the same as in the SIR model. In this section we'll go over some of the ways that modelers adapted the SIR model to apply to a worldwide pandemic rather than a localized outbreak.

Scaled models and the replication number R_0. One problem with the SIR model is that the parameters depend on the size of the city where the outbreak is happening. Typically, the larger the population, the smaller a fraction of the population that each person knows or meets and the smaller the transmission coefficient a will be. However, the product of a and the total population tends to be the same in different cities. For this reason, it's useful to write a scaled version of the model.

Let $T = S + I + R$ be the total population. This number is of course constant, or at least approximately constant. We let $s = S/T$, $i = I/T$, and $r = R/T$ be the *fractions* of the population that are susceptible, infected, or recovered, respectively, so that $s + i + r = 1$. Since $s' = S'/T$, $i' = I'/T$, and $r' = R'/T$, we can write rate equations

---

[4] After a while, people dropped the "19", stopped capitalizing the whole word, and just wrote "Covid" or even "covid".

for these quantities.

$$s' = -aSI/T = -aTsi,$$
$$i' = (aSI - bI)/T = aTsi - bi,$$
(2.15)
$$r' = bI/T = bi.$$

Modelers usually use the Greek letters β and γ in this scaled model, with

(2.16) $$\beta = aT; \quad \gamma = b.$$

This makes the scaled SIR equations:

$$s' = -\beta si,$$
$$i' = \beta si - \gamma i,$$
(2.17)
$$r' = \gamma i.$$

Conditions do vary from place to place, with β being larger in cities that are more crowded and where people are less careful, but β varies much less than a and T vary separately. As a result, it's sensible to talk about "the" values of β and γ for each disease.

In the early stages of an epidemic, when s is close to 1, people are getting sick at rate $\beta si \approx \beta i$ and recovering at rate γi. That is, there are β/γ people getting sick for every person who recovers. This ratio is called the **basic replication number**,

(2.18) $$R_0 = \frac{\beta}{\gamma},$$

and it represents the average number of new people that each sick person infects. If $R_0 > 1$, then the epidemic grows. If $R_0 < 1$, then the epidemic fizzles out.[5] For Covid, R_0 was originally estimated to be between 2 or 3. However, the speed at which it spread through the USA and Europe suggests that R_0 was actually higher. Later variants evolved to be even more infectious, with values of R_0 around or even above 10.

Controlling an epidemic then amounts to getting s and R_0 as low as possible. In the short term, public health measures, such as wearing masks, closing schools, and staying home as much as possible, can reduce β, and so can reduce R_0. Testing helps, too. Sick people don't have to infect others. If they can be identified and quarantined (or **removed**) from the general population, they can be infect**ed** without being infect**ious**.

However, such efforts can't last forever. Sooner or later people want to return to normal, they start behaving as they did prepandemic, and R_0 goes back up. The only long-term solution is reducing the fraction s of susceptibles to below $1/R_0$, either through vaccination (best case) or natural infection (worst case).

The SEIR model. If Alice is sick and coughs on Bob, then Bob might get sick. If Bob then coughs on Carol, then Carol might get sick. However, if Bob coughs on Carol immediately after meeting Alice, then Carol is safe. It takes time for Alice's viruses to grow in Bob's body to the point that he can infect Carol.

---

[5] It's unfortunate that the basic replication number uses the same letter as the number of Recovered individuals. However, since we're looking at r rather than R and mostly care about s and i, this isn't really a problem.

2.8. Exercises

- Mathematical models help us to understand real-world problems.
  - Take what we know about our problem.
  - Throw out the unimportant details and keep the main features.
  - Express those features mathematically.
  - Use math to **analyze** the model.
  - **Interpret** the results. Turn numbers and equations into real-world conclusions. Without this last step, the rest is useless.
- *Track the changes.* Many useful models involve **rate equations** that describe how fast something is changing in terms of its current state. Many of the same equations show up in different settings. The SIR equations model market penetration. The same equations also model the spread of disease.
- Once you know how fast something is changing, you can use a **linear approximation** to predict its future or explore its past. This is accurate for a short time but can't be trusted over long time intervals.
- *One step at a time.* If you don't trust a linear approximation to predict what will happen next year, just use it to predict what will happen tomorrow. Then use those results to predict what will happen the day after tomorrow, then the day after that, and so on.
- *What goes up has to stop before it comes down.* You can learn a lot by studying the point when something stops moving; that is, when the rate of change equals 0.

Expectations. You should be able to:

- Model market penetration with a particular set of rate equations (the SIR model).
- Explain how the parameters in the SIR model relate to properties of the product and market being studied.
- Use the SIR equations, together with a linear approximation, to predict future usage rates.
- Iterate this process to generate a table of values for several times.
- Relate the sign of I' to whether your product is gaining or losing market share, and relate the size of I' to how fast this is happening.
- Use the same tools to analyze the trajectory of an epidemic.

2.8. Exercises

Rate of Change

2.1. Suppose that the fees collected by a consulting firm in March and May are $1.2 million and $1.25 million, respectively. Let $F(t)$ be the fees collected t months after March. (That is, March is $t = 0$.)
  (a) What is the change in fees between March and May? Call your answer ΔF. What are the units for ΔF?
  (b) What is the change in time between these two observations? Call your answer Δt. What are the units for Δt?

(c) At what *rate* did the monthly fees change from March to May? Call your answer m_1. What are the units for m_1?

(d) Assuming that fees are a linear function of time, how much can the business expect to collect in August? What value of Δt are you using to find your answer? Use the notation and answers from parts (a), (b), and (c).

(e) Again assuming a linear relationship, how much did the business collect in January of the same year? What value of Δt are you using to find your answer?

(f) Find a linear formula for $F(t)$.

2.2. In Exercise 2.1 you developed a formula for the fees collected t months after March. Suppose that you wish to update your model using additional data. You observe that model you developed in Exercise 2.1 worked well through August, but that the *change* in the fees collected between the months of August and September was $0.1 million. That is, the firm collected $0.1 million more in September than in August.

(a) What is the monthly rate of change between August and September? Call your answer m_2.

(b) How does m_2 compare to the value of m_1 you found in Exercise 2.1c? From a business perspective, is this an improvement?

(c) Use m_2 to find an updated linear formula for predicting the fees collected for October through December. (This is similar to the formula you found in Exercise 2.1f.) State the correct domain for this expression.

(d) Use the information given in Exercise 2.1 and the result of Exercise 2.1c to write down a piecewise function that expresses the fees collected each month for the entire year (January through December), letting $t = 0$ represent the month of March.

(e) Use MATLAB to graph the piecewise function you found in part (d). Label your axes and give your graph an appropriate title. Use different colors for the two pieces of your function.

(f) Explain why using the piecewise function to express monthly consulting fees billed for the entire year is better than simply using the linear function you found in Exercise 2.1f for the entire year.

2.3. Explain the term "Rate Equation." How is a rate *equation* different from the rate of *change* m associated with a linear function?

2.4. Suppose we have a rate equation $F' = \cdots$ for a quantity $F(t)$. At a starting time t_0, we have $F(t_0) = F_0$, and the rate equation tells us that $F'(t_0) = R_0$.

(a) Explain, using proper notation, how we can approximate the change in $F(t)$ between times $t = t_0$ and $t = t_0 + a$ using a linear approximation.

(b) Explain, using proper notation, how we can approximate $F(t_0 + a)$.

(c) If $F(t)$ is a linear function, would your answers in parts (a) and (b) still be approximations? Explain.

Interpreting SIR Models

2.5. Give approximate answers to these questions about the following graph of SIR model behavior:

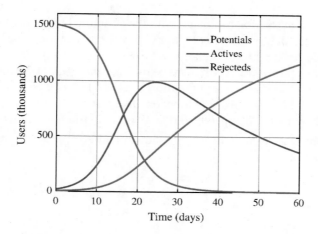

(a) When does the number of Actives reach its peak? How many people are Active at that time?
(b) Initially, how many Potentials are there? How many days does it take to cut the Potential population in half?
(c) How many days does it take the Rejected population to reach 500,000? How many people have rejected the product by day 60?
(d) On what day is the size of the Active population increasing most rapidly? When is it decreasing most rapidly? How can you tell?
(e) How many people became Active at some time during the first 30 days? (Note that this is not the same as the number of Active people on day 30!) Explain how you found this information.

2.6. Below are two graphs depicting the marketing of a product according to the usual SIR model. The initial values $S(0)$, $I(0)$, and $R(0)$ and the transmission coefficient a are the same for both graphs. However, the two graphs correspond to different attrition coefficients b.

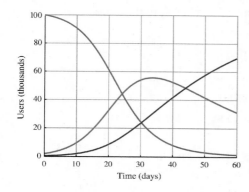
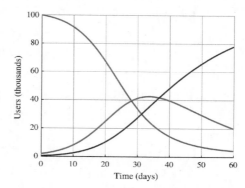

(a) For each graph, indicate which curve is S, which is I, and which is R.
(b) Which graph corresponds to the *larger* value of b? Explain.

2.7. Consider a marketing campaign with the simplified rate equation $I' = 35{,}000$ persons per day. (This is not the usual SIR model!) Suppose that there are 2,100,000 Actives on December 15.
   (a) How many Actives will there be on Christmas Day (December 25)?
   (b) When will the Active population reach 3,000,000?
   (c) How many Actives were there on December 1?
   (d) When did the campaign start? (When were there no Actives?)

2.8. A new product is launched in Smalltown and evolves according to the usual SIR equations

$$S' = -aSI,$$
$$I' = aSI - bI,$$
$$R' = bI.$$

It appears that this product keeps customers active for 20 days, and that the initial populations of Potentials, Actives, and Rejecteds was

$$S(0) = 10{,}000,$$
$$I(0) = 100,$$
$$R(0) = 0.$$

One day later, it was observed that there were 20 *new* customers.
   (a) Use the given information to complete the model. That is, find the transmission and attrition coefficients a and b.
   (b) Using $\Delta t = 2$, estimate $S(4)$, $I(4)$, and $R(4)$.
   (c) Find the threshold and explain its significance.
   (d) Suppose that upon becoming active, customers are given personal customer support and continual product updates that extend the time they stay active to an average of 30 days. What effect will this have on the threshold for S? Do you think the improvement in the threshold is likely to be worth the extra cost of the intervention?

2.9. A company has released a new game app and has created a SIR model to predict the evolution of the market for this game. The SIR equations are

$$S' = -.0000001 SI,$$
$$I' = .0000001 SI - I/30,$$
$$R' = I/30.$$

The initial values at $t = 0$ are thought to be

$$S(0) = 2{,}000{,}000, \qquad I(0) = 21{,}000, \qquad R(0) = 15{,}000.$$

   (a) Using the initial rates, estimate $S(1)$, $I(1)$, and $R(1)$.
   (b) Using the SIR equations, calculate the rates of change $S'(1)$, $I'(1)$, and $R'(1)$, and then use these values to estimate $S(2)$, $I(2)$, and $R(2)$.
   (c) Using the values of $S(2)$, $I(2)$, and $R(2)$ that you computed in part (b), calculate the rates of change $S'(2)$, $I'(2)$, and $R'(2)$. Then estimate $S(3)$, $I(3)$, and $R(3)$.

(d) Go back to the starting time $t = 0$ and to the initial values

$$S(0) = 2{,}000{,}000, \qquad I(0) = 21{,}000, \qquad R(0) = 15{,}000.$$

Recalculate the values of $S(2)$, $I(2)$, and $R(2)$ by using a time step of $\Delta t = 2$. This only requires a single round of calculations, using $S'(0)$, $I'(0)$, and $R'(0)$. How do your answers compare to those computed in part (b)? Which estimates do you think are most accurate, those in part (b) or those in part (d)? Why?

A **product bust** is when a product never catches on, with the number of active users never rising above the initial number. This can happen if the transmission coefficient is too small or if the attrition coefficient is too large.

2.10. In this exercise we consider variations of the situation in Exercise 2.8.
  (a) Suppose that the product only keeps customers active for an average of 4 days instead of 20. Will a product bust occur?
  (b) Suppose, in addition to the change in b, that $a = 0.00005$. Now will a product bust occur?
  (c) In an SIR model where $S(0) = 40{,}000$ and $b = 1/14$, how big does a need to be to avoid a product bust?
  (d) In a product bust, are there any new customers? Explain.

2.11. (a) Construct a SIR model for a product that keeps people's interest for an average of 20 days, where an average potential customer meets about 0.01% of the active population each day, and where it takes eight contacts with an Active before a Potential becomes Active.
  (b) How many Potentials are needed to avoid a product bust?

2.12. One way to create more interest in a product is to provide rewards for Actives who refer Potentials. By creating an incentive for Actives to reach out to Potentials, a reward program can increase the transmission coefficient.
  (a) Suppose, in the setting of Exercise 2.9, that a reward program is put into effect that doubles the chance that a Potential will become Active. What is the new transmission coefficient?
  (b) Compute the new threshold if the reward program in part (a) is put into effect, and compare it to the original threshold in Exercise 2.9.
  (c) Suppose that, for a different product, $S(0) = 20{,}000$, $b = 1/14$, and $a = 0.000003$. In order to be profitable, this product must have a threshold under 15,000. Will this product be profitable?
  (d) Suppose that, for the product in part (c), the company initiates a reward program, hoping to bring the threshold under 10,000. By what factor must the reward program alter the transmission coefficient to achieve this goal?

2.13. Consider two products with the same transmission coefficient a and the same initial conditions. They differ only in the average length of time that someone stays Active. Which product has the higher threshold level: the one with the higher average usage time or the lower average usage time? Explain.

2.14. **Product Trolls**. Imagine the evolution of a product in a population with the following categories of people: Potential users (S), Active users (I), Rejected

users (R), and Product Trolls (T). Product Trolls are individuals who continually give negative feedback about the product. Assume the following.
- Trolls never become Actives.
- Once a user becomes Rejected, that user never becomes Potential again.
- Potentials become Active through contact with Actives, as in the SIR model. The rate of transmission is α.
- Actives become Rejected through contact with a Troll (instead of with the passage of time). The rate of contact of the average Active with the average Troll is β.
- Potentials and Rejecteds can become Trolls through the passage of time. This happens with rate constants γ for Potentials and δ for Rejecteds.
- The system is closed. Changes in population only happen when a person moves from one class to another. This also means that the total population $N = S + I + R + T$ is constant.

(a) Draw a diagram describing the interactions between these different classes of people.
(b) Write down a system of rate equations that describes these interactions.

2.15. **Herd Immunity.** Suppose there is a measles outbreak in a small community of 40,000 people, almost all of them Susceptible, with $b = 1/(14 \text{ days})$ and $a = 0.000003$.

(a) What is the threshold in this problem? How many people will get sick before the epidemic reaches its peak?
(b) Now suppose that 10,000 people (out of 40,000 total) are immune to measles, either because they already had the disease or because they were vaccinated. How many will get sick before the epidemic reaches its peak?
(c) If enough people are immunized, the epidemic will fizzle from the start, just like a product bust. How many people would have to be immunized for that to happen?
(d) When enough people are immunized to keep outbreaks from spreading, the population is said to have **herd immunity**. Explain how the number of vaccinations needed for herd immunity is related to the threshold b/a.
(e) If you were a public health official, what would you say to somebody who doesn't want to get vaccinated and who insists that their (not) getting vaccinated is none of your business?

2.16. **Replication Number.** In an epidemic, the ratio of how many people get sick to how many recover is $aSI/(bI) = aS/b$. At the beginning of an epidemic, when S is almost the entire population, this is the same as the **basic replication number** $R_0 = \beta/\gamma$ that we saw in our scaled SIR model. If $R_0 > 1$, then the epidemic will grow, at least at first. If $R_0 < 1$, it will fizzle out.

(a) Suppose that the total population is 40,000. If $R_0 = 1.2$, what is the threshold? How many people will get sick before the epidemic reaches its peak? What if $R_0 = 10$? In general, how are R_0 (the total population) and the threshold related?
(b) In a town with total population T and a disease with basic replication number R_0, how many people would need to become immune to achieve herd immunity?

Food for thought: The number of people needed to reach herd immunity is the same as the number of people who have to get sick *before* the epidemic reaches its peak. However, a substantial number of people can still get sick *after* the peak. For instance, suppose that $R_0 = 2$. The peak will be reached when half the population has gotten sick, but nearly 30% of the population will get sick later. Only about 20% will avoid the disease altogether. This is one reason why it is much better to immunize people before an epidemic than to wait for herd immunity during an epidemic.

2.17. Consider a product evolution that progresses according to the usual SIR model, *except* that Rejecteds become Potentials again after an average of c days. (*"I haven't played that game in a while. Maybe I should play it again."*) After that, they can become Active, just like the other Potentials. Modify the usual SIR equations to take this new feature into account. Your equations should involve three unspecified parameters: a, b, and c.

Exponential Growth. In many cases, the rate of change of a quantity is proportional to the quantity itself. We will see many different examples of that in Chapter 6, but the two simplest are money earning interest and population growth. We will examine these in Exercises 2.18–2.20.

If a bank account containing $10,000 earns $150/year in interest, then an account at the same bank containing $20,000 should earn $300/year in interest. Either way, the interest is 1.5%/year of the bank balance.

If a city of 100,000 is growing at a rate of 3000 people per year, then we would expect a similar city of 200,000 (or even the same city a few years later) to grow by 6000 people per year. That is, if $P(t)$ is the population at time t, then the **net growth rate** P' is proportional to P:

$$P' = kP,$$

where k is a constant. In this example, $k = .03$/year, or 3%/year.

2.18. In the equation $P' = kP$ for the population of a city, the number k is called the **per capita growth rate**, with the Latin phrase "per capita" literally meaning "per head". Explain why the units for k are 1/year.

2.19. In 2020, the population of Australia was 25 million and was growing at 1.36%/year, while the population of the United States was 330 million and was growing at 0.71%/year. Assume that both countries' populations keep growing at these per capita rates.
   (a) Let $A(t)$ and $U(t)$ be the populations of Australia and the United States, respectively. Write down the rate equations that govern the growth of these functions.
   (b) What were the net growth rates $A'(2020)$ and $U'(2020)$ in 2020?
   (c) In general, does a country with a higher per capita growth rate necessarily have a higher net growth rate?
   (d) On average, how many seconds did it take for the population of Australia to increase by 1 in 2020? For the population of the United States? (A year is about 31.6 million seconds.)

Food for Thought: At those per capita growth rates, the populations of Australia and the United States would become equal in 2417, at which time each country would have about 4.3 *billion* people. Exponential growth is great when it comes to money, but it can be scary in other cases.

2.20. A bank account is earning interest at a constant percentage rate. When the account had $3200, it was growing at a rate of $79/year.
   (a) Write an equation that links the net growth rate (meaning P', not the interest rate) to the bank balance.
   (b) At a later time, the bank balance is growing at $100/year. What is the balance at that time?

Scaling the Time Variable. In a rate equation, the quantity Y' has units, namely whatever units we use for Y divided by whatever units we use for time. If Y is people and t is measured in days, then Y' is measured in people/day. In Section 2.6, we generated a scaled model, where S, I, and R were replaced with dimensionless quantities s, i, and r that are proportional to S, I, and R. In Exercises 2.21–2.23, we will see what happens when we similarly replace the time t with a dimensionless variable.

2.21. Suppose that $i' = 0.014$/week. That is, i is increasing at a rate of 0.014 per week.
   (a) How fast is i increasing per *day*? How fast is i increasing per *year*? Which is bigger, the growth per year or the growth per day?
   (b) Let t_d, t_w, and t_y be the number of days, weeks, and years since a fixed starting time. How are the variables related? Which is bigger, t_d or t_y?

To make a model dimensionless, we pick a fixed unit of time T. T might be a second, an hour, a day, a year, or 3723.849 years. We then measure time in units of T. That is, we define a new dimensionless variable $\tau = t/T$ that counts how many T's have elapsed since $t = 0$. If T is small, then τ will be big, while if T is big, then τ is small. A very *large* number of nanoseconds equals a very *small* number of centuries. For variables Y and P that change in time, we let $\dot{Y}$ and $\dot{P}$ denote the changes in Y and P *per change in τ*.

2.22. How is $\dot{Y}$ related to Y'? (*Hint*: If Y' is constant, how much will Y change by the time that $\tau = 1$?)

2.23. Suppose that $P' = rP$. Let $T = r^{-1}$.
   (a) Compute $\dot{P}$ in terms of P.
   (b) Compute T when $r = 1\%$/year, when $r = 3\%$/month, and when $r = 0.06$/minute.

Food for Thought: You should discover that the trajectory of $P' = rP$ looks essentially the same for all of these values of r, only we trace out that trajectory at different speeds. If we can understand the solution to $P' = rP$ for one value of the parameter r (say, for $r = 1$), then we can understand the solution to $P' = rP$ for all values of r.

2.24. Consider the scaled SIR equations (2.17). Let $T = \gamma^{-1}$. Write down a system of equations for $\dot{s}$, $\dot{i}$, and $\dot{r}$.

Food for Thought: The resulting equations still have one dimensionless parameter, namely the basic replication number R_0. The details of a product launch or epidemic depend on a lot of different parameters and initial conditions, but the general shape of the curve only depends on R_0.

Chapter 3

Close is Good Enough!

In Chapter 2, we used a lot of approximations to understand product adoption. First, we made approximations in building the SIR model, pretending that all Potentials are the same, all Actives are the same, and all Rejecteds are the same. Then, when we couldn't find a formula for the solution, we used linear approximations and baby steps to get (approximate) numerical solutions. And that was good enough.

In this chapter we're going to flesh out the ideas of linear approximation that we began in Chapter 2. Before we can do that, we need to remember how linear functions behave. Before we can do *that*, we need to understand how functions behave in general. And before we can do *that*, we need to get comfortable with the whole idea of making approximations. So let's get started!

3.1. The Idea of Approximation

Back in algebra class, you learned how to solve cut-and-dried problems such as

$$\text{Find all solutions to } x^2 - 3x + 2 = 0.$$

There are several ways to do this. We begin by defining $f(x) = x^2 - 3x + 2$.

(1) We can factor $f(x)$ as $(x-1)(x-2)$. If $(x-1)(x-2) = 0$, then either $x - 1 = 0$ or $x - 2 = 0$, so our solutions are $x = 1$ and $x = 2$.

(2) We can apply the quadratic formula.

$$(3.1) \qquad x = \frac{-b \pm \sqrt{b^2 - 4ac}}{2a} = \frac{3 \pm \sqrt{(-3)^2 - 4(1)(2)}}{2} = \frac{3 \pm 1}{2} = 1 \text{ or } 2.$$

(3) We can plot a few points to get a feel for how f behaves.

x	$f(x)$
0	2
1	0
2	0
3	2

Since $f(1) = f(2) = 0$, $x = 1$ and $x = 2$ are solutions. A quadratic function can only have two roots, so there are no other solutions, and we're done.

All of these strategies worked because *the problem was very simple*. Unfortunately, the real world rarely gives us such simple problems. It gives us complicated problems with messy numbers that we have to make sense of anyway. Most of the time, there isn't a method like factoring that gives us the answers. There isn't a formula that gives us the answers. The answers are messy, so there's almost no hope of just guessing them.

For instance, let's tackle the problem

Find all solutions to $x^5 + x - 31 = 0$.

We can't factor the polynomial. There is no formula for the roots of a fifth-order polynomial. There aren't any simple roots. So what can we do?

The answer is that we can still plot a few points. As before, we define $f(x) = x^5 + x - 31$, and we get a table of values:

x	$f(x)$
0	−31
1	−29
2	2
3	215

Since $f(1)$ is negative and $f(2)$ is positive, there must be a root between 1 and 2. What's more, $f(2)$ is a lot closer to 0 than $f(1)$ is, so that root is likely to be close to 2.

So let's try 1.9. We compute $f(1.9) = 1.9^5 + 1.9 - 31 = -4.33901$ and add it to our table.

x	$f(x)$
0	−31
1	−29
1.9	−4.33901
2	2
3	215

3.1. The Idea of Approximation

Now we know that our root is somewhere between 1.9 and 2, and probably closer to 2 than 1.9 (Why?) So we try 1.96 and 1.97 and get

x	$f(x)$
0	-31
1	-29
1.9	-4.33901
1.96	-0.11453
1.97	0.64093
2	2
3	215

Now we know that our root is somewhere between 1.96 and 1.97, and probably a lot closer to 1.96 than 1.97.

Graphing our results helps us to visualize what is going on. By plotting points between $x = 0$ and $x = 2$, as in the left-hand graph in Figure 3.1, we see that our root is between 1.8 and 2, and looks to be between 1.9 and 2. By plotting points between 1.9 and 2, as in the right-hand graph in Figure 3.1, we see that our root is between 1.96 and 1.97. Some would say that we still haven't solved the problem, because we still don't know what the answer is *exactly*. But we've actually found the answer to two decimal places. It's 1.96! By plugging in a few more numbers, we can figure out that the answer is between 1.961 and 1.962, in other words that to three decimal places it's 1.961. Then that to four decimal places it's 1.9615. Then that to five decimal places it's 1.96153. No matter how much accuracy we need, this simple method can get it for us.

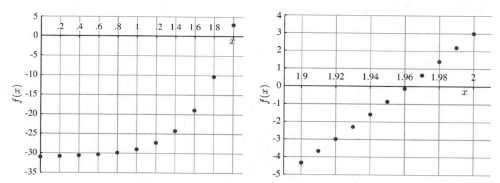

Figure 3.1. The graph of $f(x)$ on two different scales

Lessons from this example

- We didn't use any fancy mathematical tools. Having good tools is useful, of course, and we're going to develop some great tools in the rest of the book, but most math can be done with the tools you learned in middle school.

- We didn't do anything especially clever or elegant. Cleverness and elegance are overrated! In time we'll learn some clever methods that give us more accuracy with less effort, but you don't need to be clever to do good math.
- Having the right mindset *is* important. *Close is good enough!* If you can't find an exact answer, find a way to get an approximate answer. Then find a way to improve that answer, and improve it again, until you have as much accuracy as you need.
- The most important thing is *perseverance*. In this problem, there was no way to find the answer in one step, or even in several steps, but we didn't let that stop us. Instead, we solved the problem to several decimal places, one step at a time.
- Computers aren't clever, and the mathematical tools they have aren't fancy. However, they have incredible perseverance, and will do millions of calculations for us every second. We just have to understand the procedure well enough to ask them to do it. That is, we have to write a program. We'll have a lot to say about using computers to solve rate equations later in this chapter.

The same lessons apply to almost everything in calculus. We almost never can get an answer directly, so we almost always look for an approximate solution. Then we look for ways to improve that approximation and home in on the true solution. More accuracy usually means more work, but with enough perseverance we can obtain as much accuracy as we need. If we're too lazy to do all that work ourselves (and who isn't?), we can ask computers to do it for us, but only after we've understood the procedure well enough to explain it to them.

3.2. Functions

A **function** is a rule that assigns one and only one output to every possible input, as shown in Figure 3.2.

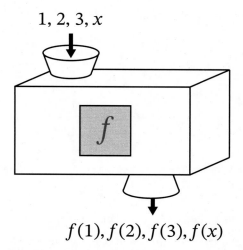

Figure 3.2. A function viewed as a machine for turning inputs into outputs.

3.2. Functions

The inputs can be, well, anything, and so can the outputs. For instance, the function *SC* has states as inputs and cities as outputs, and thus

$$SC(\text{Alabama}) = \text{Montgomery}$$
$$SC(\text{Alaska}) = \text{Juneau}$$
$$SC(\text{Arizona}) = \text{Phoenix}$$
$$\vdots \quad \vdots$$
(3.2) $$SC(\text{Wyoming}) = \text{Cheyenne}$$

You may have already guessed that $SC(\text{New York}) = \text{Albany}$, that $SC(\text{Texas}) = \text{Austin}$, and that *SC* stands for "state capital".

Most of the time, however, our inputs and outputs will be ordinary numbers. (Occasionally the inputs and outputs will be pairs or triples of numbers.) Such a numerical function can be described in four ways:

(1) **As a procedure for computing the output in terms of the input.** For instance, there is a function *SQ* where the output is what you get when you multiply the input by itself.

There is also a function *DJ* whose input is a date and whose output is the closing value of the Dow Jones Industrial Average on that date. Computing *DJ* for a date in the past is easy: just look up the records of the New York Stock Exchange. Computing *DJ* for a future date is a lot harder and involves waiting for that date!

(2) **As a formula.** We give the input a name (usually x or t), and we try to find a formula for the output. In the case of *SQ*,

$$SQ(x) = \text{What you get when you multiply } x \text{ by itself}$$
(3.3) $$= x^2.$$

Of course, we could just as well have called the input variable something else and written $SQ(y) = y^2$ or $SQ(t) = t^2$ or $SQ(\text{George}) = \text{George}^2$. (We don't usually use human names for variables, but there's no reason why we couldn't.) It's all the same function that takes the input and squares it.

Warning! Not every function has a formula. There simply isn't any formula for $DJ(t)$, neither for the past nor the future. Even when looking at the past, the fluctuations of the stock market are too complicated to be described by mathematical formulas.

(3) **As a table.** We just make a list of inputs and outputs, as with *SC*, and write something like

x	−3	−2	−1	0	1	2	3
$SQ(x)$	9	4	1	0	1	4	9

In this case, the table doesn't describe the function completely, since it doesn't tell us what $SQ(7)$ or $SQ(\pi)$ or $SQ(-24)$ are. There are infinitely many possible inputs, and we can only fit a finite number of rows in the table! We have to use our imagination to figure out what $SQ(x)$ is when x isn't on the list.

In many real-world applications, however, we don't know the formula for the function, but we have measured data. For instance, the function might give the annual profit of a certain business as a function of what year it is, or the high temperature in a certain city as a function of the date. Or we could be talking about $DJ(t)$. In those cases, the table *is* the function.

(4) **As a graph.** The graph of a function is the set of all ordered pairs (x, y), where x is an input and y is the corresponding output. This gives a curve in the plane that passes the **vertical line test**, which says that a vertical line can only hit the curve once. It contains a lot more information than a table, since it shows what happens for every possible input x, at least within the interval that we are graphing. For instance, Figure 3.3 shows what SQ does when the input x is between -5 and 5. If we want to know what $SQ(3.728)$ is, we just have to look closely to see where the vertical line $x = 3.728$ hits the graph.

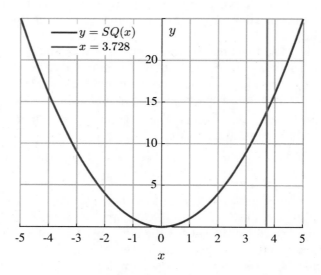

Figure 3.3. The graph of SQ

If we have two ordinary functions—let's call them f and g—that take numbers as inputs and outputs, then we can combine them by feeding the output of one function to the other function. We use the symbol $f \circ g$ to mean the function "apply g to the input, and then apply f to that". In other words,

(3.4) $$(f \circ g)(x) = f(g(x)).$$

Be careful with the order. **We read from right to left, applying first g and then f.** Likewise, if we had a function $f \circ g \circ h$, then we would first apply h, then g, then f.

3.3. Linear Functions and Their Graphs

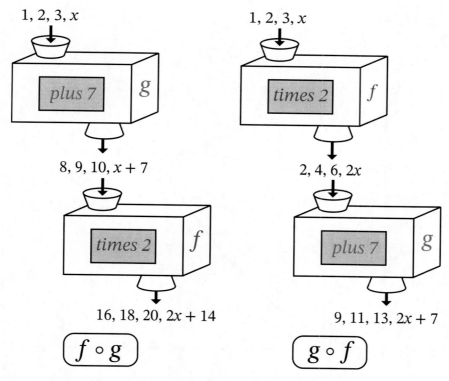

Figure 3.4. The functions $f \circ g$ and $g \circ f$ for $f(x) = 2x$ and $g(x) = x + 7$

The order matters, as you can see in Figure 3.4. If f is the "multiply by 2" function, described by the formula $f(x) = 2x$, and if g is the "add 7" function, described by the formula $g(x) = x + 7$, then

(3.5)
$$\begin{aligned} f \circ g(x) &= f(x+7) &= 2(x+7) = 2x + 14, \\ g \circ f(x) &= g(2x) &= 2x + 7. \end{aligned}$$

Those are different functions!

3.3. Linear Functions and Their Graphs

If f is a function and x_1 and x_2 are two possible inputs, we are often interested in how fast the output changes as the input goes from x_1 to x_2. Let $y_1 = f(x_1)$, and let $y_2 = f(x_2)$, so (x_1, y_1) and (x_2, y_2) are two points on the curve $y = f(x)$. We use the Greek letter Δ to indicate the change in a quantity, so

(3.6)
$$\begin{aligned} \Delta y &= y_2 - y_1 &= \text{Change in } y, \\ \Delta x &= x_2 - x_1 &= \text{Change in } x. \end{aligned}$$

We often call Δy the **rise** and Δx the **run**, even though these quantities may in fact be negative.

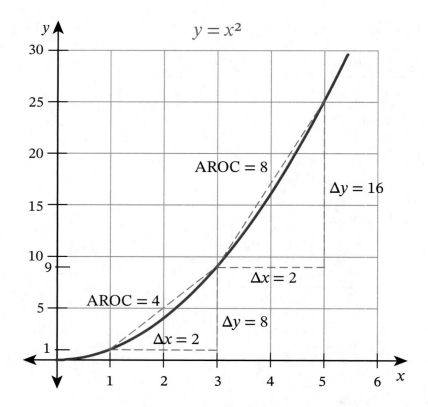

Figure 3.5. The rate of change of a typical function is not constant.

The ratio $\Delta y/\Delta x$ measures how fast our function is changing. We call it the **average rate of change**, or AROC, of the function between x_1 and x_2. For instance, in Figure 3.5 the AROC between 1 and 3 is

$$(3.7) \qquad \frac{\Delta y}{\Delta x} = \frac{9-1}{3-1} = 4,$$

while the AROC between 3 and 5 is

$$(3.8) \qquad \frac{\Delta y}{\Delta x} = \frac{25-9}{3-1} = 8.$$

The function is increasing faster between 3 and 5 than between 1 and 3.

A **linear function** is a function whose rate of change is constant, as in Figure 3.6. That is, there is a number m such that, for any two points, $\Delta y/\Delta x = m$. For instance, if $y = f(x) = 4x$, then y always changes four times as much as x changes. The same is true if $y = f(x) = 4x + 5$. The number m goes by many names: it is the **slope** of the line $y = f(x)$ or the **rate of change** or the **multiplier** of the function f.

3.3. Linear Functions and Their Graphs

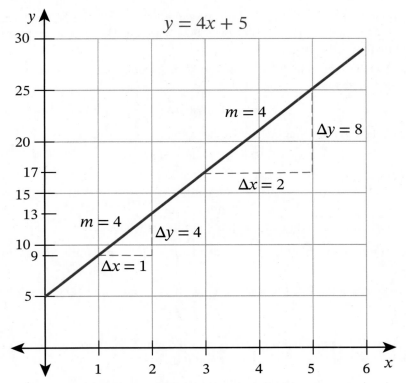

Figure 3.6. The rate of change of a linear function *is* constant.

Most questions about linear functions have two parts. First, we want to figure out the multiplier from data. Second, we want to use that multiplier to find the equation of the line.

For instance, suppose that we are driving along a highway at constant speed. At 1:00 pm, we reach mile marker 132. At 3:00 pm we reach mile marker 262. How fast are we going? Where will we be at 4:00 pm? Where will we be at time t? If our destination is at mile marker 457, when will we get there?

Our input variable is the time t, measured in hours, and $f(t)$ is our position at time t, measured in miles. We are told that $f(1) = 132$ and $f(3) = 262$, so our rate of change is

$$(3.9) \qquad m = \frac{\Delta y}{\Delta t} = \frac{262 \text{ mi} - 132 \text{ mi}}{3 \text{ hr} - 1 \text{ hr}} = \frac{130 \text{ mi}}{2 \text{ hr}} = 65 \text{ miles/hour}.$$

To figure out where we'll be at 4:00, we compare 4:00 to 1:00. We have $\Delta t = 4 - 1 = 3$ (hours), so $\Delta y = 65(3) = 195$ (miles). That is, at 65 miles/hour we will drive 195 miles between 1:00 and 4:00, putting us at mile marker 132+195=327. At time t we will have gone $65(t-1)$ miles since 1:00, so we will be at mile marker $132 + 65(t-1)$. That's the equation of our position function,

$$(3.10) \qquad f(t) = 132 + 65(t-1),$$

where $f(t)$ is measured in miles and t is measured in hours. If we prefer, we can do a little algebra to rewrite $f(t)$ as $65t+67$ or as $262+65(t-3)$. The formulas look different, but the function is the same.

So far we have been using m to convert known values of Δt into values of Δy. We can also do the reverse:

$$\text{(3.11)} \qquad \Delta y = m\Delta t, \quad \text{so } \Delta t = \frac{\Delta y}{m}.$$

When we reach our destination, we will have gone

$$\text{(3.12)} \qquad \Delta y = (457 - 132) \text{ miles} = 325 \text{ miles}.$$

That takes us

$$\text{(3.13)} \qquad \Delta t = \frac{\Delta y}{m} = \frac{325 \text{ miles}}{65 \text{ miles/hour}} = 5 \text{ hours},$$

so we'll reach our destination at 1:00 pm + 5 hours = 6:00 pm.

Let's summarize what we've learned, and apply it to an arbitrary linear function $f(x)$.

- Two points determine the slope of a line. If $f(x_1) = y_1$ and $f(x_2) = y_2$, then

$$\text{(3.14)} \qquad m = \frac{y_2 - y_1}{x_2 - x_1} = \frac{\text{rise}}{\text{run}}.$$

- A point and a slope give you the equation of a line. If our point is (x_1, y_1) and the slope is m, and if (x, y) is any other point, then

$$\text{(3.15)} \qquad y - y_1 = \Delta y = m\Delta x = m(x - x_1)$$

so

$$\text{(3.16)} \qquad y = y_1 + m(x - x_1).$$

This is called the **point-slope** form of the equation of the line. See Figure 3.7.

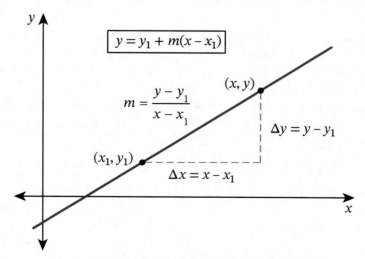

Figure 3.7. The point-slope form of the equation of a line

3.4. Linear Approximations and Microscopes

- Some people prefer to use a little algebra to rewrite the equation in **slope-intercept** form, as

$$(3.17) \qquad y = mx + b,$$

where $b = y_1 - mx_1$ is the value of y when $x = 0$, also known as the y-**intercept**. In other words, slope-intercept form is what you get when you write point-slope form with $x_1 = 0$. Most algebra classes stress slope-intercept form, but you should get used to using point-slope form, *which is much more useful for calculus*.

- The multiplier m is a conversion factor for getting Δy from Δx, but *it's also a conversion factor for getting Δx from Δy*.

$$(3.18) \qquad \Delta y = m \Delta x, \qquad \Delta x = \frac{\Delta y}{m}.$$

To figure out when a linear function reaches a certain value y, first compute $\Delta y = y - y_1$, then $\Delta x = \Delta y / m$, and finally $x = x_1 + \Delta x$.

3.4. Linear Approximations and Microscopes

Linear functions are perfect for describing quantities that change at a constant rate. They are also great for *approximating* functions that change at a *nearly* constant rate. If a curve doesn't curve much, then a line through any two points will do a good job of approximating the entire curve. This allows us to **extrapolate** and **interpolate** data.

For instance, consider the graph shown in Figure 3.8. As you can see, it is pretty close to a straight line. If I tell you that $f(1) = 1$ and $f(1.01) = 1.0201$, you can figure

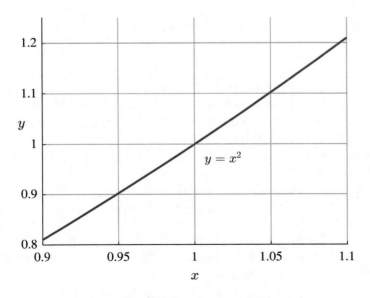

Figure 3.8. Zoomed in, most graphs look linear.

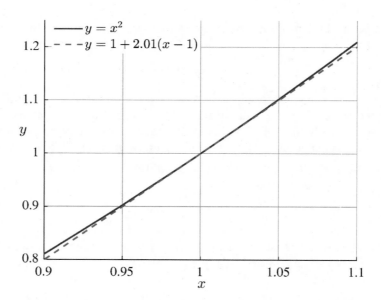

Figure 3.9. Linear approximations work well over short distances.

out the slope of that line.

(3.19) $$m = \frac{f(1.01) - f(1)}{1.01 - 1} = \frac{.0201}{.01} = 2.01.$$

Once you have the slope, you can write down the equation of the line in point-slope form:

(3.20) $$y = 1 + 2.01(x - 1).$$

Once you have the equation of the line, you can use it to approximate any point along the curve:

(3.21) $$f(x) \approx 1 + 2.01(x - 1),$$

as shown in Figure 3.9. For example, $f(1.005) \approx 1 + 2.01(0.005) = 1.01005$. Since 1.005 is halfway from 1 to 1.01, we approximate $f(1.005)$ as the value halfway from $f(1)$ to $f(1.01)$. Using known values of a function at two points x_1 and x_2 to estimate $f(x)$ for x between x_1 and x_2 is called **interpolation**.

We can use the same reasoning to estimate $f(x)$ when $x < x_1$ or when $x > x_2$. For instance, $f(1.03) \approx 1 + 2.01(.03) = 1.0603$ while $f(0.98) \approx 1 + 2.01(-.02) = 0.9598$. Extending our line past x_1 and past x_2 is called **extrapolation**.

Notice the advantage of using point-slope form instead of slope-intercept form. In point-slope form, the numbers $x - 1$ are small and easy to multiply by 2.01, and you can see what is going on. The equation $y = 2.01x - 1.01$ describes the same line in slope-intercept form, but it requires you to multiply 2.01 by 1.005 or 1.03 or 0.98, which is more complicated and much more likely to lead to arithmetic errors.

Linear approximations are not exact, of course. In this example, the function was our old friend $f(x) = x^2$. The exact value of $f(1.005)$ was 1.010025, not 1.01005, the

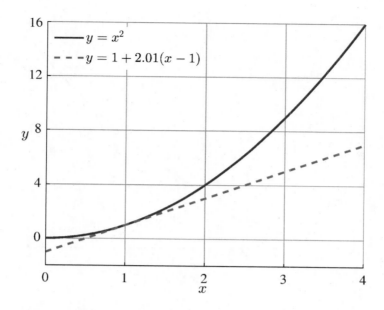

Figure 3.10. Over longer distances, linear approximations aren't nearly as good.

exact value of $f(1.03)$ was 1.0609 rather than 1.0603, and the exact value of $f(0.98)$ was 0.9604 instead of 9.9598. Not exact, but very, very close.

So did we just get lucky by studying a function whose graph didn't curve much? Not at all! As you can see in Figure 3.10, the graph of $y = x^2$ curves quite a bit, and is much steeper at $x = 3$ than at $x = 1$. The curve hugs the line near $x = 1$, but eventually veers away. As a result, the linear approximation $f(3) \approx 1 + 2(2.01) = 5.02$ doesn't come anywhere close to the true value of $f(3) = 9$.

However, $y = x^2$ doesn't curve much *between* $x = 0.9$ *and* $x = 1.1$. Most curves have the same property: if we zoom in on them enough, they begin to look straight. Linear approximations can be used for all sorts of functions, but we should *only use them over short distances*.

3.5. Euler's Method and Compound Interest

In the 1700s, the mathematician Leonhard Euler (pronounced "Oiler") invented a method for solving rate equations. We already used his method on the SIR equations in Chapter 2. Next we're going to apply Euler's method to a financial example, and then we'll develop the method in general.

Suppose that we invest $1000 in a fund that pays 6% interest per year. How much money will we have after t years?

Our contract with the bank says that our investment is supposed to grow at a rate that is 6% of whatever is there. In other words, our rate equation is

(3.22) $$y' = 0.06y.$$

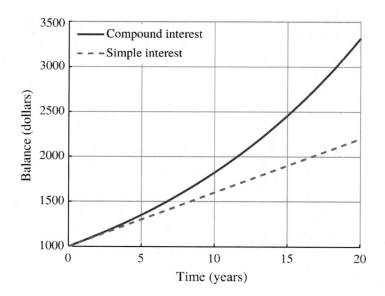

Figure 3.11. The actual solution to $y' = 0.06y$ and the linear approximation.

Since we started with $1000, our initial condition is $y(0) = 1000$. Putting these together, we have $y'(0) = 0.06(1000) = 60$ dollars/year. The equation of the line through $(0, 1000)$ with slope 60 is

(3.23) $$y(t) = 1000 + 60t.$$

This equation says that the bank gives us $60/year forever, which is what would happen if the account earned **simple interest**. According to this **linear approximation** shown by the dashed line in Figure 3.11, we should have $1060 after one year, $1120 after 2 years, $1600 after 10 years, and $2800 after 30 years.

But that's not how investments actually work. In the second year, we will have more than $1000 in our fund, since we will have the $1000 that we started with, plus the interest we earned in the first year. That means that we should earn more than $60 in interest in the second year. We should earn even more the third year, and so on. This is called **compound interest**.

To take that into account, we can update our estimate of y' every year. That is, we can:

(1) Use $y(0) = 1000$ and the rate equation $y' = 0.06y$ to compute $y'(0) = 60$.
(2) Use the linear approximation $y(t) \approx y(0) + y'(0)t$ to approximate $y(1)$, getting $y(1) \approx \$1060$.
(3) Use $y(1) \approx 1060$ and the rate equation to estimate $y'(1) \approx 63.6$.
(4) Use the linear approximation $y(t) \approx y(1) + y'(1)(t-1)$ to approximate $y(2) \approx 1123.6$.
(5) Repeat the process, one year at a time, for as many years as we hold the investment.

3.5. Euler's Method and Compound Interest

Applying this for 5 years generates the data shown in Table 3.1. The time step is $\Delta t = 1$ year. The first column is the time, and the second column is our bank balance. The third is the rate at which we are earning interest, which is 0.06 times the second column. The fourth is how much interest we earn in one time step (which looks like the third column because $\Delta t = 1$, but if we were working 2 years at a time, this column would be twice the third column.) The fifth is our balance at the end of the year, which also gives us our starting balance the next year.

Table 3.1. Computing our balance after 5 years of annual compounding

t	y	y'	y'Δt	y + y'Δt
0	1000.0	60.0	60.0	1060.0
1	1060.0	63.6	63.6	1123.6
2	1123.6	67.4	67.4	1191.0
3	1191.0	71.5	71.5	1262.5
4	1262.5	75.7	75.7	1338.2
5	1338.2			

In fact, this is how banks used to compute interest, with updates done yearly. This interest is said to be **compounded annually** or **compounded yearly**. If instead we calculated interest every six months, our interest would be **compounded semi-annually**. As you can see from Table 3.2, semi-annual compounding is better than annual compounding, but the difference is fairly small. Monthly compounding is a little bit better than semi-annual compounding, and daily compounding is a tiny bit better than monthly compounding. These are all approximations to the solution of the rate equation, which is called **continuous compounding**. But they're very good approximations! If we used our procedure with a time step of one month to estimate the result of 20 years of continuously compounded interest, we would only be off by $9.90.

Table 3.2. 20-year earnings from $1000 invested at 6% interest with simple, annual, semi-annual, monthly, and continuous interest

Year	Simple	Annual	Semi-annual	Monthly	Continuous
0	1000.0	1000.0	1000.0	1000.0	1000.0
5	1300.0	1338.2	1343.9	1348.9	1349.9
10	1600.0	1790.8	1806.1	1819.4	1822.1
15	1900.0	2356.6	2427.3	2454.1	2459.6
20	2200.0	3207.1	3262.0	3310.2	3320.1

More generally, suppose that we have a rate equation

(3.24) $$y' = f(y, t),$$

where $f(y, t)$ is some fixed function of y and t. (In the banking example, we had $f(y, t) = 0.06y$, but it could just as well have been $f(y, t) = 0.06y + t$.) We start at a particular time t_0, with initial condition $y = y_0$. There is a later time t_f, and we want to estimate $y(t_f)$.

(1) Pick a number of steps N, which makes our **time step** $\Delta t = (t_f - t_0)/N$.

(2) Make a table similar to Table 3.1 with $N + 1$ rows and with at least three columns, labeled t, y, and y'. It's convenient (but not necessary) to have two more columns, labeled $y'\Delta t$ and $y + y'\Delta t$. Enter t_0 and y_0 in the first row.

(3) Fill out the table one row at a time. Once we have t and y, the rate equation tells us y', and we can then figure out $y'\Delta t$ and $y + y'\Delta t$.

(4) Start on the next row. The new t is the previous t plus Δt. The new y is the previous y plus the previous $y'\Delta t$. This is our linear approximation.

(5) Repeat until the entire table is filled out. Then read off our final answer $y(t_f)$. If we want to see the whole progression, we can plot the y column against the t column.

(6) If we want more accuracy, we can do it all over again with N larger and Δt smaller.

3.6. The SIR Model by Computer

We now revisit the SIR model that we discussed in Chapter 2.

If this book were written in 1940, there would be nothing more to say. We have a nice model describing market penetration and the spread of disease. In principle we know how to solve rate equations by Euler's method, but in practice it takes a lot of work. Computing ten years of compound interest with $\Delta t = 1$ requires ten turns of the crank. Doing it with $\Delta t = 1/2$ requires 20 turns, and doing it with $\Delta t = 0.1$ requires 100 turns. Solving the SIR equations, with three different variables to keep track of, is even messier. Who has the patience for all that?

Fortunately, we live in the 21st century, and we have computers to do our chores for us. With the right instructions, a computer can do the calculation with $\Delta t = 0.1$ or even $\Delta t = 0.01$ in the blink of an eye. We just need to understand the SIR rate equations, and Euler's method for solving them, well enough to give the computer its instructions—in other words, we need to write a program.

We begin with a clear step-by-step description of the algorithm that we can turn into computer code.

(1) Specify the parameters of our model, namely the transmission coefficient a and the attrition coefficient b.

(2) Specify our initial conditions, that is, the values of S, I, and R at time $t = 0$.

(3) Decide how far into the future we are going to run. We'll call the final time t_f. It's also possible to set a negative value of t_f, in which case we're using the SIR model to understand the past.

(4) Decide how many steps we are going to take. Call this number N. Our **step size** will then be $\Delta t = t_f/N$.

(5) Generate a list of times $t_0 = 0$, $t_1 = \Delta t$, $t_2 = 2\Delta t$, up through $t_N = N\Delta t = t_f$. Note that there are $N + 1$ different times that we care about, not N.

3.6. The SIR Model by Computer

(6) Make a table with seven columns and $N+1$ rows. Label the columns t, S, I, R, S', I', and R'. (Some people like to add three more columns labeled $S'\Delta t$, $I'\Delta t$, and $R'\Delta t$, and maybe even three after that for $S + S'\Delta t$, $I + I'\Delta t$, and $R + R'\Delta t$.) Put the list of times in the first column. Fill in the second, third, and fourth entries of the first row with $S(0)$, $I(0)$, and $R(0)$.

(7) Use the SIR equations to fill out the rest of the first row. That is, compute $S'(0) = -aS(0)I(0)$, $I'(0) = aS(0)I(0) - bI(0)$, and finally $R'(0) = bI(0)$.

(8) Use a linear approximation to figure out the values of S, I, and R at the next time. We'll call these S_{new}, I_{new}, and R_{new}, and call the previous values S_{old}, etc. This involves adding $S'_{old}\Delta t$, $I'_{old}\Delta t$, and $R'_{old}\Delta t$ to S_{old}, I_{old}, and R_{old}. Write down the new values on the next row.

If we aren't interested in keeping track of S', I', and R', we can combine steps (7) and (8), writing

$$\begin{aligned} S_{new} &= S_{old} - aS_{old}I_{old}\Delta t, \\ I_{new} &= I_{old} + (aS_{old}I_{old} - bI_{old})\Delta t, \\ R_{new} &= R_{old} + bI_{old}\Delta t. \end{aligned}$$

(9) Repeat steps (7) and (8) (or the combined step) N times. At this point, the first N rows of our table will be filled in completely, and the last row will feature t, S, I, and R. Congratulations! We're done!

```
function [S,I,R,t] = SIREulers(tmax,N,a,b,S0,I0,R0)
dt = tmax/N;                                  % Compute Δt
S = zeros(1,N+1);                             % Start with empty
I = zeros(1,N+1);                             % vectors for
R = zeros(1,N+1);                             % S, I and R.
S(1) = S0;                                    % Initialize S
I(1) = I0;                                    % Initialize I
R(1) = R0;                                    % Initialize R
for n = 1:N                                   % Start the loop
    S(n+1) = S(n) - a*S(n)*I(n)*dt;           % Rate eq.for S
    I(n+1) = I(n) + (a*S(n)*I(n) - b*I(n))*dt; % Rate eq. for I
    R(n+1) = R(n) + b*I(n)*dt;                % Rate eq. for R
end                                           % End the loop
t = linspace(0,tmax,N+1);                     % A vector of times
plot(t,S,'red', t,I,'blue', t,R,'green')      % Graph the results
title('Plot of SIR model behavior')           % Add a title
ylabel('Individuals'), xlabel('Days')         % Add labels
legend('Potentials','Actives','Rejecteds')    % Add a legend.
```

In the program code above we see how that looks with an actual program, called SIREulers. The programs in this book are all written in MATLAB, which includes some simple graphing routines for seeing our output. However, if you prefer to work with another programming language, that's fine, too.

There are a few differences between this program and the algorithm above.

(1) The program is written to be flexible. Instead of having lines that specify t_f, N, a, b, $S(0)$, $I(0)$, and $R(0)$, we type in these parameters when we use the program. In fact, most of the data in Tables 2.1 and 2.2, where we studied the SIR model with $a = 0.000002$, $b = 1/30$, $S(0) = 40,000$, $I(0) = 2100$, and $R(0) = 7900$, were generated with SIREulers.[1] Most of the data for Table 2.1 was generated by typing

SIREulers(10,5,0.000002, 1/30, 40000, 2100, 7900),

and the data for Table 2.2 was generated by typing

SIREulers(-30,15,0.000002, 1/30, 40000, 2100, 7900).

To run it again with $N = 10$ or $N = 100$ or $N = 10{,}000$ steps, just replace "5" or "15" with "10" or "100" or "10000". By picking a big enough value of N, we can get as much accuracy as we want.

(2) The program doesn't generate tables directly. Rather, it generates arrays. In MATLAB, arrays are indexed starting at 1, so we have to make our index n run from 1 to $N + 1$ instead of from 0 to N. We just have to remember that $S(n)$, $I(n)$, and $R(n)$ represent the values of S, I, and R at time $t_{n-1} = (n-1)\Delta t$ and not at time t_n. In particular, our final answers are $S(N + 1)$, $I(N + 1)$, and $R(N + 1)$.

(3) The program combines steps (7) and (8) and doesn't actually keep track of S', I', and R'. (To create Table 2.1, those values were calculated after the fact.)

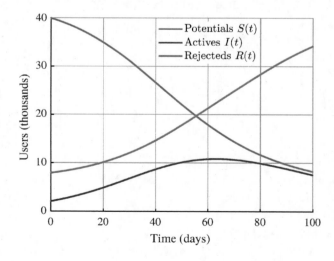

Figure 3.12. Graphs generated by SIREulers

---

[1] To use this program we must work in a directory that contains the file SIREulers.m. You are welcome to copy and use SIREulers.m, which is not a built-in part of MATLAB.

(4) The program doesn't just compute numbers. The last five lines of the program *graph* the output, with S, I, and R shown as differently colored lines on the same chart. Since SIREulers saves the vectors S, I, R, and t, all of the graphing can be done after we have run the program. For instance, if we only wanted to graph S we could just type `plot(t,S)`, and similarly for I and R. Graphing after the fact also allows us to vary the colors and format of the graph, for instance generating the graphs of Figure 3.12.

3.7. Solving Algebraic Equations

We close this chapter with two approximation methods for computing roots of algebraic equations. One is an elaboration of what we did at the beginning of the chapter to solve $x^5 + x - 31 = 0$. The other is a very efficient algorithm for computing square roots.

The bisection method. There is a common guessing game where one person thinks of a whole number between 1 and 100 and the second person tries to guess it. After each guess, the first person says whether the number is higher or lower than the guess. Here is a typical run.

> Alice: Guess my number.
> Bob: 50
> Alice: Lower!
> Bob: 25
> Alice: Lower!
> Bob: 12
> Alice: Higher!
> Bob: 18
> Alice: Higher!
> Bob: 22
> Alice: Lower!
> Bob: 20
> Alice: Right!

At each step, Bob guessed a number in the middle of the range of possible answers (sometimes rounding up or down). When Alice told him whether he needed to go higher or lower, that range was cut in half and he guessed in the middle again. 50 is in the middle of 1–100, 25 is in the middle of 1–49, 12 is in the middle of 1–24, 18 is in the middle of 13–24, 22 is in the middle of 19–24, and 20 is in the middle of 19–21. By shrinking the interval each time, Bob homed in on Alice's number. No matter what number Alice picked, Bob was guaranteed to get the answer in at most seven guesses.

Now suppose that we have a polynomial $p(x)$ and points a and b such that $p(a)$ and $p(b)$ have opposite signs. Polynomials are continuous and can't jump from positive to negative, so there has to be a **root** (that is, a point where $p(x) = 0$) somewhere between a and b. We then guess the midpoint, $c = (a + b)/2$, and compute $p(c)$. From the sign of $p(c)$ we know whether to go higher or lower. Like Bob, we narrow our interval and keep guessing in the middle until our interval is so small that any point is close enough.

Table 3.3. Solving $x^5 + x - 31 = 0$ by the bisection method

Lower	Upper	Guess	p(Guess)	Adjustment
1.00000	2.00000	1.50000	−21.906	Higher!
1.50000	2.00000	1.75000	−12.837	Higher!
1.75000	2.00000	1.87500	−5.9507	Higher!
1.87500	2.00000	1.93750	−1.7596	Higher!
1.93750	2.00000	1.96875	0.54566	Lower!
1.93750	1.96875	1.95312	−0.2655	Higher!
1.95312	1.96875	1.96093	−0.0449	Higher!
1.96093	1.96875	1.96484	0.24921	Lower!

For instance, suppose that $p(x) = x^5 + x - 31$. We already know that $p(1) = -29 < 0$ and $p(2) = 3 > 0$, which gives us a starting point. The result of our game is shown in Table 3.3.

In eight steps, we went from "somewhere between 1 and 2" to "somewhere between 1.96093 and 1.96484", an interval that is a factor of 256 smaller. Every ten steps improves our accuracy by a factor of $2^{10} = 1024$, in other words by about three decimal places. This is true not only when we solve $p(x) = x^5 + x - 31 = 0$, but when we find a root of any polynomial.

The Babylonian Method. More than 3000 years ago (scholars differ on the exact date), the ancient Babylonians invented a very efficient method for computing square roots using only addition, subtraction, multiplication, and division. Suppose they were trying to find the square root of a number a. They started with a guess, which we'll call x. Since

(3.25) $$x \cdot \frac{a}{x} = a = \sqrt{a} \cdot \sqrt{a},$$

one of x and $\frac{a}{x}$ is bigger than $\sqrt{a}$ and the other is smaller. Either way, the average of x and $\frac{a}{x}$ is likely to be closer to $\sqrt{a}$ than x was. They then replaced x with the average of x and $\frac{a}{x}$ and repeated the process as many times as necessary to get the accuracy they wanted.

In fact, this procedure homes in on the true square root amazingly quickly. Table 3.4 shows what happens when we use this method to find $\sqrt{5}$ with the initial guess $x = 2$.

3.8. Chapter Summary

Table 3.4. Computing $\sqrt{5}$ with the Babylonian method gives 12 decimal places of accuracy in just four steps.

x	$5/x$	Average
2	2.5	2.25
2.25	2.222222222222	2.236111111111
2.236111111111	2.236024844720	2.236067977916
2.236067977916	2.23606797084	2.236067977500
2.236067977500	2.236067977500	2.236067977500

The numbers in Table 3.4 were generated with a very simple MATLAB program:

```
a=5           % We're finding the square root of 5
x=2           % That's our initial guess
for j=1:4     % Run the loop four times
   y=a/x      % Compute a/x
   x=(x+y)/2  % Replace x with the average
end           % Complete the loop
```

To compute the square root of a different number, just change the first two lines.

You might think that the Babylonian method is obsolete, since nowadays we have calculators and computers with built-in square root functions. But how do you think the square root function on your calculator actually works? It uses the Babylonian method!

Several thousand years after the Babylonians, Isaac Newton modified the Babylonian method to find roots of all sorts of functions, not just quadratic polynomials. We'll learn about Newton's method in Section 5.6.

3.8. Chapter Summary

The Main Ideas.

- Only very easy problems can be solved quickly and exactly. Most of the time we need to find approximate answers and then improve the approximations.
- A function is a rule for turning an input into an output. It can be understood as
  - an algorithm,
  - a formula,
  - a table, or
  - a graph.
- Linear functions are functions that change at a constant rate m, called the *slope*, *multiplier*, or *rate of change*. The formulas

$$m = \frac{\Delta y}{\Delta x}, \quad \Delta y = m\Delta x, \quad \Delta x = \frac{\Delta y}{m}$$

are mathematically equivalent, but they carry three different lessons:
  - You can compute the slope from two points as rise/run.
  - You can predict Δy from Δx. This gives us the equation of our line in *point-slope* form: $y = y_1 + m(x - x_1)$.
  - You can predict Δx from Δy.
- You can use lines to approximate curves, *interpolating* and *extrapolating* from given data to estimate other values.
- Euler's method is a "follow your nose" method for finding an approximate solution to a rate equation. More accuracy requires more work.
- With a simple program, we can get a computer to do that work for us.
- For a very, very long time, people have been using approximations to solve algebraic equations. The bisection method and the Babylonian method are simple and accurate.

Expectations. You should be able to:

- Translate back and forth between different representations of a function.
- Find the slope of a line from the coordinates of two points.
- Find the equation of a line in point-slope form from a slope and the coordinates of one point.
- Manipulate that equation to answer questions such as "where will you be at such-and-such a time?" and "when will you reach such-and-such a place?"
- Use linear approximations to approximate functions. Extrapolate and interpolate from nearby data points.
- Implement Euler's method by hand, generating a table of (approximate) values of y and y' as a function of time.
- Explain the steps in a computer program that uses Euler's method for a certain rate equation, and modify the program to handle a slightly different rate equation.

3.9. Exercises

Understanding Functions and Composite Functions

Exercises 3.1 and 3.2 refer to the following functions:

$$h(s) = \frac{2s}{s+3},$$
$$f(x) = x^2,$$
$$g(u, v) = u + 2v,$$
$$q(x, y) = 3,$$
$$r(z) = \begin{cases} 2 & \text{if } z < 0, \\ z^2 + 3 & \text{if } 0 \leq z < 5, \\ 5 - z & \text{if } 5 \leq z. \end{cases}$$

3.9. Exercises

3.1. Evaluate each function.
   (a) $h(3)$
   (b) $(f \circ h)(2)$
   (c) $g(3, 5)$
   (d) $q(\pi, -3)$
   (e) $h(f(3))$
   (f) $r(r(r(-2)))$

3.2. **True or False?** If true, explain why. If false, provide a counterexample.
   (a) If $a > b$, then $f(a) > f(b)$.
   (b) For all real numbers a and b, $f(a + b) = f(a) + f(b)$.
   (c) For all real numbers c and d, $g(c, d) > 0$.
   (d) For all real numbers c, $(f \circ h)(c) = (h \circ f)(c)$.

3.3. Graph $y = |x|$. Does this relationship between x and y describe a function? Why or why not?

3.4. **Sales Tax.** The Texas state sales tax rate is 6.25%. That is, the sales tax T (in dollars) is proportional to the price P of an object, also in dollars, and the constant of proportionality is $k = 6.25\% = 0.0625$. Write a formula that expresses the sales tax as a linear function of the price. Use your formula to compute the tax on a computer that costs $1200.00.

3.5. Suppose you work at a cell phone store. Your weekly salary is $400 plus a 4% commission on total sales over $8000. Let x represent your total sales (in dollars) in a particular week.
   (a) In words, briefly describe what the function $f(x) = 0.04x$ is computing.
   (b) In words, briefly describe what the function $g(x) = x - 8000$ is computing.
   (c) Suppose that you sold enough cell phones that week to earn a commission. Which of the following would correctly compute your earnings for that week:

   $$S_1(x) = 400 + (f \circ g)(x) \quad \text{or} \quad S_2(x) = 400 + (g \circ f)(x)?$$

   Explain.

3.6. Suppose that you want to grow your own vegetables in raised garden beds. You go to the local farm supply store and buy wood, fertilizer, soil, and other supplies for the beds. The store will deliver your purchases to your house for a fee. You pay for your purchases, the sales tax, and the delivery fee. The cost of the supplies, before taxes, is x. The delivery fee is $30. Sales tax is 7.5%.
   (a) Write a function $g(x)$ for the total cost, *after* taxes, *not* counting the delivery fee. (Remember that x is the price of your purchases.)
   (b) Write a function $f(x)$ for the total cost, *before* taxes, *including* the delivery fee.
   (c) Calculate and interpret $(f \circ g)(x)$ and $(g \circ f)(x)$.
   (d) If sales taxes are *not* charged on delivery fees, which of the compositions in part (c) represents your actual cost? Explain.

Linear Functions and Slope

3.7. Graph the function $f(x) = 0.8x - 2$ on the interval $-5 \leq x \leq 5$.
   (a) Read from your graph the value of $f(x)$ when $x = -2$ and when $x = 2$. What is the difference between these function values? Call this difference Δy. What is the difference between the two x-values? Call this difference Δx. Compute the slope using Δx and Δy. According to the equation of the line, which was given in slope-intercept form, what is the slope? Does your computed slope agree with this value? Why or why not?
   (b) Rewrite the equation of the line in point-slope form, using the point $(1, -1.2)$ as your reference.

3.8. Graph the function $f(x) = 0.8(x + 2) - 3.6$ on the interval $-5 \leq x \leq 5$.
   (a) Read from your graph the value of $f(x)$ when $x = -2$ and when $x = 2$. What is the difference between these function values? Call this difference Δy. What is the difference between the two x-values? Call this difference Δx. Compute the slope using Δx and Δy. According to the equation for the line, what is the slope? Does your computed slope agree with this value?
   (b) Rewrite the equation of the line in slope-intercept form.

3.9. A car is driving at a constant speed of 64 miles/hour. At 2:00 it passes mile marker 150.
   (a) How far will the car travel in the next half hour? Where will it be at 2:30?
   (b) How far will the car travel between 2:00 and t o'clock? Where will it be at t o'clock?
   (c) Express the position of the car at t o'clock as a linear function *in point-slope form*. Then plug in $t = 2.25$ to figure out the location of the car at 2:15.
   (d) Rewrite this function in *slope-intercept form*. Then plug in $t = 2.25$ to figure out the location of the car at 2:15.

Food for Thought: Which was easier, plugging in $t = 2.25$ to point-slope or slope-intercept form? If you did your calculations with a calculator, it probably didn't matter much. However, if you did them by hand, you probably found multiplying the slope in part (c) by 1/4 to be a lot easier than multiplying the slope in part (d) by $2\frac{1}{4}$. In general, point-slope is much better for understanding a linear function close to your reference point, which is what we usually care about, while slope-intercept is better for understanding that function near $t = 0$ (noon in this case).

3.10. A car is driving at constant speed. At 3:00 it is at mile marker 145. At 4:00 it is at mile marker 201.
   (a) How fast is the car moving?
   (b) Write a function, *in point-slope form*, for the location $x(t)$ of the car at t o'clock, using 3:00 as your reference time.
   (c) Use this function to determine where the car is at 3:15.
   (d) Now rewrite your function in *slope-intercept form*. Use this function to determine where the car was at 2:45.

3.11. You should be able to answer all parts of this exercise without ever finding the equations of the functions involved.
   (a) Suppose that $f(x)$ is a linear function with multiplier $m = 1/2$. If $f(-2) = 2$, what is $f(-2.1)$? $f(-2.016)$? $f(10)$?
   (b) Suppose that $h(x)$ is a linear function with multiplier $m = -3$. If $h(1) = 4$, for what value of x is $h(x) = -5$? $h(x) = 19$? $h(x) = 97$?
   (c) Suppose that $g(x)$ is a linear function with $g(3) = 3$ and $g(6) = 4.5$. What is $g(3.1)$? What is $g(3 + a)$?

3.12. Suppose that $w = 150 - 12s$. How much does w change when s changes from 4 to 10? From 4 to 4.5? From 4 to 4.1? Let Δs denote a change in s, and let Δw denote the corresponding change in w. Is $\Delta w = m \Delta s$ for some constant m? If so, what is m?

3.13. Consider two quantities x and y.
   (a) Explain what it means for y to be a linear function of x.
   (b) If y is a linear function of x, is x necessarily a linear function of y? Either provide an argument for why it always is, or an example in which it isn't.
   (c) In the following table, x and y are linear functions of each other. Fill in the rest of the table.

x	−5	−2			3	6
y	8		−0.5	−4		−10

   (d) Based on the table in part (c), express Δy as a function of Δx. Then find a formula for y in terms of x.

3.14. **Thermometers**. Temperatures are usually measured in degrees Fahrenheit in the United States and in degrees Celsius elsewhere. The Fahrenheit reading F and the Celsius reading C are linear functions of each other, based on two reference points:
   - 0 degrees Celsius is the freezing point of water, which is 32 degrees Fahrenheit.
   - 100 degrees Celsius is the boiling point of water, which is 212 degrees Fahrenheit.

   (a) Use the freezing and boiling points of water to write F as a linear function of C.
   (b) Express C as a linear function of F.
   (c) Which represents a larger change in temperature—a degree Celsius or a degree Fahrenheit?
   (d) Suppose that an item heats up or cools down. Express ΔF (the change in the Fahrenheit reading) as a linear function of ΔC, and vice versa.
   (e) Find a temperature that has the same reading on both scales.

A True Story: On a spring day in 1975, a young boy from Nebraska was riding with his father to Canada. As they neared the border, they picked up a radio station from Winnipeg. The weather report came on and said that the temperature was 5 degrees. The boy and his father were shocked that it could be so cold in the spring. After a commercial break, the disc jockey came on and said, "That's 5 degrees Celsius, or 41 degrees Fahrenheit." Canada had switched its weather forecasts a few weeks earlier from Fahrenheit to Celsius. It took a while for Canadians (not to mention Americans) to get used to the change.

Falling Bodies. Ignoring air resistance, the velocity of a falling body increases in proportion to the increase in the time that the body has been falling. The constant of proportionality is approximately 32 feet per second squared, or 32 ft/s^2.

3.15. A ball is falling at a velocity of 48 feet/second after 1 second.
   (a) How fast is it falling after 3 seconds?
   (b) Express the change in the ball's velocity Δv as a linear function of the change in time Δt.
   (c) Express the ball's velocity $v(t)$ as a linear function of time.

3.16. If d and v denote the distance fallen (in feet) and the velocity (in feet per second) of a falling body, then the motion can be described by the following equations:
$$d' = v, \quad v' = 32.$$
The first equation says that velocity is the rate of change of distance. The second equation is what we discussed in Exercise 3.15.
   Suppose that $d(0) = 0$ and $v(0) = 16$. Using a step size of 1 second, estimate $d(3)$ and $v(3)$.

3.17. **Depreciation.** A farmer buys a new combine (used for harvesting crops) for $380,000. He assumes that the value of the combine will drop, or **depreciate**, at a constant rate, and that it will be worth $100,000 in ten years. (We will examine other depreciation methods in the Chapter 6 exercises.) Let $V(t)$ represent the value of the combine t years after it was purchased.
   (a) Find an expression relating ΔV to Δt. Then find an expression relating V to t.
   (b) What will the combine be worth after 7 years?
   (c) When will the value of the combine fall below $80,000?
   (d) Graph V for $0 \leq t \leq 20$ and indicate on your graph the answers for parts (b) and (c).
   (e) This method of estimating the value of a used piece of equipment is called **straight-line depreciation**. Explain why this is a sensible name.

3.18. Using MATLAB, graph the function $f(x) = 1/x$ on the interval $2 \leq x \leq 4$.
   (a) Determine the equation of a line between the points $(2, f(2))$ and $(4, f(4))$. Write this equation in the form $h(x) = m(x - 2) + b$.
   (b) Using the MATLAB "hold on" command, graph $h(x)$ on the same set of axes that you graphed $f(x)$ on.
   (c) Compare the values $f(2)$ and $h(2)$. Do you think $h(x)$ would be a good approximation for $f(x)$ at $x = 3$? Regardless of your answer, how could you create a line that would provide a better approximation for $f(x)$ at $x = 3$?

3.19. **Euler's Method.** For the rate equation $y'(t, y) = 2t + y - y^2$ and initial condition $y(0) = 2$, use Euler's method and given step size to answer the following.
   (a) For step size $\Delta t = 2$, approximate $y(2)$.
   (b) For step size $\Delta t = 1$, approximate $y(2)$.

(c) For step size $\Delta t = 0.5$, approximate $y(1)$.

(d) How does the $y(1)$ computed in part (c) compare to the $y(1)$ computed in the intermediate step for part (b)? Which of the two estimates for $y(1)$ is more accurate? Explain why.

The Solow Growth Model. Some companies use a lot of equipment and employ relatively few workers. Others employ a lot of workers but use relatively little equipment. The **capital-to-labor ratio** is the ratio of a company's capital expenses to its labor costs. Over time, a company's capital-to-labor ratio may change, so we express it as a function of time $k(t)$.

The **Solow growth model** is often used to predict the change in $k(t)$ as a company evolves. In this model, the rate equation is
$$\frac{dk}{dt} = sk^\alpha - \lambda k,$$
where s, α, and λ are parameters. k^α is called the **production function**.

3.20. Suppose that the Solow growth model for a particular company has parameters $\alpha = 1/3$ and $s = \lambda = 1$, and initial value $k(0) = 0.5$.
  (a) Use Euler's method with time step $\Delta t = 0.5$ to estimate $k(1)$.
  (b) Do you think the company is automating its operations (increasing capital expenses but reducing labor costs) or relying on aging machinery (saving on capital expenses but incurring higher labor costs)?
  (c) In the long run, what number will $k(t)$ approach? (*Hint*: At equilibrium, what is $k'(t)$?)

3.21. **Radioactivity**. During the 1986 disaster at the Chernobyl nuclear power plant, an estimated 27 kg of radioactive Cesium-137 was released into the environment. Cesium-137 has a half-life of 30 years, meaning that the amount $C(t)$ in the environment satisfies the rate equation
$$C' = -0.0231C,$$
where time is measured in years. Using a step size of 10 years, estimate how much Cesium-137 will remain in 2036, on the 50th anniversary of the disaster.

Food for Thought: Not understanding radioactivity can be deadly. During Russia's 2022 invasion of Ukraine, Russian soldiers who had never heard of the Chernobyl disaster were sent to camp and dig trenches in the most contaminated part of the Chernobyl Exclusion Zone, the so-called Red Forest. Their digging unearthed contaminated soil that had previously been buried. Many of those soldiers died of radiation poisoning.

The Bisection and Babylonian Methods

3.22. Use the bisection method to solve $x^3 + x^2 = 20$ to at least two decimal places, starting with the fact that $2^3 + 2^2 < 20 < 3^3 + 3^2$.

3.23. Use five steps of the bisection method to estimate the positive solution to $x^2 = 10$ (or estimate $\sqrt{10}$), starting with the fact that $3^2 < 10 < 3.2^2$.

3.24. In the setup of Exercise 3.23, how many steps would it take to get eight digit accuracy?

3.25. Use the Babylonian method to estimate $\sqrt{10}$ to at least eight digit accuracy, starting with the initial guess $x_0 = 3$. (This should only take three steps.)

Chapter 4

Track the Changes!

So far, we've talked a lot about rates of change, but we never really defined what *rate of change* actually means. We tiptoed around the concept, which was the best we could do with the tools we had. But now we have the tools to take a closer look at the **derivative** of a function. That's the precise mathematical object that corresponds to the vague idea of "rate of change".

4.1. The National Debt

The United States government, like almost every government in the world, finances much of its operations through bonds and other borrowing. Also like almost every other government, it takes in less in taxes than it spends, so it gets deeper in debt every year. As long as the debt grows reasonably slowly, this isn't a problem, both because inflation erodes the value of the debt and because the economy keeps growing. The $250 billion that the government owed in 1945 was a much bigger drain on the economy than the $5.7 *trillion* that it owed in 2000. However, a debt that grows faster than the economy for an extended period of time can cause very serious problems.

So, how fast has the debt actually been growing? We'll look at three time periods, one in the 1990s, one in the 2000s, and one in the 2010s.

Year	Debt (in $B)	Year	Debt (in $B)	Year	Debt (in $B)
1997	5413	2005	7933	2016	19,573
1998	5526	2006	8507	2017	20,245
1999	5656	2007	9008	2018	21,516

In 2017, the national debt was $20,245 billion = $20.245 trillion. If we let $D(t)$ be the debt (in billions of dollars) at time t (in years), then $D(2017) = 20,245$. There are three ways to estimate the rate $D'(2017)$ at which that debt was growing, as shown in Figure 4.1.

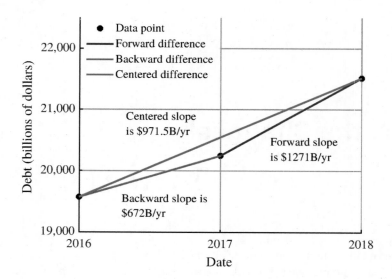

Figure 4.1. Three ways to estimate $D'(2017)$

(1) We can compare the debt in 2017 to the debt in 2018. The debt went up by $\Delta D = 21{,}516 - 20{,}245 = 1271$ in time $\Delta t = 2018 - 2017 = 1$, so we estimate $D'(2017) \approx 1271$. ΔD is called a **forward difference**, and the ratio $\Delta D/\Delta t$ is called a (forward) **difference quotient**.

(2) We can compare the debt in 2017 to the debt in 2016. As before, we compute $\Delta D = D(2017 + \Delta t) - D(2017)$ and divide by Δt, only now $\Delta t = -1$. Then $\Delta D = 19{,}573 - 20{,}245 = -672$, and we estimate that $D'(2017) \approx 672$. Here ΔD is called a **backward difference**.

(3) We can compare the debt in 2016 to the debt in 2018. That is,

(4.1)
$$D'(2017) \approx \frac{D(2017 + \Delta t) - D(2017 - \Delta t)}{2\Delta t}$$
$$= \frac{21{,}516 - 19{,}573}{2} = 971.5.$$

ΔD is called a **centered difference** or a **central difference**. Most of the time, centered difference quotients are much better estimates than forward or backward difference quotients. However, lack of data sometimes forces us to use forward or backward differences.

We can do similar computations to compute $D'(1998)$ and $D'(2006)$. Using centered differences, we get

(4.2) $\qquad D'(1998) \approx 121.5, \qquad D'(2006) \approx 537.5.$

The debt grew much faster in the 2000s than in the 1990s, and even faster in the 2010s.

4.1. The National Debt

> **Historical Note**: The budget **deficit** is the difference in any year between what the government takes in and what it spends. If the government used the same accounting methods for debt and deficit, then the deficit in any year would be exactly the same as the change in the debt that year.
>
> However, there is a technical difference in the way that Social Security and Medicare are treated. From the beginning of Social Security in 1935 until 2020, employees have paid more in Social Security and Medicare taxes every year than has been spent on retiree benefits. (This is expected to change as the population ages and more people retire.) The Social Security and Medicare surpluses go into two trust funds that are supposed to pay for future benefits, and the value of these trust funds grew to more than $5 trillion by 2020. In computing the national debt, these trust funds are *not* considered part of the government. If they were, they would offset about a quarter of the national debt.
>
> However, when the government reports the annual budget deficit, these funds *are* considered part of the government! This leads to a mismatch between the deficit and the change in the debt. For instance, between 1997 and 1998 the debt grew by $113 billion, but the trust funds grew by $182 billion, for a net surplus of $69 billion. Between 1999 and 2000 there was officially a budget surplus of $236 billion, but the debt still grew by $18 billion (and the trust funds grew by $254 billion).
>
> So which is more important—D' or the official budget deficit? That question has more to do with politics than with calculus or economics. Politicians almost always talk about the deficit (which looks better) when their party is in power and about the change in the debt (which looks worse) when the opposing party is in power!

Now that we have $D'(1998)$, we can use it to do a linear approximation. The equation of the line through (1998, 5526) with slope 121.5 is

$$(4.3) \qquad D(t) \approx 5526 + 121.5(t - 1998).$$

This linear approximation[1] predicts a debt of $5526 + 2(121.5) = 5769$ in 2000, which is reasonably close to the actual value of 5674. However, it also predicts a debt of $5526 + 121.5(7) = 6376.5$ in 2005, which is nowhere near the actual value of 7933. What happened?!

The answer is that historical events changed the trajectory. In the year 2000 it looked like federal budget surpluses would continue and even grow, and that the entire national debt might be paid off by 2020. But then 9/11 happened. The Bush tax cuts happened. The Iraq war happened. All of these pushed spending up and revenue down and increased the debt. In 2008 the Great Recession happened, which sharply decreased tax revenues and increased the demand for social programs like unemployment insurance. The bank bailouts (under Bush) and the stimulus package (under Obama) led to even more red ink, and D' reached a peak of over $1.7 trillion in 2009.

---

[1] Note that this **tangent line** is not the line through (1997, 5413) and (1999, 5656). It has the same slope, since we used a centered difference to estimate $D'(1998)$, but it touches the actual graph at one point, not two.

Things improved in the mid-2010s, but D' increased to record levels in the late 2010s and early 2020s, driven first by the Trump tax cuts and then by the Covid pandemic.

4.2. Marginal Cost, Revenue, and Profit

In the previous example, our input variable was time and D' was the rate at which D changed per unit of time. There are many other examples where the variable is time. We study the rate of change of position with respect to time and call that **velocity**. We study the rate of change of velocity with respect to time and call that **acceleration**. We study the rate at which the population is growing, or the concentration of CO_2 in the atmosphere is growing, or the polar ice caps are melting. However, the concept of **rate of change** applies even when the input variable has nothing to do with time.

Consider a factory that makes widgets. It costs $22,000/month just to keep the factory open, plus $7 for every widget that it produces. If it makes 10,000 widgets per month, how much does it cost to produce a widget?

Alice says that the cost is $9.20/widget, because it costs $92,000/month to produce 10,000 widgets/month and $\frac{\$92,000/\text{month}}{10,000\,\text{widgets/month}} = \$9.20/\text{widget}$. Bob says that the cost is $7/widget, because the $22,000/month is a fixed cost. Once the factory is open, it only costs $7 to produce each additional widget. Carol says that the cost is $22,007/widget, because that's what it takes to produce exactly one widget. Who is right?

The answer is that the three of them are talking about three different things. Alice is talking about the **average cost**. Bob is talking about the **marginal cost**. And Carol is just being a troublemaker.

In general, suppose that $C(x)$ is the cost of making x objects. In the case of our widget factory,

$$(4.4) \qquad C(x) = 22,000 + 7x.$$

The **average** cost if we make x widgets is

$$(4.5) \qquad \overline{C}(x) = \frac{C(x)}{x} = \frac{22,000}{x} + 7.$$

The **marginal** cost if we make x widgets is

$$(4.6) \qquad C'(x) = 7.$$

The marginal cost is often described as the extra cost of making one more widget since, using a forward difference with $\Delta x = 1$,

$$(4.7) \qquad C'(x) \approx \frac{C(x+1) - C(x)}{1} = C(x+1) - C(x).$$

Meanwhile, Carol's answer is just $C(1)$.

The graph of the cost function $C(x)$ is called the **cost curve**, even if (as in this case) it happens to be a straight line. The marginal cost is the slope of the cost curve. The average cost is the slope of a line from the origin to a spot on the cost curve. See Figure 4.2.

4.2. Marginal Cost, Revenue, and Profit

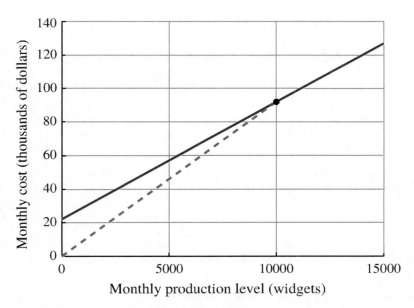

Figure 4.2. The marginal cost $C'(10,000)$ is the slope of the cost curve at $x = 10,000$. The average cost $\overline{C}(10,000)$ is the slope of the dotted line.

These quantities have different uses in business. Suppose that we can sell our widgets for $9 each. Alice points out that the price is lower than our average cost, so the factory is losing money. In her opinion, we should consider shutting it down altogether. Bob argues that each additional widget costs us $7 but brings in $9, for a profit of $2/widget. He thinks we should increase production, so that our increased profits can pay for the fixed cost of keeping the factory open.

Alice and Bob are looking at two sides of the same problem and both raise valid points. The marginal cost C' is the right thing to look at when considering small changes. If each additional widget costs us $7 and brings in $9, it makes sense to make more widgets. But the average cost $\overline{C}$ is the right thing to look at when evaluating the factory as a whole. Bob is comparing making 10,000 widgets/month to making 9,999 or 10,001, and he sees that 10,001 is better than 10,000, which is better than 9,999. Alice is comparing 10,000 to 0, and sees that 0 is better than 10,000. Together they can prepare a report that says "We are losing money making 10,000 widgets/month. We need to either increase production or go out of business."

Unfortunately, Bob's plan to increase production has a flaw. By the law of supply and demand, more widgets on the market would cause the price of widgets to drop. The marketing team estimates that the price (in dollars per widget) would be

$$(4.8) \qquad p(x) = 10 - \frac{x}{10,000}.$$

As a result, the monthly revenue from making and selling x widgets would be

$$(4.9) \qquad R(x) = xp(x) = 10x - \frac{x^2}{10,000},$$

and the monthly profit (which may be positive or negative) would be

(4.10) $$P(x) = R(x) - C(x) = 3x - \frac{x^2}{10,000} - 22,000.$$

To analyze the situation, Bob looks at the **marginal revenue** $R'(x)$ and the **marginal profit** $P'(x) = R'(x) - C'(x)$. Comparing $R(10,001)$ and $R(9999)$ and doing a centered difference, he estimates that

(4.11) $$R'(10,000) \approx \frac{R(10,001) - R(9999)}{2} = \frac{90,007.9999 - 89,991.9999}{2} = 8,$$

and $P'(10,000) = R'(10,000) - C'(10,000) = 8 - 7 = 1$. Each additional widget brings in an additional \$8 in revenue and increases the factory's cost by \$7, and so brings in one extra dollar of profit.[2] Despite the falling price of widgets, it makes sense to increase production.

In Chapter 5 we will revisit this factory and figure out how much to increase production to maximize profit, as well as whether it's worthwhile to keep the factory open.

4.3. Local Linearity and Microscopes

In the last chapter we saw a linear approximation work very well on the function $SQ(x) = x^2$, approximating x^2 as $1 + 2(x - 1)$ for x between 0.9 and 1.1. Earlier in this chapter, we saw a linear approximation do a very poor job of extrapolating the national debt from the late 1990s into the mid-2000s. In this section we examine the question of when a linear approximation is justified, and when it isn't. The general rule is:

> **The Microscope Equation:** *If $f'(a)$ exists, then the linear approximation*
>
> (4.12) $$f(x) \approx f(a) + f'(a)(x - a)$$
>
> *is good whenever x is close to a. If x is far from a or if $f'(a)$ does not exist, then all bets are off.*

To get a feel for how close is close enough, let's look at the function $f(x) = \sin(x)$, shown at four different scales in Figure 4.3, with x measured in radians. On the interval $[-9, 11]$, the graph of $f(x)$ is nowhere near straight. In fact, it undergoes a little more than three full oscillations! Trying to approximate all those wiggles with a straight line is ridiculous.

On the interval $[0, 2]$, however, it only undergoes about a third of a cycle. It mostly goes up, although it heads downhill a little at the end. Approximating this curve with a straight line isn't as absurd as trying it on $[-9, 11]$, but it still isn't a very good fit.

On the interval $[-.9, 1.1]$, however, the sine function only undergoes a 30th of a cycle. In that short stretch, it doesn't have room to change direction very much. For any two numbers x_1 and x_2 between 0.9 and 1.1, $\Delta y/\Delta x$ is somewhere between 0.45

---

[2]The 10,001st widget sells for about \$9 and costs \$7 to make, so you might think that the marginal profit should be \$2. However, making that last widget also lowers the price of widgets by \$.0001, meaning that the revenue from selling the other 10,000 widgets is \$1 less than before, for a net gain of only \$1.

4.3. Local Linearity and Microscopes

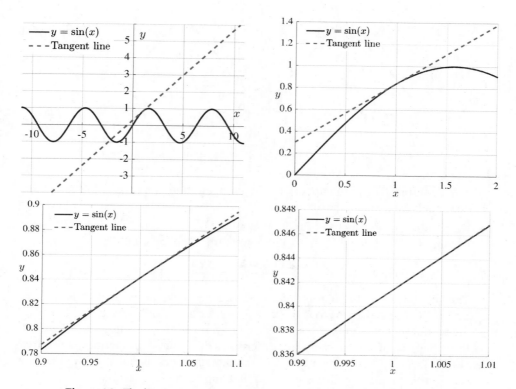

Figure 4.3. The line tangent to $y = \sin(x)$ at $x = 1$ is a terrible approximation on $[-9, 11]$, is poor on $[0, 2]$, is good on $[0.9, 1.1]$, and is indistinguishable from the curve on $[0.99, 1.01]$.

and 0.62. On the interval $[-0.99, 1.01]$ the fit is even better, and $\Delta y / \Delta x$ only varies between 0.53 and 0.55. Since $f(1) = \sin(1)$ is close to 0.8415, the linear approximation

(4.13) $$\sin(x) \approx 0.8415 + 0.54(x - 1),$$

while only reasonably accurate when x is between 0.9 and 1.1, is very accurate when x is between 0.99 and 1.01.

The upshot is that as you zoom in on a smooth curve—*any* smooth curve—the curve looks straighter and straighter, and the linear approximation (4.12) gets better and better. We call equation (4.12) the **microscope equation**, since it describes what our graph looks like through a microscope.

Before moving on, let's consider a few functions whose graphs are not smooth, and for which linear approximations don't work.

- Consider the absolute value function $f(x) = |x| = \begin{cases} x & x \geq 0, \\ -x & x < 0, \end{cases}$ shown in Figure 4.4. The graph of this function between -1 and 1 looks like a V. There is a line $y = x$ that works for $x \geq 0$, and a line $y = -x$ that works for $x \leq 0$, but there isn't any line that works on both sides. No matter how much we zoom in, we still see

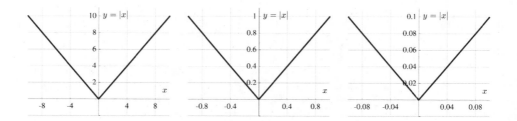

Figure 4.4. The graph of $f(x) = |x|$ is V-shaped at all length scales.

a V, not a line. The graph does not have a tangent line at $x = 0$, and the function does not have a linear approximation.

- Next consider the cube-root function $g(x) = x^{1/3}$, shown in Figure 4.5. If we try to compute the slope m at $x = 0$ with a forward difference, we get

$$(4.14) \quad m \approx \frac{\Delta y}{\Delta x} = \frac{g(0 + \Delta x) - g(0)}{\Delta x} = \frac{\Delta x^{1/3}}{\Delta x} = \Delta x^{-2/3}.$$

The smaller Δx is, the bigger $\Delta y/\Delta x$ is. If $\Delta x = 0.001$, then $\Delta y/\Delta x = 100$. If $\Delta x = 0.000001$, then $\Delta y/\Delta x = 10,000$. As Δx gets closer and closer to 0, our approximating line gets steeper and steeper, approaching a vertical line. The same thing happens if we use backward differences or centered differences. The curve simply has a vertical tangent at $x = 0$. A vertical line is not the graph of any function, so there is no linear function that approximates g near $x = 0$.

- Imagine flipping a fair coin over and over again. For each whole number n, let $h(n)$ be the number of heads in the first n tosses minus the number of tails. This is called a **random walk**. $h(n+1) - h(n)$ will either be 1 (if the $(n+1)$-st toss comes up heads) or -1 (if it comes up tails). Since there is no way of predicting what the

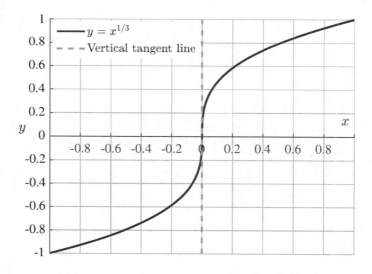

Figure 4.5. The graph of $y = x^{1/3}$ has a vertical tangent at $x = 0$.

4.4. The Derivative

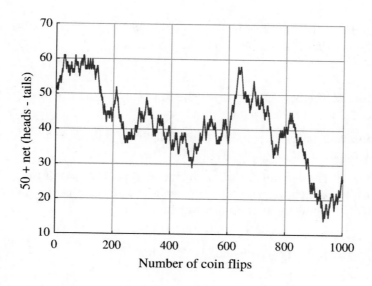

Figure 4.6. A random walk with 1000 coin flips, starting at $h = 50$. The resemblance to a typical stock chart is *not* coincidental.

$(n+1)$-st toss will be, it is absolutely impossible to use data of $h(1), h(2), \ldots, h(n)$ to predict whether h will go up or down next. The graph of $h(x)$ (with straight lines connecting consecutive integer points) shown in Figure 4.6 is like the graph of $|x|$, only with corners all over the place. It's impossible to find a line that reliably approximates the graph of h near any integer.

That won't stop people from trying, of course. After a run of heads (which happens by chance every once in a while), some people will say "the coin is lucky" and bet on the next toss being heads. Others will say "we've had so many heads that it's bound to be tails next" and bet the other way. Casinos and con artists make a lot of money exploiting that kind of woolly thinking.

The price of a stock from day to day, or a stock average like the Dow Jones, has many of the same mathematical properties as a random walk. There is a slow long-term pattern of growth, a few percent per year, but the short-term fluctuations are completely unpredictable.

Besides pointing out their existence, we don't have much to say about functions with the wild behavior of the last three examples. We are interested in functions that *do* have good linear approximations, and whose graphs *do* have (nonvertical) tangent lines. Such functions are called **differentiable**, and there are plenty of them to keep us busy.

4.4. The Derivative

After several chapters of exploration, we are finally ready to define the **derivative** $f'(a)$ of the function $f(x)$ at $x = a$. There are several different definitions, and it's important to understand why they all describe the exact same thing. If $f'(a)$ satisfies any one of these definitions, then it also satisfies all of the others.

> **Definition 1.** The derivative $f'(a)$ of the function $f(x)$ at $x = a$ is the rate of change of the function at $x = a$.
>
> **Definition 2.** The derivative $f'(a)$ of the function $f(x)$ at $x = a$ is the slope of the line tangent to the graph $y = f(x)$ at $(a, f(a))$.
>
> **Definition 3.** The derivative $f'(a)$ of the function $f(x)$ at $x = a$ is the multiplier in the best linear approximation to $f(x)$ near $x = a$. That is, for any constant $c > 0$, no matter how small, the approximation
>
> (4.15) $$f(x) \approx f(a) + f'(a)(x - a)$$
>
> is accurate to within $c|x - a|$ when x is close enough to a.
>
> **Definition 4.**
>
> (4.16) $$f'(a) = \lim_{x \to a} \frac{f(x) - f(a)}{x - a}.$$
>
> **Definition 5.**
>
> (4.17) $$f'(a) = \lim_{h \to 0} \frac{f(a+h) - f(a)}{h}.$$

Definitions 4 and 5 use the language and notation of **limits**, and that deserves an explanation. We use the term **limit** to describe what a certain expression is homing in on. The expression "$x \to a$" is read as "x approaches a", while "$\lim_{x \to a}$" (also written "$\lim_{x \to a}$") is "the limit as x approaches a." The statement $f'(a) = \lim_{x \to a} \frac{f(x) - f(a)}{x - a}$ means that $\frac{f(x) - f(a)}{x - a}$ gets closer and closer to $f'(a)$ as x gets closer and closer to a. The statement $f'(a) = \lim_{h \to 0} \frac{f(a+h) - f(a)}{h}$ means that $\frac{f(a+h) - f(a)}{h}$ gets closer and closer to $f'(a)$ as h gets closer and closer to 0.

As we saw in the last section, not every function has a derivative. If $f'(a)$ exists, then we say that f is **differentiable at** a. If $f'(a)$ exists for every number a, we say that $f(x)$ is **differentiable**, period. The study of differentiable functions is called **differential calculus**, and that's what the first half of this book is about. (The second half is about **integral calculus**.) For the next few chapters, almost all of the functions that we encounter will be differentiable.

Let's see why all the definitions are really talking about the same thing.

If the function $f(x)$ is changing at rate $m = f'(a)$ near $x = a$, then for any two points x_1 and x_2 near a, $f(x_2) - f(x_1)$ is approximately $m(x_2 - x_1)$. But that means that the ratio

(4.18) $$\frac{\Delta f}{\Delta x} = \frac{f(x_2) - f(x_1)}{x_2 - x_1}$$

4.4. The Derivative

is close to m, and so

(4.19) $$f(x_2) - f(x_1) \approx m(x_2 - x_1),$$

with an error that gets smaller and smaller, as a fraction of $x_2 - x_1$, as x_2 and x_1 get closer and closer to a. In particular, if $x_1 = a$ and $x_2 = x$, then

(4.20) $$f(x) \approx f(a) + f'(a)(x - a),$$

which is Definition 3.

If Definition 3 applies, then we have a good linear approximation whose graph has slope $f'(a)$. This graph is the line tangent to the graph of $f(x)$ at $x = a$ (Definition 2). If a line with slope $f'(a)$ is tangent to the graph, then the rate of change of the linear approximation is $f'(a)$, so the rate of change of the original function $f(x)$ must also be $f'(a)$. That is,

(4.21) $$\text{Definition 1} \Rightarrow \text{Definition 3} \Rightarrow \text{Definition 2} \Rightarrow \text{Definition 1},$$

so the first three definitions are all equivalent.

Now suppose that $f'(a)$ satisfies Definition 3. Then, for x close enough to a, $f(x) - f(a)$ must be within $c|x - a|$ of $f'(a)(x - a)$, so $\frac{f(x)-f(a)}{x-a}$ must be within c of $f'(a)$. As x gets closer and closer to a, we can take c smaller and smaller, so the ratio $\frac{f(x)-f(a)}{x-a}$ must be getting closer and closer to $f'(a)$. That is exactly Definition 4.

The reasoning also runs the other way. If $\frac{f(x)-f(a)}{x-a}$ is close to $f'(a)$, then $f(x) - f(a)$ is close (within a small fraction of $x - a$) to $f'(a)(x - a)$, so $f(x) \approx f(a) + f'(a)(x - a)$ is a good linear approximation. The upshot is that Definitions 1–4 are all equivalent.

Finally, Definition 5 is just Definition 4 in slightly different notation. If we define $h = x - a$ (what we previously called Δx), then $x = a + h$, and looking at what happens as $x \to a$ is the same as looking at what happens when $h \to 0$. In particular,

(4.22) $$\lim_{x \to a} \frac{f(x) - f(a)}{x - a} = \lim_{h \to 0} \frac{f(a+h) - f(a)}{a+h-a} = \lim_{h \to 0} \frac{f(a+h) - f(a)}{h}.$$

Definitions 4 and 5 look like forward difference quotients, but it's a little more complicated than that. When we say $\lim_{x \to a}$ we are considering what happens for *all* values of x that are close to a. x might be slightly greater than a, in which case we're looking at a forward difference quotient, or x might be slightly less than a, in which case we're looking at a backward difference quotient. Likewise, Definition 5 involves the possibility that h is slightly positive, and also the possibility that h is slightly negative. For the function $f(x)$ to be differentiable at a, both the forward difference quotients and the backward difference quotients have to approach the *same* number $f'(a)$.

What about centered differences? The centered difference quotient

(4.23) $$\frac{f(a+h) - f(a-h)}{2h} = \frac{1}{2}\left(\frac{f(a+h) - f(a)}{h} + \frac{f(a) - f(a-h)}{h}\right)$$

is the average of a forward difference quotient and a backward difference quotient. If $f(x)$ is differentiable at $x = a$, then both the forward and backward difference quotients will be close to $f'(a)$, so their average will also be close to $f'(a)$. Using centered differences is a perfectly good way to compute a derivative.

The trouble is that centered difference quotients sometimes give us an answer even when the function isn't differentiable! Consider the function $f(x) = |x|$ near $x = 0$. The forward difference quotient $f(h)/h$ with $h > 0$ is always 1. The backward difference quotient $f(-h)/(-h)$ (with $h > 0$) is always -1. The centered difference quotient

$$(4.24) \qquad \frac{f(h) - f(-h))}{2h} = \frac{|h| - |-h|}{2h} = \frac{0}{2h}$$

is always 0. It's tempting to just say that $f'(0) = 0$, but that's wrong. The line $y = 0$ is *not* tangent to the curve and is *not* the graph of a good linear approximation. This glitch is why we don't have a Definition 6 involving centered differences.

Examples of Derivatives. Now that we know what a derivative *is*, let's compute the derivative of the function $f(x) = x^2$ at $x = 1$. Using Definition 5, we have to find the difference quotient $\frac{f(1+h)-f(1)}{h}$ and compute its limit as $h \to 0$. We could plug in numbers like $h = 1, -1, 0.1, -0.1, 0.01, -0.01$ into a calculator, but we can do a lot better using algebra:

$$(4.25) \qquad \frac{f(1+h) - f(1)}{h} = \frac{(1+h)^2 - 1^2}{h} = \frac{h^2 + 2h + 1 - 1}{h} = \frac{h^2 + 2h}{h} = h + 2.$$

See Figure 4.7. As h gets closer and closer to 0, the difference quotient gets closer and closer to 2, no matter which way we do the limit. The result is that

$$(4.26) \qquad f'(1) = \lim_{h \to 0} \frac{f(1+h) - f(1)}{h} = \lim_{h \to 0} h + 2 = 2,$$

and we're done.

Note that using forward differences ($h > 0$) gives estimated slopes $h + 2$ that are a little too high. That's because the graph of $f(x) = x^2$ is curving up, so it is steeper to the right of $x = 1$ than at $x = 1$ itself. (That's why the slope of our linear approximation

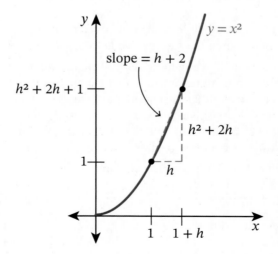

Figure 4.7. Computing the derivative of $f(x) = x^2$ at $x = 1$.

4.4. The Derivative

in Chapter 3 was 2.01 instead of 2. We estimated $f'(1)$ using a forward difference with $h = 0.01$.) Using backward differences ($h < 0$) gives estimated slopes that are a little too low. If we use centered differences, those errors happen to cancel out:

$$(4.27) \qquad \frac{f(1+h) - f(1-h)}{2h} = \frac{(h^2 + 2h + 1) - (h^2 - 2h + 1)}{2h} = \frac{4h}{2h} = 2.$$

In this case we got lucky and the errors canceled completely. Most of the time, using centered differences just makes the error a lot smaller. Forward or backward difference quotients typically have errors that are roughly proportional to h, with one difference quotient being too high and the other being too low. Centered difference quotients are typically off by something proportional to h^2. When h is much less than 1, h^2 is a lot smaller than h! If we use $h = 0.001$, we might get an answer accurate to three decimal places using forward or backward difference quotients, while getting six decimal places of accuracy using centered difference quotients. That's why we should use centered differences with numerical data whenever possible.

Now let's compute the derivative of $f(x) = 1/x$ at $x = 3$. We compute a difference quotient

$$\begin{aligned}
\frac{f(3+h) - f(3)}{h} &= \frac{1}{h}\left(\frac{1}{3+h} - \frac{1}{3}\right) \\
&= \frac{1}{h}\left(\frac{3}{3(3+h)} - \frac{3+h}{3(3+h)}\right) \\
&= \frac{1}{h}\left(\frac{-h}{3(3+h)}\right) \\
&= \frac{-1}{3(3+h)}.
\end{aligned}$$
(4.28)

As $h \to 0$, this gets closer and closer to $-1/9$, so $f'(3) = -1/9$.

If we had used centered differences, we would have computed

$$\begin{aligned}
\frac{f(3+h) - f(3-h)}{2h} &= \frac{1}{2h}\left(\frac{1}{3+h} - \frac{1}{3-h}\right) \\
&= \frac{1}{2h}\left(\frac{3-h}{(3+h)(3-h)} - \frac{3+h}{(3+h)(3-h)}\right) \\
&= \frac{1}{2h}\left(\frac{-2h}{9-h^2}\right) \\
&= \frac{-1}{9-h^2}.
\end{aligned}$$
(4.29)

This also approaches $-1/9$ as $h \to 0$. If we're doing things algebraically and taking a limit, it doesn't matter whether we use forward/backward differences or centered differences. Most of the time, the algebra is a little easier using Definition 5 as written, so we'll stick with that.

However, if we're working from data and have to stop at a not-so-small value of h, which method we use can make a big difference, as you can see in Table 4.1.

Table 4.1. Estimating $f'(3)$ with one-sided and centered differences

h	$\frac{-1}{3(3+h)}$	$\frac{-1}{9-h^2}$
-1	-0.1666667	-0.125
-0.1	-0.114943	-0.111234
-0.01	-0.111483	-0.111112
0.01	-0.110742	-0.111112
0.1	-0.107527	-0.111234
1	-0.083333	-0.125

Both $\frac{-1}{3(3+h)}$ and $\frac{-1}{9-h^2}$ approach $-1/9 \approx -0.111111$ as $h \to 0$. However, $\frac{-1}{9-h^2}$ gets there a *lot* faster! Estimating $f'(1)$ by using $\frac{-1}{9-h^2}$ with $h = \pm 0.01$ is about 300 times more accurate than using $\frac{-1}{3(3+h)}$ with $h = \pm 0.01$. Even using $\frac{-1}{9-h^2}$ with $h = \pm 0.1$ is more accurate than using $\frac{-1}{3(3+h)}$ with $h = \pm 0.01$. *When crunching numerical data, use centered differences whenever possible.*

4.5. A Global View

So far we have been looking at $f'(a)$, the derivative of f at a single point $x = a$. However, f' is also a function in its own right. After all, it's a rule for converting an input a into an output $f'(a)$. In this section we'll look at the derivative function $f'(x)$ and its graph, and see how that relates to the graph of the original function $f(x)$.

Consider the function $f(x) = x^2$, shown in Figure 4.8. At $x = -2$, the graph of $f(x)$ is heading downhill, and the slope of the tangent line is negative, so $f'(-2) < 0$. When $x = -1$, the graph is still heading downhill, but not as steeply. That is, $f'(-2) < f'(-1) < 0$. When $x = 0$ the graph is horizontal, so $f'(0) = 0$. When $x = 1$ the graph is heading uphill, so $f'(1) > 0$. When $x = 2$, the graph is heading uphill more steeply, so $f'(2) > f'(1)$.

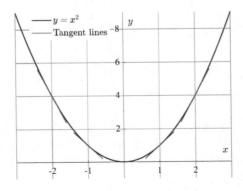

Figure 4.8. The graph of $f(x)$ has a slope that is negative when $x < 0$ and is positive when $x > 0$.

4.5. A Global View

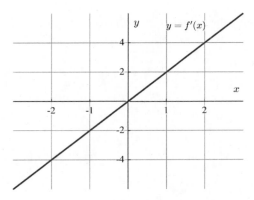

Figure 4.9. The *value* of $f'(x)$ gives the *slope* of the graph of $f(x)$.

Putting this all together, we have

(4.30) $$f'(-2) < f'(-1) < f'(0) = 0 < f'(1) < f'(2).$$

Without doing a single computation, we know that the graph of $f'(x)$ looks more-or-less like the line shown in Figure 4.9.

Notice that:

- The sign of $f'(x)$ tells you which way the graph of $f(x)$ is heading. When $f'(x) > 0$, the graph of $f(x)$ is heading uphill. When $f'(x) < 0$, the graph of $f(x)$ is heading downhill.
- The size of $f'(x)$ tells you how steeply $f(x)$ is heading uphill or downhill. Steep uphills have $f'(x) \gg 0$, while steep downhills have $f'(x) \ll 0$.
- When $f'(x)$ changes sign from negative to positive, it means that $f(x)$ has stopped going downhill and is about to go uphill. In other words, we're at the bottom of a valley, also called a **local minimum**. At the top of a hill (a **local maximum**, not shown in this example), $f'(x)$ goes from positive to negative.
- If $f(x)$ is differentiable at a local maximum (or minimum), we must have $f'(x) = 0$ there. This is the 3rd Pillar of Calculus:

> *What goes up has to stop before it comes down.*

- The value of $f'(x)$ says absolutely nothing about the value of $f(x)$. It only says how fast the value of $f(x)$ is changing.

Now that we've studied our example qualitatively, let's do some calculations. We can compute $f'(-2)$ from the definition, using algebra:

$$f'(-2) = \lim_{h \to 0} \frac{(-2+h)^2 - (-2)^2}{h}$$

(4.31) $$= \lim_{h \to 0} \frac{4 - 4h + h^2 - 4}{h} = \lim_{h \to 0} -4 + h = -4.$$

Similar calculations give the values of $f'(-1)$, $f'(0)$, $f'(1)$, and $f'(2)$:

$$f'(-1) = \lim_{h \to 0} \frac{(-1+h)^2 - (-1)^2}{h}$$
$$= \lim_{h \to 0} \frac{1 - 2h + h^2 - 1}{h} = \lim_{h \to 0} -2 + h = -2,$$
$$f'(0) = \lim_{h \to 0} \frac{(0+h)^2 - (0)^2}{h}$$
$$= \lim_{h \to 0} \frac{h^2}{h} = \lim_{h \to 0} h = 0,$$
$$f'(1) = \lim_{h \to 0} \frac{(1+h)^2 - (1)^2}{h}$$
$$= \lim_{h \to 0} \frac{1 + 2h + h^2 - 1}{h} = \lim_{h \to 0} 2 + h = 2,$$
$$f'(2) = \lim_{h \to 0} \frac{(2+h)^2 - (2)^2}{h}$$
(4.32)
$$= \lim_{h \to 0} \frac{4 + 4h + h^2 - 4}{h} = \lim_{h \to 0} 4 + h = 4.$$

Notice that we did essentially the same calculation each time. Instead of repeating it for each value of x that we care about, we can use algebra to do the calculation for every value of x, once and for all, as in Figure 4.10:

$$\begin{aligned} f'(x) &= \lim_{h \to 0} \frac{(x+h)^2 - (x)^2}{h} \\ &= \lim_{h \to 0} \frac{x^2 + 2xh + h^2 - x^2}{h} \\ &= \lim_{h \to 0} 2x + h \\ &= 2x. \end{aligned}$$
(4.33)

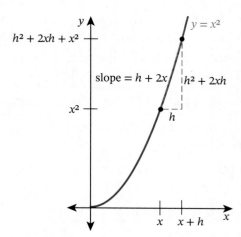

Figure 4.10. The calculation of $f'(x)$ is just like the calculation of $f'(1)$.

4.5. A Global View

We have now explored the derivative function $f'(x)$ (of $f(x) = x^2$) from all four perspectives:

(1) *As an algorithm.* For each value of a, compute the slope $f'(a)$ of the graph $y = x^2$ using the methods of the previous section.

(2) *As a table.*

x	$f'(x)$
-2	-4
-1	-2
0	0
1	2
2	4

(3) *As a graph.* See Figure 4.9 on p. 73.

(4) *As a formula.* $f'(x) = 2x$.

There are several different notations for the derivative function. The most common are

$$(4.34) \qquad f'(x) \quad \text{and} \quad \frac{df(x)}{dx}, \quad \text{or} \quad \frac{df}{dx} \text{ for short.}$$

If there is a variable y and $y = f(x)$, we sometimes write dy/dx instead of df/dx. The letter d stands for "difference", and the letter Δ that we previously used for the change in a variable is just the Greek equivalent of d. When computing difference quotients we write $\Delta y/\Delta x$, but when we talk about the limit of that quotient we write dy/dx.

We also sometimes use a dot to mean a derivative with respect to time. That is, if $y = f(t)$, then

$$(4.35) \qquad \dot{y} = \frac{df}{dt} = f'(t).$$

The dot notation is especially useful in physics, where many quantities depend both on position (x) and time (t), and we use a dot to denote a derivative with respect to time. Finally, if $f(x)$ is given by a formula, say $f(x) = x^2 \sin(x)$, we might write $(x^2 \sin(x))'$ instead of $f'(x)$ and $d(x^2 \sin(x))/dx$ instead of df/dx.

So why do we have all these different notations? One reason is historical (and stupid). The two primary inventors of calculus were Isaac Newton and Gottfried Leibniz, and there was a big controversy over who should get credit for what. Newton used f' and $\dot{f}$, so generations of British scientists followed his example. Leibniz used dy/dx and df/dx and df/dt, so generations of continental European scientists followed *his* example. There was a big rivalry between British and continental scientists, neither side would budge, and both notations stuck.

A better reason is that there are real advantages to both notations. We'll see that when we get to the chain rule in Chapter 5. Leibniz's notation gets to the heart of what the chain rule means and is much better for doing related rates problems, while

Newton's notation is much better for computing the derivative of $\sin(x^2 - 7)$. It's worth your while to learn all of these notations.

We close this section with a worked problem:

> **Problem:** Use a linear approximation to estimate 9.975^2 and 10.05^2.
>
> **Solution:** We are studying the function $f(x) = x^2$. Since $f'(x) = 2x$, $f'(10) = 20$. The microscope equation tells us that, for x close to 10,
>
> $$x^2 = f(x) \approx f(10) + f'(10)(x - 10)$$
> (4.36)
> $$= 100 + 20(x - 10).$$
>
> Thus
>
> $$9.975^2 \approx 100 + 20(-0.025) = 99.5, \text{ and}$$
> (4.37)
> $$10.05^2 \approx 100 + 20(0.05) \quad = 101.$$
>
> The exact values turn out to be $9.975^2 = 99.500625$ and $10.05^2 = 101.0025$. Our linear approximation is both simple and accurate.

4.6. Chapter Summary

The Main Ideas.

- We can estimate rates of change from data using forward, backward, or centered difference quotients. If $h > 0$, those are

(4.38)
$$\frac{f(x+h) - f(x)}{h}, \quad \frac{f(x) - f(x-h)}{h}, \quad \text{and} \quad \frac{f(x+h) - f(x-h)}{2h},$$

respectively. Sometimes there is only enough data to do it one of the three ways, but when we have a choice we should generally use centered differences, which are more accurate.

- Rates of change aren't just for things that vary in time. If a factory makes x items, then the rate of change with respect to x of the cost, revenue, and profit are called the **marginal cost**, **marginal revenue**, and **marginal profit**. Decisions on whether to increase or decrease production should be based on whether the marginal profit is positive or negative, or equivalently on whether the marginal revenue is greater than or less than the marginal cost.

- All smooth functions look linear when you zoom in enough, so it makes sense to look for linear approximations.

- If a function $f(x)$ changes at rate $f'(a)$ when $x = a$, then the **linear approximation**

(4.39)
$$f(x) \approx f(a) + f'(a)(x - a),$$

also called the **microscope equation**, is very accurate whenever x is sufficiently close to a.

- The **derivative** $f'(a)$ of the function $f(x)$ at $x = a$ can be understood as:
  - the rate of change of the function when $x = a$,

- the slope of the line tangent to the graph at $x = a$,
- the multiplier in the best linear approximation,
- the limit $\lim_{x \to a} \frac{f(x)-f(a)}{x-a}$, and
- the limit $\lim_{h \to 0} \frac{f(a+h)-f(a)}{h}$.

- We don't need tables of values to compute derivatives. Most of the time, it's easier to use algebra.

- The **derivative function** is the function $f'(x)$ whose value at each point is the derivative of $f(x)$ at that point.

- The *value* of $f'(x)$ gives the **rate of change** of $f(x)$, but does not say anything about the value of $f(x)$. Equivalently, the *height* of the graph of $f'(x)$ gives the **slope** of the graph of $f(x)$, but does not say anything about the height of the graph of $f(x)$.

- The sign of $f'(x)$ tells whether $f(x)$ is increasing.
    - When $f'(x) > 0$, $f(x)$ is **increasing**.
    - When $f'(x) < 0$, $f(x)$ is **decreasing**.
    - When $f(x)$ hits a maximum or a minimum, then $f'(x) = 0$ (or does not exist).

- It is often possible to find a formula for $f'(x)$ by using algebra to compute the limit

(4.40) $$f'(x) = \lim_{h \to 0} \frac{f(x+h) - f(x)}{h}.$$

Expectations. You should be able to:

- Use tables of data to compute rates of change.
- Use rates of change to make linear approximations (the microscope equation).
- Use marginal costs, revenues, and profits to improve the profitability of a factory.
- Explain how the different definitions of $f'(a)$ are related.
- Compute the derivatives of simple functions such as $f(x) = x^2$ and $f(x) = 1/x$ using limits and algebra.
- Relate the derivative of a function to the shape of its graph.
- Relate the shape of the graph of a function to the graph of its derivative.

4.7. Exercises

MATLAB tip: A number of the following exercises involve graphing functions with MATLAB and zooming in and out of the picture. The graphs are most easily generated using the "plot" or "fplot" commands. The drop-down menus for the figure itself contain commands for zooming in and out. You can generate multiple graphs on the same axes directly using the "plot" command (type "help plot" at the MATLAB prompt for details), or by adding new graphs one at a time with the help of the "hold on" command.

Population Growth. The following table gives the populations of Texas and Puerto Rico (in millions, to two decimal places) between 2001 and 2020. Use this data for Exercises 4.1–4.3.

Year	TX	PR	Year	TX	PR
2001	21.33	3.67	2011	25.67	3.56
2002	21.72	3.66	2012	26.06	3.54
2003	22.10	3.65	2013	26.45	3.50
2004	22.49	3.64	2014	26.96	3.45
2005	22.93	3.63	2015	27.47	3.38
2006	23.51	3.62	2016	27.86	3.28
2007	23.90	3.62	2017	28.30	3.16
2008	24.33	3.61	2018	28.63	3.04
2009	24.78	3.60	2019	29.00	2.93
2010	25.25	3.58	2020	29.90	2.86

4.1. We are going to compare forward, backward, and centered differences for understanding how fast Texas was growing in 2006. Let $TX(t)$ be the population of Texas in year t.
  (a) Using a forward difference with $\Delta t = 5$ years, estimate the rate of change $TX'(2006)$. Repeat with $\Delta t = 2$. Repeat again with $\Delta t = 1$.
  (b) Now use backward differences with $\Delta t = 5$ years, 2 years, and 1 year to estimate $TX'(2006)$.
  (c) Finally, use centered differences with $\Delta t = 5$ years, 2 years, and 1 year to estimate $TX'(2006)$.
  (d) Which of these nine estimates do you think is the most accurate? Why?

4.2. Now that we know the population of Texas in 2006 and how fast it was changing, let's see what that says about the population of Texas in other years.
  (a) Using your best estimated value of $TX'(2006)$ from Exercise 4.1, write the microscope equation for $TX(t)$.
  (b) Use the linear approximation from part (a) to estimate the population of Texas in 2010, 2015, and 2020.
  (c) Using MATLAB, generate a graph that shows both the actual population and the linear approximation from $t = 2001$ through $t = 2020$ on the same set of axes. Would you say that the linear approximation does a good job of describing Texas's population growth?

4.3. Now let's study the population $PR(t)$ of Puerto Rico in year t.
  (a) Using a forward difference with $\Delta t = 2$ years, estimate $PR'(2006)$. Repeat with $\Delta t = 1$.
  (b) Now use backward differences with $\Delta t = 2$ years and 1 year to estimate $PR'(2006)$.
  (c) Finally, use centered differences with $\Delta t = 2$ years and 1 year to estimate $PR'(2006)$.
  (d) Which of these six estimates do you think is the most accurate? Why?

4.7. Exercises

4.4. The problem with using a small Δt to estimate $PR'(2006)$ is that, after rounding to the nearest 10,000, we don't see any change at all between 2006 and 2007! To get accurate results, we have to make sure that the change in our function is bigger than our round-off errors, or our actual errors in measurement.

(a) Here is Puerto Rico's population (in millions) to three decimals instead of two:

Year	PR
2005	3.631
2006	3.623
2007	3.616
2008	3.608

Using this more accurate data set, estimate $PR'(2006)$.

(b) Explain why we needed more accurate data to estimate $PR'(2006)$, but we didn't need more accurate data to estimate $TX'(2006)$.

Food for Thought: In general, using the smallest possible Δt is a good idea when our data is exact, as with a mathematical function like x^2 or e^x. However, when there are errors in the data, it's sometimes necessary to use a bigger baseline to average out random fluctuations and rounding errors.

4.5. The following table shows the net annual revenue for Dell Technologies, in billions of dollars, between 1996 and 2020.

Year	$B	Year	$B	Year	$B
		2001	31.9	2011	61.5
		2002	31.2	2012	62.1
		2003	35.3	2013	56.9
		2004	41.3	2014	55.6
		2005	49.1	2015	54.1
1996	5.3	2006	55.8	2016	50.9
1997	7.8	2007	57.4	2017	61.6
1998	12.3	2008	61.1	2018	79.0
1999	18.2	2009	61.1	2019	90.6
2000	25.3	2010	52.9	2020	92.2

(a) Using MATLAB, graph the revenue as a function of time from 1996 through 2020.

(b) This graph is not at all close to a straight line, so it doesn't make sense to approximate the *entire* revenue curve with a single linear function. However, there are shorter periods of time when Dell's revenue grew (or shrank) at a more-or-less constant rate. Identify at least two such time periods.

(c) For one of those time periods, draw a straight line (either electronically or with a pencil) that comes close to tracking the curve. What is the slope of that line? Does that correspond to the actual difference quotient $\Delta f/\Delta t$ for an interval?

Functions that Repeat

4.6. For this exercise, look the graph of the function $f(x) = \sin(x)$ between $x = -1$ and $x = 12$. While any graph will do, a graph generated with MATLAB is best, since then you can zoom in on different portions of the graph.
   (a) What are the x-intercepts of $\sin(x)$? That is, where is the function 0?
   (b) On what interval(s) is the function $f(x)$ increasing? Decreasing?
   (c) At what x value(s) is $f(x)$ increasing most rapidly? Approximately how fast is $f(x)$ increasing at this value of x?
   (d) What is the largest value of $f(x)$ on the interval $-1 \leq x \leq 12$?
   (e) Zoom in on the portion of the graph near $x = 0$ until it looks like a straight line. What is the slope of this line?
   (f) Use the slope you found in part (e) and the point $(0, f(0))$ to find a linear approximation to $\sin(x)$ for x close to 0.

4.7. For this exercise, look at the graphs of $h(x) = \cos(x)$ and $g(x) = \cos(2x)$ between $x = -2\pi$ and $x = 2\pi$.
   (a) How far apart are the x-intercepts of $h(x)$? How far apart are the x-intercepts of $g(x)$?
   (b) The graph of $g(x)$ has a repeating pattern. How often does it repeat? The graph of $h(x)$ also has a repeating pattern. How often does it repeat?
   (c) Compare the graphs of $h(x)$ and $g(x)$. Can you get one of them by compressing or stretching the other vertically? If so, how? Can you get one of them by compressing or stretching the other horizontally? If so, how?
   (d) How fast is $h(x)$ increasing at its steepest point? How fast is $g(x)$ increasing at its steepest point?

4.8. In Exercises 4.6 and 4.7, the functions involved had a repeating pattern. Give three examples of applications where a function that repeats would be useful for modeling a behavior.

Approximate Derivatives and Microscopes

4.9. The price at which a manufacturer can sell x units of a certain product is $p(x) = 300 - x$ for $0 \leq x \leq 300$.
   (a) The function $p(x)$ is obviously a decreasing function of x. Explain why price functions are generally decreasing functions of quantity.
   (b) Compute the revenue function $R(x)$.
   (c) Let $Q(x) = \dfrac{R(x) - R(100)}{x - 100}$. Using a rough graph for $R(x)$, draw a schematic diagram explaining the geometric meaning of this difference quotient.
   (d) Use algebra to find a simplified formula for $Q(x)$.

4.7. Exercises

(e) Using your expression from part (d), compute $Q(102)$, $Q(101)$, $Q(100.5)$, and $Q(100.1)$. What value does $Q(x)$ appear to be approaching as x approaches 100? What is your best estimate of $R'(100)$?

(f) Using the value of $R'(100)$ that you found in part (e), approximate the increase in revenue that would be expected if production increased from 100 units to 101 units. Use good notation and fully explain your solution.

(g) Compute the actual increase in revenue that would occur if production increased from 100 units to 101 units. How does that compare to your answer in part (f)?

4.10. Approximate the derivative of each function, at the given value $x = a$, in two different ways.
- Use a computer microscope (e.g., MATLAB) to view the graph of f near $x = a$. Zoom in until the graph looks straight and find its slope.
- Use a calculator to find the value of the centered difference quotient

$$\frac{f(a+h) - f(a-h)}{2h}$$

for $h = 0.1, 0.01, 0.001$, and 0.0001. Based on these quotients, give your best estimate for $f'(a)$.

(a) $f(x) = 1/x^2$ at $x = 1$.
(b) $f(x) = x^3$ at $x = 2$.

4.11. Write the microscope equation for each function at the indicated point. You will want to use your results from Exercise 4.10.
(a) $f(x) = 1/x^2$ at $x = 1$.
(b) $f(x) = x^3$ at $x = 2$.

4.12. This exercise uses the microscope equation for $f(x) = x^3$ at $x = 2$ that you constructed in Exercise 4.11(b).
(a) What is Δx when $x = 2.05$? What estimate does the microscope equation give for $f(2.05)$? How does that compare to the exact value of $f(2.05)$, namely 8.615125?
(b) What estimate does the microscope equation give for $f(2.02)$? How far is that from the exact value of 8.242408?
(c) What estimate does the microscope equation give for $f(1.99)$? How far is this from the exact value of 7.880599?
(d) The error in parts (b)–(d) is roughly proportional to what power of Δx?

4.13. Suppose $h(x)$ is a function for which $h(3) = 10$ and $h'(3) = 1.5$.
(a) Write the microscope equation for h at $x = 3$.
(b) Draw the graph of what you would see in a microscope. Do you need a formula for $h(x)$ in order to do this?
(c) If $x = 3.2$, what is Δx in the microscope equation? What estimate does the microscope equation give for Δh? What estimate does the microscope equation then give for $h(3.2)$?
(d) What estimates does the microscope equation give for $h(3.25)$, $h(2.9)$, and $h(5)$? Do you think these estimates are equally reliable? Why or why not?

4.14. Suppose $f(t)$ is a function for which $f(-2) = 5$ and $f'(-2) = 2$.
   (a) Write the microscope equation for f at $t = -2$.
   (b) Draw the graph of what you see in the microscope.
   (c) Estimate $f(-2.5)$ and $f(-2.75)$.
   (d) For what value of t near -2 would you estimate that $f(t) = 6$? For what value of t would you estimate $f(t) = 5.5$?

4.15. Suppose that $f(a) = b$, $f'(a) = -2$, and k is a small number. Which of the following is the estimate for $f(a+k)$ that comes from the microscope equation? Explain.

$$a + 2k \qquad a - 2k \qquad b + 2k \qquad b - 2k$$
$$2a + 2b \qquad 2a - 2b \qquad 2b - k \qquad f'(a - 2k)$$

4.16. Suppose that f is differentiable at a, and that h is a very small number. Explain why each expression either is or isn't a reasonable approximation of $f'(a)$.

(a) $\dfrac{f(a - h) - f(a)}{h}$ \qquad (b) $\dfrac{f(a + 2h) - f(a - 2h)}{4h}$

(c) $\dfrac{f(a + h) - f(h)}{h}$ \qquad (d) $\dfrac{f(a + h) - f(a - h)}{h}$

Functions and Their Derivatives

4.17. If a and b are any two numbers, then
$$(a + b)^3 = a^3 + 3a^2 b + 3ab^2 + b^3.$$
Using this identity and the formula $f'(1) = \lim_{h \to 0} \dfrac{f(1+h) - f(1)}{h}$, compute the derivative of $f(x) = x^3$ at $x = 1$.

4.18. As in Exercise 4.17, let $f(x) = x^3$. Using the formula
$$f'(a) = \lim_{h \to 0} \dfrac{f(a+h) - f(a)}{h}$$
for the derivative of $f(x)$ at an arbitrary point $x = a$, compute the following derivatives.

(a) $f'(-1)$ \qquad (b) $f'(0)$ \qquad (c) $f'(2)$

(d) Repeating the calculation, only with x instead of -1, 0, or 2, find the formula for $f'(x)$.

4.19. If a and b are any two numbers and if $a \neq b$, then
$$\dfrac{b^3 - a^3}{b - a} = b^2 + ab + a^2.$$
Let $f(x) = x^3$. Using the identity above and the formula
$$f'(a) = \lim_{x \to a} \dfrac{f(x) - f(a)}{x - a},$$

4.7. Exercises

compute the following derivatives.

(a) $f'(0)$ (b) $f'(1)$ (c) $f'(2)$

(d) What is the problem with using this approach to find the formula for $f'(x)$?

4.20. Consider the graph of a mystery function $f(x)$ between $x = 0$ and $x = 3$. We know that $f(x)$ is differentiable and that $f(0) = 1$. We are also told that

$$0.25 \leq f'(x) \leq 1 \quad \text{when } 0 \leq x \leq 1,$$
$$0 \leq f'(x) \leq 2 \quad \text{when } 1 \leq x \leq 2,$$
$$-0.5 \leq f'(x) \leq 0 \quad \text{when } 2 \leq x \leq 3.$$

(a) First consider $f(x)$ for x between 0 and 1. Write an expression (in terms of x) for the largest possible value of $f(x)$. Write an expression for the smallest possible value of $f(x)$. What are the possible values of $f(1)$?

(b) Repeat part (a) for x between 1 and 2, using what you just learned about $f(1)$. What are the possible values of $f(2)$?

(c) Repeat part (a) for x between 2 and 3.

(d) Draw and shade the smallest region in the x-y plane that the graph of $f(x)$ must lie in.

4.21. Suppose that for a function $g(x)$, $g(0) = 0$, and that $0 \leq g'(x) \leq 2$ for all values of x between 0 and 4.

(a) Explain why the graph of g must lie entirely in the shaded region below:

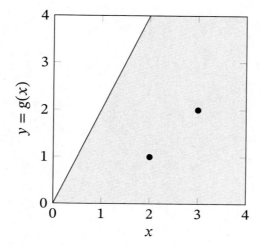

(b) Suppose that $g(2) = 1$ and $g(3) = 2$, as indicated in the figure. Draw the smallest region in which the graph of g must lie. (*Hint:* How do the conditions on $g'(x)$ constrain the values of $g(x)$ to the right *and to the left* of the indicated points?)

4.22. For each function graphed below, sketch the graph of its derivative.

(a)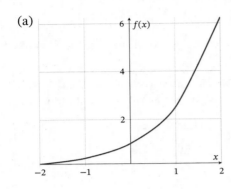

(b)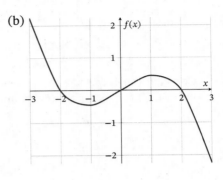

(c)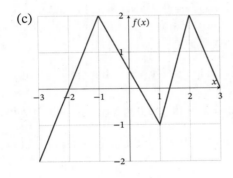

(d)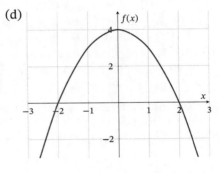

Chapter 5

Computing and Using Derivatives (What Goes Up Has to Stop Before it Comes Down)

There are three questions you should ask about every mathematical concept that you encounter:

- What is it?
- How do you compute it?
- What is it good for?

For derivatives, Chapter 4 was all about the first question. We learned that derivatives are limits of difference quotients and are closely connected to linear approximations. In this chapter, we'll tackle the second and third questions.

In the first half of the chapter, we'll develop some rules and formulas for computing derivatives. In Chapter 4 we saw that the derivative of $f(x) = x^2$ is $f'(x) = 2x$, but how can we find the derivative of $g(x) = \dfrac{5x^3 - 3x + \sin(x^2)}{\cos(2^x + 17)}$? Working that out directly from the definition $g'(x) = \lim\limits_{h \to 0} \dfrac{g(x+h) - g(x)}{h}$ is ridiculously hard, but with the methods we are going to develop, computing derivatives of even the ugliest functions will become easy.

In the second half of the chapter, we'll see what derivatives are good for. The biggest application is finding the values of a variable that maximize or minimize a function like profit, allowing us to **optimize** our business strategy. We consider two kinds of business optimization problems, one where we control the production level and the market determines the price, and one where we control the price and the market determines the demand.

Another important application is to use properties of $f'(x)$ to sketch the graph of a function $f(x)$, thereby visualizing how the function behaves. "A picture is worth a thousand words!" We'll take the derivative of $f'(x)$ (also called the **second derivative** of $f(x)$, denoted $f''(x)$ or d^2f/dx^2) and use it to tell us even more about the graph of $f(x)$. Finally, we will see how derivatives can help us solve complicated equations.

5.1. Building Blocks

Most functions that you'll see are built by combining several very basic functions:

- *Powers of x.* That is, $f(x) = x^n$. This includes not only functions like x^2 and x^3, but also functions like $\frac{1}{x} = x^{-1}$, $\sqrt{x} = x^{1/2}$ and $1 = x^0$.
- *The trigonometric functions* $\sin(x)$ *and* $\cos(x)$. Later we'll add $\tan(x)$, $\cot(x)$, $\sec(x)$ and $\csc(x)$ to the list, but those can be built from $\sin(x)$ and $\cos(x)$.
- *The exponential function* $f(x) = a^x$, where a is a fixed positive number. Eventually we'll add logarithms, which are inverse exponentials, to the list.

So let's figure out the derivatives of these functions, starting with simple powers of x.

The Power Law (Newton's Hammer). We already saw that the derivative of x^2 is $2x$. The derivative of x^3 is

$$
\begin{aligned}
(x^3)' &= \lim_{h \to 0} \frac{(x+h)^3 - x^3}{h} \\
&= \lim_{h \to 0} \frac{(x^3 + 3x^2h + 3xh^2 + h^3) - x^3}{h} \\
&= \lim_{h \to 0} \frac{3x^2h + 3xh^2 + h^3}{h} \\
&= \lim_{h \to 0} 3x^2 + 3xh + h^2 \\
&= 3x^2.
\end{aligned}
$$
(5.1)

The details of the expansion of $(x+h)^3$ aren't very important. What's important is that

(5.2) $\qquad (x+h)^3 = x^3 + 3x^2h +$ terms with 2 or more powers of h,

meaning that the extra terms involve h^2 or h^3, but not h to a power that is less than 2. Subtracting x^3 and dividing by h, we get that

(5.3) $\qquad \dfrac{(x+h)^3 - x^3}{h} = 3x^2 +$ terms with 1 or more power of h.

5.1. Building Blocks

However, all terms with 1 or more power of h go to 0 as $h \to 0$, so

$$\lim_{h \to 0} \frac{(x+h)^3 - x^3}{h} = \lim_{h \to 0} \frac{3x^2 h + \text{terms with 2 or more powers of } h}{h}$$

$$= \lim_{h \to 0} (3x^2 + \text{terms with 1 or more power of } h)$$

(5.4)
$$= 3x^2.$$

We will use the notation $O(h^2)$ to mean "terms with 2 or more powers of h", and $O(h)$ to mean "terms with 1 or more power of h". This simplifies our algebra, since when we add two $O(h^2)$ terms or when we multiply an $O(h^2)$ term by x, we still get $O(h^2)$. The first few powers of $x + h$ are then as follows.

$$(x+h)^2 = \quad x^2 + 2xh + h^2 \quad = x^2 + 2xh + O(h^2),$$

$$(x+h)^3 = \quad (x+h)^2(x+h)$$
$$= \quad (x^2 + 2xh + O(h^2))(x+h)$$
$$= \quad x^3 + x^2h + 2x^2h + 2xh^2 + O(h^2) \quad = x^3 + 3x^2h + O(h^2),$$

$$(x+h)^4 = \quad (x+h)^3(x+h)$$
(5.5)
$$= \quad (x^3 + 3x^2h + O(h^2))(x+h) \quad = x^4 + 4x^3h + O(h^2).$$

Do you see the pattern? No matter how many powers of $(x + h)$ we take, we always have

(5.6)
$$(x+h)^n = x^n + nx^{n-1}h + O(h^2).$$

To make absolutely sure that equation (5.6) works for all powers n, we apply a recursive argument called **mathematical induction**. Once we know that equation (5.6) works for a particular value of n (say, $n = k$), we can check it for the next value of n:

$$(x+h)^{k+1} = (x+h)^k(x+h)$$
$$= (x^k + kx^{k-1}h + O(h^2))(x+h)$$
$$= x^{k+1} + kx^k h + x^k h + O(h^2)$$
(5.7)
$$= x^{k+1} + (k+1)x^k h + O(h^2),$$

which is equation (5.6) with $n = k + 1$. Since we already know that equation (5.6) works for $n = 1$, the recursion tells us that it must also work for $n = 2$, so it must also work for $n = 3$, $n = 4$, and so on forever.

Equation (5.6) tells us that

$$(x^n)' = \frac{dx^n}{dt} = \lim_{h \to 0} \frac{(x+h)^n - x^n}{h}$$
$$= \lim_{h \to 0} \frac{x^n + nx^{n-1}h + O(h^2) - x^n}{h}$$
$$= \lim_{h \to 0} \left(nx^{n-1} + O(h) \right)$$
(5.8)
$$= nx^{n-1}.$$

That's it! We have just figured out the derivative of infinitely many powers of x.

The **power law** formula

(5.9) $$(x^n)' = nx^{n-1}$$

is so useful that it even has a nickname: **Newton's hammer**. Strictly speaking, we have only proven Newton's hammer when n is a whole number, but the same formula also works when n is a fraction or is negative. Once we develop the product, quotient, and chain rules, we'll see why.

There is an important special case of this formula. When $n = 0$, then $x^n = x^0 = 1$. By definition, constants don't change, so the derivative of a constant is 0. This agrees with our formula, since when $n = 0$, $nx^{n-1} = 0x^{-1} = 0$.

Sines and Cosines. Next we turn to the trigonometric functions $\sin(x)$ and $\cos(x)$. We'll go into these in much more detail in the next chapter, but for now just look at their graphs. The graph of $\sin(x)$, shown in Figure 5.1, shows that $\sin(x)$ is increasing whenever $\cos(x)$ is positive and is decreasing whenever $\cos(x)$ is negative. This suggests that the derivative of $\sin(x)$ has something to do with $\cos(x)$. Similarly, $\cos(x)$ is increasing whenever $\sin(x)$ is *negative* and is decreasing whenever $\sin(x)$ is *positive*, so the derivative of $\cos(x)$ has something to do with $-\sin(x)$.

In fact, as long as we work in radians,[1] the derivatives of $\sin(x)$ and $\cos(x)$ are simply as shown in (5.10).

(5.10) $$\frac{d \sin(x)}{dx} = \cos(x) \quad \text{and} \quad \frac{d \cos(x)}{dx} = -\sin(x)$$

Equation (5.10) *looks* right, but how do we know that it's really true? To actually compute the derivatives of $\sin(x)$ and $\cos(x)$, we need the addition-of-angle formulas from trigonometry. (See the supplemental materials in Section 6.7 for a derivation of these formulas, as well as for a general review of trigonometry.) If A and B are any angles, then

$$\sin(A + B) = \sin(A)\cos(B) + \cos(A)\sin(B),$$
(5.11) $$\cos(A + B) = \cos(A)\cos(B) - \sin(A)\sin(B).$$

---

[1] If we work in degrees, then the derivative of $\sin(x)$ is $\frac{\pi}{180}\cos(x)$, which is ugly. When doing calculus, we always use radians.

5.1. Building Blocks

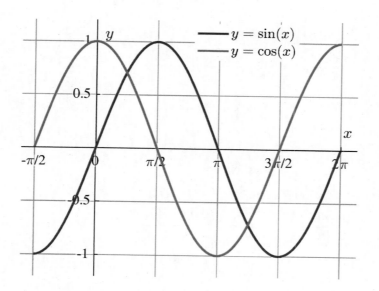

Figure 5.1. The graphs of $\sin(x)$ and $\cos(x)$

This means that

$$\frac{\sin(x+h)-\sin(x)}{h} = \frac{\sin(x)\cos(h)+\cos(x)\sin(h)-\sin(x)}{h}$$

$$= \sin(x)\left(\frac{\cos(h)-1}{h}\right)+\cos(x)\frac{\sin(h)}{h},$$

$$\frac{\cos(x+h)-\cos(x)}{h} = \frac{\cos(x)\cos(h)-\sin(x)\sin(h)-\cos(x)}{h}$$

(5.12)
$$= \cos(x)\left(\frac{\cos(h)-1}{h}\right)-\sin(x)\frac{\sin(h)}{h}.$$

To understand these limits as $h \to 0$, we have to understand how the quotients $\frac{\cos(h)-1}{h}$ and $\frac{\sin(h)}{h}$ behave as $h \to 0$. As you can see from Table 5.1, the closer h gets to

Table 5.1. Behavior of $\frac{\sin(h)}{h}$ and $\frac{\cos(h)-1}{h}$ as $h \to 0$

h	$\sin(h)$	$\frac{\sin(h)}{h}$	$\cos(h)-1$	$\frac{\cos(h)-1}{h}$
-1.00	-0.8414700	0.84147	-0.459700	0.45970
-0.10	-0.0998330	0.99833	-0.004996	0.04996
-0.01	-0.0099998	0.99998	-0.000050	0.00500
0.01	0.0099998	0.99998	-0.000050	-0.00500
0.10	0.0998330	0.99833	-0.004996	-0.04996
1.00	0.8414700	0.84147	-0.459700	-0.45970

0, the closer $\frac{\sin(h)}{h}$ gets to 1 and the closer $\frac{\cos(h)-1}{h}$ gets to 0.[2] Thus

$$\frac{d\sin(x)}{dx} = \lim_{h \to 0} \frac{\sin(x+h) - \sin(x)}{h}$$

$$= \lim_{h \to 0} \left(\sin(x) \left(\frac{\cos(h) - 1}{h} \right) + \cos(x) \frac{\sin(h)}{h} \right)$$

$$= 0 \sin(x) + 1 \cos(x)$$

(5.13) $$= \cos(x).$$

The derivative of $\cos(x)$ is similar:

$$\frac{d\cos(x)}{dx} = \lim_{h \to 0} \frac{\cos(x+h) - \cos(x)}{h}$$

$$= \lim_{h \to 0} \left(\cos(x) \left(\frac{\cos(h) - 1}{h} \right) - \sin(x) \frac{\sin(h)}{h} \right)$$

$$= 0 \cos(x) - 1 \sin(x)$$

(5.14) $$= -\sin(x).$$

Exponential Functions. Finally, we consider exponential functions. We will go over exponentials and logarithms systematically in Section 6.2, but for now we just need the basic fact that

(5.15) $$a^{b+c} = a^b a^c.$$

In particular, $a^{x+h} = a^x a^h$, so

$$\frac{da^x}{dx} = \lim_{h \to 0} \frac{a^{x+h} - a^x}{h}$$

$$= \lim_{h \to 0} \frac{a^h a^x - a^x}{h}$$

$$= \lim_{h \to 0} \frac{a^h - 1}{h} a^x$$

(5.16) $$= k_a a^x,$$

where

(5.17) $$k_a = \lim_{h \to 0} \frac{a^h - 1}{h}.$$

That is, the derivative of a^x is proportional to a^x, and the proportionality constant depends on a. From the data in Table 5.2, we see that k_2 is around 0.69, k_3 is around 1.10, and k_4 is around 1.38.

---

[2]If you don't find the table convincing, look at the supplemental materials in Section 5.8, where we provide rigorous computations of these limits.

5.1. Building Blocks

Table 5.2. Estimating k_2, k_3, and k_4 numerically

h	$\frac{2^h-1}{h}$	$\frac{3^h-1}{h}$	$\frac{4^h-1}{h}$
-1.000	0.50000	0.66667	0.75000
-0.100	0.66967	1.04042	1.29449
-0.010	0.69075	1.09260	1.37673
-0.001	0.69291	1.09801	1.38533
0.001	0.69339	1.09922	1.38726
0.010	0.69555	1.10467	1.39595
0.100	0.71773	1.16123	1.48698
1.000	1.00000	2.00000	3.00000

Since $k_2 < 1 < k_3$, there has to be a number, somewhere between 2 and 3, for which the constant is exactly 1. We call this number **Euler's number** or **Napier's constant** (but not *Euler's constant*, which is a different number!), and we denote it with the letter e. Numerically, $e \approx 2.71828$.

This makes $f(x) = e^x$ a very special function. Since $k_e = 1$, we have

$$(5.18) \qquad \frac{de^x}{dx} = k_e e^x = e^x,$$

so e^x is its own derivative! We also have $e^0 = 1$, since $a^0 = 1$ for any value of a. This means that $y = e^x$ is the solution to the rate equation

$$(5.19) \qquad y' = y$$

with initial condition $y(0) = 1$. We previously saw this rate equation (with an extra factor of r) in the context of compound interest. If you borrow a dollar from a loan shark at 100% annual interest (compounded continuously), you will owe him e^t dollars after t years. Be sure to pay him back quickly, since if you wait 10 years you'll wind up owing him over \$22,000!

Summary. The results of this section are summarized in Table 5.3. Be sure to *memorize it!*

Table 5.3. Derivatives of basic functions

$f(x)$	$f'(x)$
x^n	nx^{n-1}
$\sin(x)$	$\cos(x)$
$\cos(x)$	$-\sin(x)$
a^x	$k_a a^x$
e^x	e^x

5.2. Adding, Subtracting, Multiplying, and Dividing Functions

So far we've figured out the derivatives of functions like x^3, $\sin(x)$, and e^x. But what about functions like $x^3 + 2\sin(x) + 17e^x$? Or $5x^3 \sin(x)$? Or $\frac{e^x}{x^3 + \sin(x)}$? It turns out that if we know the derivatives of two functions $f(x)$ and $g(x)$, then it's fairly easy to find the derivatives of $f(x) + g(x)$ and $f(x) - g(x)$ and $cf(x)$, where c is any constant. Finding the derivatives of the product $f(x)g(x)$ and the quotient $f(x)/g(x)$ is a little more complicated, requiring the **product rule** and the **quotient rule**, which we will learn in this section. With those rules, we'll be able to handle all of the examples listed at the beginning of this paragraph.

The Trivial Rules: Adding, Subtracting, and Multiplying by Constants.
Imagine that a company makes green and blue widgets and stores them for later sale. At time t, the total number of green widgets in the warehouse is $W_1(t)$ and the number of blue widgets is $W_2(t)$. That is, the company is producing green widgets at rate $W_1'(t)$ and blue widgets at rate $W_2'(t)$. At a certain time, $W_1'(t) = 10{,}000$ widgets/month and $W_2'(t) = 5000$ widgets/month. At what rate is the company's total widget inventory $W_1 + W_2$ increasing?

The answer, of course, is 15,000 widgets/month. After a month, there are 10,000 new green widgets and 5000 new blue widgets, or 15,000 new widgets in all. That is,

$$(5.20) \qquad \frac{d(W_1 + W_2)}{dt} = \frac{dW_1}{dt} + \frac{dW_2}{dt}.$$

If there's a competition between green and blue teams to see who can produce the most widgets, then we might ask how fast green's lead over blue is growing. The answer, of course, is

$$(5.21) \qquad \frac{d(W_1 - W_2)}{dt} = \frac{dW_1}{dt} - \frac{dW_2}{dt}.$$

Now suppose that each green widget is worth \$9. How fast is the *value* of the green widgets growing? Since the value of that inventory is $9W_1$, we're asking for the derivative of $9W_1$. On the other hand, this is the value per unit of time of the newly produced green widgets, namely $9W_1'$. Of course these are the same, and we have the rule $(cW)' = c(W')$.

In terms of functions $f(x)$ and $g(x)$ and a constant c, we have the following.

The basic rules:

$$\frac{d}{dx}(f(x) + g(x)) = \frac{df}{dx} + \frac{dg}{dx},$$

$$\frac{d}{dx}(f(x) - g(x)) = \frac{df}{dx} - \frac{dg}{dx},$$

$$(5.22) \qquad \frac{d}{dx}(cf(x)) = c\frac{df}{dx}.$$

5.2. Multiplying and Dividing Functions

These can be combined into a single rule. If a and b are constants, then

$$\frac{d}{dx}(af(x)+bg(x)) = a\frac{df}{dx} + b\frac{dg}{dx}, \tag{5.23}$$

or in Newton's notation,

$$(af(x)+bg(x))' = af'(x) + bg'(x). \tag{5.24}$$

The same rule applies to combinations of three, four, or more functions. Thus

$$\begin{aligned}(x^3 + 2\sin(x) + 17e^x)' &= (x^3)' + 2(\sin(x))' + 17(e^x)' \\ &= 3x^2 + 2\cos(x) + 17e^x.\end{aligned} \tag{5.25}$$

The combined rule (5.24) can also be proven directly, as follows:

$$\begin{aligned}(af(x)+bg(x))' &= \lim_{h\to 0}\frac{(af(x+h)+bg(x+h))-(af(x)+bg(x))}{h} \\ &= \lim_{h\to 0} a\left(\frac{f(x+h)-f(x)}{h}\right) + b\left(\frac{g(x+h)-g(x)}{h}\right).\end{aligned} \tag{5.26}$$

As $h \to 0$, $\frac{f(x+h)-f(x)}{h}$ gets closer and closer to $f'(x)$, while $\frac{g(x+h)-g(x)}{h}$ gets closer and closer to $g'(x)$, so $a\frac{f(x+h)-f(x)}{h} + b\frac{g(x+h)-g(x)}{h}$ gets closer and closer to $af'(x)+bg'(x)$.

The Product Rule. If we know the derivative of $f(x)$ and the derivative of $g(x)$, then what is the derivative of $f(x)g(x)$? An obvious guess, which almost every calculus student has believed at one time or another, is

$$(f(x)g(x))' = f'(x)g'(x). \qquad \text{(Wrong!)} \tag{5.27}$$

Let's test whether it works using functions that we already understand.

If $f(x) = x^2$ and $g(x) = x^3$, then $f(x)g(x) = x^5$. The derivative of x^2 is $2x$, the derivative of x^3 is $3x^2$, and $f'(x)g'(x) = (2x)(3x^2) = 6x^3$. However, the derivative of x^5 is $5x^4$, not $6x^3$. The naive rule (5.27) *does not work*.

Let's try another example. If $f(x)=1$ and $g(x)=e^x$, then $f'(x)=0$, so $f'(x)g'(x) = 0$. But $f(x)g(x) = e^x$, and the derivative of e^x is not 0. Again, rule (5.27) fails.

A good way to check whether a formula is plausible is to see if the units line up. Suppose we have a rectangle of length $L(t)$ and width $W(t)$, and both the length and the width are changing with time. Then $L(t)$ has units of length (say, measured in meters), as does $W(t)$, t has units of time (say, measured in seconds), and both $L'(t)$ and $W'(t)$ have units of meters/second. Their product has units of square meters/square second. But the area $A(t) = L(t)W(t)$ has units of square meters, so A' has units of square meters/second. Something with units of m²/s can't possibly be equal to something with units of m²/s². Formula (5.27) isn't just wrong, it's nonsensical.

So if that isn't the right formula, what is? We'll work that out with an example about an oil well.

Suppose that an oil well produces oil at a rate Q, and that oil sells for a price p. Right now, $Q = 200$ barrels/year, and $p = \$70$/barrel. So the revenue $R = pQ$ generated by the well is \$14,000/year. What happens if the production increases to 205 barrels/year?

What happens if the price increases to $73/barrel? What happens if the production increases to 205 barrels/year *and* the price increases to $73/barrel?

If the production goes up and the price stays the same, then we can treat p as a constant. We have $R = 70Q$, so the change in R is 70 times the change in Q. That is,

(5.28) $$\Delta R = 70\Delta Q = 70(205 - 200) = 350.$$

Five more barrels per year at $70/barrel is worth an extra $350/year.

If the price goes up and the production stays the same, we can treat Q as a constant and write

(5.29) $$\Delta R = 200\Delta p = 200(73 - 70) = 600.$$

Three more dollars for each of the 200 barrels gives us an extra $600/year.

If both the production and price go up, then our new revenue (in dollars/year) is

(5.30) $$\begin{aligned}(70 + 3)(200 + 5) &= 70(200) + 70(5) + 3(200) + 3(5) \\ &= 14{,}000 + 350 + 600 + 15.\end{aligned}$$

The first term is our old revenue, the second is the effect of raising production, and the third is the effect of raising the price. The fourth term, involving both the change in production and the change in price, is much smaller than the others. To find the *change* in revenue, we subtract our original revenue of 14,000 from this sum. More generally, if Q changes to $Q + \Delta Q$ and p changes to $p + \Delta p$, then

(5.31) $$\begin{aligned}\Delta R &= (p + \Delta p)(Q + \Delta Q) - pQ \\ &= pQ + p\Delta Q + Q\Delta p + \Delta p \Delta Q - pQ \\ &= p\Delta Q + Q\Delta p + \Delta p \Delta Q.\end{aligned}$$

On the last line, the first term is the effect of changing just Q and the second term is the effect of changing just p. If we're talking about small changes, then the cross-term $\Delta p \Delta Q$ is a lot smaller than the two main terms.

The same reasoning applies to functions in general. Suppose that $f(x)$ and $g(x)$ are functions and we change x by a small amount h, resulting in changes $\Delta f = f(x+h) - f(x)$ and $\Delta g = g(x+h) - g(x)$ to f and g, and a change $\Delta(fg) = f(x+h)g(x+h) - f(x)g(x)$ to fg. Then

(5.32) $$\Delta(fg) = f(x)\Delta g + g(x)\Delta f + \Delta f \Delta g,$$

so

(5.33) $$\frac{\Delta(fg)}{h} = f(x)\frac{\Delta g}{h} + g(x)\frac{\Delta f}{h} + h\frac{\Delta f}{h}\frac{\Delta g}{h}.$$

As $h \to 0$, $\frac{\Delta f}{h}$ approaches $f'(x)$ (since that's the definition of $f'(x)$!) and $\frac{\Delta g}{h}$ approaches $g'(x)$. The first term on the right-hand side approaches $f(x)g'(x)$, the second approaches $g(x)f'(x)$, and the third goes to $0f'(x)g'(x) = 0$. Thus, in both Newton's notation and Leibniz's, we have the following.

5.2. Multiplying and Dividing Functions

> **The product rule:**
> $$(f(x)g(x))' = f(x)g'(x) + f'(x)g(x),$$
> (5.34) $$\frac{d(f(x)g(x))}{dx} = f(x)\frac{dg(x)}{dx} + \frac{df(x)}{dx}g(x).$$

In both versions, the first term is the effect of g changing and the second term is the effect of f changing.

Let's see how that works with some examples, including the ones that equation (5.27) failed at:

- If $f(x) = x^2$ and $g(x) = x^3$, then

(5.35) $$f(x)g'(x) + f'(x)g(x) = x^2(3x^2) + 2x(x^3) = 5x^4,$$

which is the derivative of $f(x)g(x) = x^5$.

- If $f(x) = 1$ and $g(x) = e^x$, then $f'(x) = 0$ and $g'(x) = e^x$, so $f(x)g'(x) + f'(x)g(x) = 1(e^x) + 0(e^x) = e^x$, which is the derivative of $f(x)g(x) = e^x$.
- If $f(x) = 5x^3$ and $g(x) = \sin(x)$, then $f'(x) = 15x^2$ and $g'(x) = \cos(x)$, so

(5.36) $$\left(5x^3 \sin(x)\right)' = 15x^2 \sin(x) + 5x^3 \cos(x).$$

- If $f(x) = g(x) = \sqrt{x}$, then we don't need the product rule to take the derivative of $f(x)g(x) = x$; instead, we use it to figure out the derivative of $\sqrt{x}$. Since the derivative of x is $x' = 1$, we have

$$x = \sqrt{x}\sqrt{x},$$
$$(x)' = \sqrt{x}(\sqrt{x})' + (\sqrt{x})'\sqrt{x},$$
$$1 = 2\sqrt{x}(\sqrt{x})',$$

(5.37) $$(\sqrt{x})' = \frac{1}{2\sqrt{x}},$$

in agreement with Newton's hammer: $(x^{1/2})' = \frac{1}{2}x^{-1/2}$.

The Quotient Rule. Next we turn to the ratio of two functions. Suppose that $h(x) = f(x)/g(x)$. How can we write the derivative of $h(x)$ in terms of $f'(x)$ and $g'(x)$?

Since $f(x) = g(x)h(x)$, we already have a rule that relates $f'(x)$, $g'(x)$, and $h'(x)$. We just have to apply the product rule to $f = gh$ and solve for h':

$$f(x) = g(x)h(x),$$
$$f'(x) = g(x)h'(x) + g'(x)h(x),$$
$$g(x)h'(x) = f'(x) - g'(x)h(x),$$
$$h'(x) = \frac{f'(x) - g'(x)(f(x)/g(x))}{g(x)}$$
(5.38)
$$= \frac{f'(x)g(x) - g'(x)f(x)}{(g(x))^2}.$$

In both Newton's and Leibniz's notation, we have the following.

The quotient rule:
$$\left(\frac{f}{g}\right)' = \frac{f'g - g'f}{g^2},$$
(5.39)
$$\frac{d}{dx}\left(\frac{f}{g}\right) = \frac{\frac{df}{dx}g - \frac{dg}{dx}f}{g^2}.$$

The hardest thing about remembering the quotient rule is keeping straight which term is positive and which is negative. Think of it this way: if we have a ratio $f(x)/g(x)$ of positive functions, then increasing the numerator $f(x)$ makes the ratio *bigger*. This means the term involving $f'(x)$ must be *positive*. On the other hand, increasing the denominator $g(x)$ makes the ratio *smaller*, so the term involving $g'(x)$ must be *negative*.

Here are some examples involving the quotient rule.

- If $f(x) = 1$ and $g(x) = x^n$, then $f'(x) = 0$ and $g'(x) = nx^{n-1}$, so

(5.40)
$$(x^{-n})' = \left(\frac{1}{x^n}\right)' = \frac{(1)'x^n - (x^n)'1}{x^{2n}} = \frac{-nx^{n-1}}{x^{2n}} = -nx^{-n-1}.$$

This shows that Newton's hammer works for negative powers as well as positive powers of x.

- If $f(x) = e^x$ and $g(x) = x^3 + \sin(x)$, then $f'(x) = e^x$ and $g'(x) = 3x^2 + \cos(x)$, so

$$\left(\frac{e^x}{x^3 + \sin(x)}\right)' = \frac{(x^3 + \sin(x))e^x - (3x^2 + \cos(x))e^x}{(x^3 + \sin(x))^2}$$
(5.41)
$$= \frac{(x^3 - 3x^2 + \sin(x) - \cos(x))e^x}{(x^3 + \sin(x))^2}.$$

- Besides $\sin(x)$ and $\cos(x)$, we also have the trigonometric functions $\tan(x) = \sin(x)/\cos(x)$, $\sec(x) = 1/\cos(x)$, $\cot(x) = \cos(x)/\sin(x)$, and $\csc(x) = 1/\sin(x)$.

5.3. The Chain Rule

The derivative of $\tan(x)$ is

$$
\begin{aligned}
\left(\frac{\sin(x)}{\cos(x)}\right)' &= \frac{\cos(x)(\sin(x))' - \sin(x)(\cos(x))'}{\cos^2(x)} \\
&= \frac{\cos(x)\cos(x) - \sin(x)(-\sin(x))}{\cos^2(x)} \\
&= \frac{\cos^2(x) + \sin^2(x)}{\cos^2(x)} \\
&= \frac{1}{\cos^2(x)} = \sec^2(x),
\end{aligned}
\tag{5.42}
$$

since $\sin^2(x) + \cos^2(x) = 1$.

- Similarly, you can (should!) use the quotient rule to compute the derivatives of $\sec(x)$, $\cot(x)$, and $\csc(x)$. I mean *you*, not *we*! The answers appear in Table 5.4, but don't peek until you've done the calculations yourself.

Table 5.4. More results about derivatives

Function	Derivative	Comments
$af(x)+bg(x)$	$af'(x)+bg'(x)$	Obvious
$f(x)g(x)$	$f'(x)g(x)+g'(x)f(x)$	The product rule is **not** $f'(x)g'(x)$.
$\frac{f(x)}{g(x)}$	$\frac{f'(x)g(x)-g'(x)f(x)}{g(x)^2}$	Quotient rule. Watch the signs.
x^{-n}	$-nx^{-n-1}$	Newton's hammer still works.
$\tan(x)$	$\sec^2(x)$	which equals $1/\cos^2(x)$
$\sec(x)$	$\sec(x)\tan(x)$	which equals $\sin(x)/\cos^2(x)$
$\cot(x)$	$-\csc^2(x)$	which equals $-1/\sin^2(x)$
$\csc(x)$	$-\csc(x)\cot(x)$	which equals $-\cos(x)/\sin^2(x)$

5.3. The Chain Rule

As of press time, a British pound (GBP) was worth about 1.2 American dollars (USD), and an American dollar was worth about 20 Mexican pesos (MXN).[3] At those exchange rates, how many pesos is a pound worth?

This an easy problem:

$$
\begin{aligned}
1 \text{ GBP} &= 1.2 \text{ USD}, \\
1 \text{ USD} &= 20 \text{ MXN}, \\
1 \text{ GBP} &= 1.2(20 \text{ MXN}) \\
&= 24 \text{ MXN}.
\end{aligned}
\tag{5.43}
$$

---

[3] In Mexico, the symbol $ is used to mean pesos, not dollars, so we'll refer to dollars as USD and pesos as MXN.

In a two-step chain, where pounds go to dollars and dollars go to pesos, the **conversion factor** from pounds to pesos equals the product of the conversion factors from pounds to dollars and from dollars to pesos.

Now suppose that a company does business in Britain, the United States, and Mexico. Let x be the value of their holdings in pounds, y the value in dollars, and z the value in pesos. We have

$$y = 1.2x \quad \text{so} \quad \frac{dy}{dx} = 1.2,$$

$$z = 20y \quad \text{so} \quad \frac{dz}{dy} = 20,$$

(5.44)
$$z = 24x \quad \text{so} \quad \frac{dz}{dx} = 24.$$

That is,

> **The chain rule (Leibniz's notation):**
> (5.45)
> $$\frac{dz}{dx} = \frac{dz}{dy}\frac{dy}{dx}.$$

We derived the chain rule in this simple example, where x, y, and z were all proportional to one another. However, it applies much more generally.

Suppose that x, y, and z are variables, that y is a function of x, and that z is a function of y. For instance, we might have $y = x^2$ and $z = \sin(y) = \sin(x^2)$. More generally, suppose that $y = g(x)$ and $z = f(y) = f(g(x))$. As long as f and g are differentiable, the microscope equation tells us that small changes in x result in proportional changes in y, and that small changes in y result in proportional changes in z. There are conversion factors m_1, m_2, and m_3 such that

(5.46)
$$\Delta y \approx m_1 \Delta x,$$
$$\Delta z \approx m_2 \Delta y, \quad \text{and}$$
$$\Delta z \approx m_3 \Delta x,$$

where $m_1 = \frac{dy}{dx}$, $m_2 = \frac{dz}{dy}$, and $m_3 = \frac{dz}{dx}$. Since m_3 is obviously equal to $m_1 m_2$, we conclude that $\frac{dz}{dx} = \frac{dz}{dy}\frac{dy}{dx}$.

For instance, suppose that $y = x^2 - 1$ and $z = 3e^y = 3e^{x^2-1}$. When $x = 1$, $y = 1^2 - 1 = 0$, so $z = 3e^0 = 3$. Since $dy/dx = 2x = 2$, every small change in x results in twice as big a change in y. If x becomes $1 + h$, then y will become about $0 + 2h = 2h$. Since $dz/dy = 3e^y = 3$, every change in y results in three times as big a change in z, so if y becomes $2h$, z must increase by $6h$ and become approximately $3 + 6h$. Restating

5.3. The Chain Rule

this example in equations, we have

$$y = x^2 - 1 \quad \text{so} \quad \frac{dy}{dx} = 2x,$$

$$z = 3e^y \quad \text{so} \quad \frac{dz}{dy} = 3e^y = 3e^{x^2-1},$$

(5.47) $$\frac{dz}{dx} = \frac{dz}{dy}\frac{dy}{dx} \quad \text{so} \quad \frac{dz}{dx} = 6xe^{x^2-1}.$$

Plugging in $x = 1$ gives us $\frac{dy}{dx} = 2$, $\frac{dz}{dy} = 3$, and $\frac{dz}{dx} = 6$, but the formulas also apply at other values of x. The derivative of $3e^{x^2-1}$ with respect to x equals $6xe^{x^2-1}$ everywhere.

The chain rule in equation (5.45) is very useful for **related rates** problems. For instance, suppose that we are pumping air into a large balloon at a rate of 3.7 cubic feet/minute. How fast is the *radius* of the balloon growing when the radius is 1 foot? When the radius is 5 feet?

In this problem, our three variables are the radius r, the volume V, and the time t. We imagine a chain going from time to radius to volume. We know how volume is related to radius:

(5.48) $$V = \frac{4}{3}\pi r^3, \quad \text{so} \quad \frac{dV}{dr} = 4\pi r^2.$$

We know that $dV/dt = 3.7$ cubic feet/minute. We just need to write down the chain rule and solve for dr/dt:

$$\frac{dV}{dt} = \frac{dV}{dr}\frac{dr}{dt},$$

$$3.7 \text{ ft}^3/\text{min} = 4\pi r^2 \frac{dr}{dt},$$

$$\frac{dr}{dt} = \frac{3.7 \text{ ft}^3/\text{min}}{4\pi r^2}$$

(5.49) $$= \frac{0.2944 \text{ ft}^3/\text{min}}{r^2}.$$

When $r = 1$ foot, this is 0.2944 ft/min. (r^2 has units of square feet, and cubic feet divided by square feet make feet.) When $r = 5$ feet, this is 0.01178 ft/min. We are pumping air into the balloon at a constant rate, but the radius grows much more slowly when $r = 5$ feet than when $r = 1$ foot. You may have noticed a similar effect when blowing up party balloons. At first they seem the grow quickly, but then you get out of breath and they hardly seem to grow at all.

So far we have studied the chain rule in Leibniz's notation, which is great for understanding why it works, and great for related rates problems. Now we're going to switch to Newton's notation, which is better for solving problems like "find the derivative of $\sin(e^x + x^3)$".

Instead of talking about three variables x, y, and z, we'll talk about an input variable x, two functions f and g, and the compound function $f \circ g(x) = f(g(x))$. It's sometimes useful to let $y = g(x)$ and $z = f(y) = f(g(x))$, but that isn't always necessary. Translating from Leibniz to Newton,

$$\frac{dy}{dx} \quad \text{becomes} \quad g'(x),$$

$$\frac{dz}{dy} \quad \text{becomes} \quad f'(y)$$

$$\text{which equals} \quad f'(g(x)),$$

(5.50) $$\frac{dz}{dx} \quad \text{becomes} \quad (f(g(x)))',$$

and we get the following.

The chain rule (Newton's notation):

(5.51) $$(f(g(x)))' = f'(g(x))\,g'(x).$$

Note that the derivative of f is evaluated at y, which is to say at $g(x)$, and not at x! The derivative of $\sin(x^2)$ is $2x\cos(x^2)$, not $2x\cos(x)$.

There is also a hybrid form of the chain rule, where we write u in place of $g(x)$:

(5.52) $$\frac{df(u)}{dx} = f'(u)\frac{du}{dx}.$$

In particular, if u is any function of x (such as $x^2 - 1$ or $\cos(x)$ or $e^x + 7$ or whatever you want), we have

$$\frac{d(u^n)}{dx} = nu^{n-1}\frac{du}{dx},$$

$$\frac{d\sin(u)}{dx} = \cos(u)\frac{du}{dx},$$

$$\frac{d\cos(u)}{dx} = -\sin(u)\frac{du}{dx},$$

(5.53) $$\frac{de^u}{dx} = e^u\frac{du}{dx}.$$

Here are some examples:

- The derivative of $\sin(x^2)$ with respect to x is $2x\cos(x^2)$. This is the second equation of (5.53) with $u = x^2$.
- The derivative of $\sin^2(x)$ is $2\sin(x)\cos(x)$. This is the first equation of (5.53) with $u = \sin(x)$.
- If k is a constant, then the derivative of e^{kx} is ke^{kx}. This will come up a lot when we study growth models.
- The chain rule tells us how to combine two links of a chain. If our chain is longer than two links, we can apply the rule multiple times. For instance, $\sin(e^{5x})$ is a

5.3. The Chain Rule

three-step chain: x becomes $5x$, which becomes e^{5x}, which becomes $\sin(e^{5x})$. Its derivative is

$$
\begin{aligned}
\left(\sin\left(e^{5x}\right)\right)' &= \cos\left(e^{5x}\right) \times \left(e^{5x}\right)' \\
&= \cos\left(e^{5x}\right) \times e^{5x} \times (5x)' \\
&= 5e^{5x} \cos\left(e^{5x}\right).
\end{aligned}
$$
(5.54)

- The chain rule can be combined with the product rule and the quotient rule. For instance, if we want to compute the derivative of the function $g(x) = \frac{5x^3 - 3x + \sin(x^2)}{\cos(2^x + 17)}$ that we saw at the beginning of the chapter, we first apply the quotient rule to get that $g'(x)$ equals

(5.55)
$$
\frac{\cos(2^x + 17)\left(5x^3 - 3x + \sin(x^2)\right)' - (\cos(2^x + 17))'\left(5x^3 - 3x + \sin(x^2)\right)}{\cos^2(2^x + 17)}.
$$

Next we use the sum rule and the chain rule to evaluate

(5.56)
$$
\left(5x^3 - 3x + \sin(x^2)\right)' = 15x^2 - 3 + 2x \cos(x^2).
$$

Remembering that the derivative of 2^x is $k_2 2^x$, where $k_2 \approx 0.69$, we use the chain rule to get

(5.57)
$$
(\cos(2^x + 17))' = -\sin(2^x + 17)(k_2 2^x).
$$

Combining everything, we get that the derivative of $g(x)$ is

(5.58)
$$
\frac{\cos(2^x + 17)\left(15x^2 - 3 + 2x \cos(x^2)\right) + k_2 2^x \sin(2^x + 17)\left(5x^3 - 3x + \sin(x^2)\right)}{\cos^2(2^x + 17)}.
$$

The derivatives of even the ugliest functions (and this one was pretty darn ugly!) can be evaluated patiently, *one step at a time*. We don't have to solve the whole problem by applying a single rule! We just have to turn it into a slightly easier problem, and then repeat until we're done.

- So far we have been using the chain rule to get the derivative of $f \circ g$ from the derivatives of f and g. However, it can also be used to get the derivative of f from the derivatives of $f \circ g$ and g, or the derivative of g from the derivatives of $f \circ g$ and f.

For instance, we originally obtained Newton's hammer for positive integer powers of x, and then we extended it to negative integer powers with the quotient rule. What about fractional powers? If $x^{p/q}$ is a fractional power of x (where p and q are integers and $q \neq 0$), we can write $x^p = u^q$, where $u = g(x) = x^{p/q}$. Using the first equation of (5.53),

$$
\begin{aligned}
(x^p)' &= (u^q)', \\
px^{p-1} &= qu^{q-1} u', \\
px^{p-1} &= qx^{p - \frac{p}{q}} \left(x^{p/q}\right)', \\
\left(x^{p/q}\right)' &= \frac{p}{q} x^{\frac{p}{q} - 1}.
\end{aligned}
$$
(5.59)

Newton's hammer works for fractional powers, too!

5.4. Optimization

In this section we'll see how a manufacturer can maximize revenue or profit by adjusting the amount it produces or the price it charges for its product. The price and the demand are related by market conditions, so a company can only control one of these quantities. In some settings, it sets a production level x and then sees what price $p(x)$ it can get for its product. In other settings, it sets a price p and then sees how much demand $D(p)$ there is for the product at that price. Both the **price function**[4] $p(x)$ and the **demand function** $D(p)$ express the law of supply and demand. The difference is that $p(x)$ gives the price as a function of how many items are made, while $D(p)$ gives how many items can be sold (which is usually the same as the number of items the company decides to make) as a function of the price. In both settings, we apply the 3rd Pillar of Calculus:

> *What goes up has to stop before it comes down.*

Controlling quantity. In Section 4.2 we encountered a struggling widget factory. Producing x widgets in a month costs

$$(5.60) \qquad C(x) = 22,000 + 7x,$$

in dollars. If the factory produces x widgets per month, it can sell them for

$$(5.61) \qquad p(x) = \left(10 - \frac{x}{10,000}\right) \text{ dollars/widget},$$

giving them a monthly revenue of

$$(5.62) \qquad R(x) = \left(10x - \frac{x^2}{10,000}\right) \text{ dollars}.$$

The monthly profit, in dollars, is then

$$(5.63) \qquad P(x) = R(x) - C(x) = 3x - \frac{x^2}{10,000} - 22,000.$$

The factory is currently producing 10,000 widgets/month. Alice realized that $P(10,000) = -2,000$ dollars, so it is losing money, and she argued for closing the factory. Bob computed the **marginal** cost, revenue, and profit numerically, and saw that $R'(10,000) > C'(10,000)$, so $P'(10,000) > 0$, so the factory can do better by increasing production.[5] What production level will maximize the factory's profit, and is it enough to keep the factory open?

---

[4] The price *function* is also called the demand *curve*, but for now we'll stick with the less confusing term price function.

[5] Some people like to think in terms of marginal profit, while others think in terms of marginal revenue and marginal cost. Since $P = R - C$ and $P' = R' - C'$, having a positive marginal profit is exactly the same as having a marginal revenue that is bigger than the marginal cost.

5.4. Optimization

To decide this question, we compute the marginal cost, revenue, and profit using Newton's hammer:

(5.64)
$$\begin{aligned} C'(x) &= 7, \\ R'(x) &= 10 - \frac{x}{5,000}, \quad \text{and} \\ P'(x) &= 3 - \frac{x}{5,000}. \end{aligned}$$

As long as $x < 15,000$, the marginal profit is positive, so the factory can make more money by increasing production. Once x exceeds 15,000, the marginal profit is negative, so it's better to decrease production. The ideal production level is therefore $x = 15,000$, as seen in Figure 5.2.

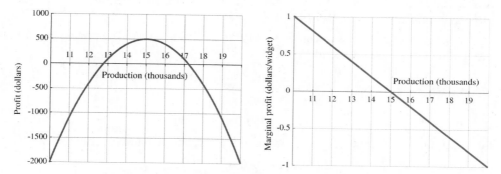

Figure 5.2. The profit $P(x)$ and marginal profit $P'(x)$ as functions of x

This calculation shows that the best production level is $x = 15,000$, but it doesn't tell us anything about how much money the factory is making or losing. For that, we have to plug $x = 15,000$ into our formula for $P(x)$:

(5.65) $$P(15,000) = 3(15,000) - \frac{(15,000)^2}{10,000} - 22,000 = 500.$$

At the new production level the factory turns a (small) profit. The factory stays open and the workers' jobs are saved.

To recap, the graph of $P'(x)$ tells us what happens as we make changes. When $P'(x) > 0$, $P(x)$ is an increasing function. When $P'(x) < 0$, $P(x)$ is a decreasing function. The **maximum** of $P(x)$ occurs when $P(x)$ stops increasing and is about to start decreasing. That's the point where $P'(x)$ goes from positive to negative and, in particular, equals 0. If we were trying to minimize a function, we would look for where the derivative stops being negative and starts being positive.

> The standard method for finding both maxima and minima is to take the derivative of the function we are studying, set it equal to 0, and solve for our input variable.

Controlling Price. In most business situations, we don't produce a certain number of items and then try to sell them at auction. Rather, we set a *price* and see how many orders we get. The **demand function** $D(p)$ gives the number of items that customers are willing to buy at price p. We then produce $x = D(p)$ items and sell them all for a total cost, revenue, and profit of

$$C = C(D(p)),$$
$$R = pD(p),$$
(5.66) $$P = R - C = pD(p) - C(D(p)),$$

where $C(x)$ is the cost of making x items.

To maximize our profit, we take the derivative of the revenue R *with respect to the price p* and compare it to the derivative of the cost C (also with respect to p) to get the derivative of profit with respect to p. If $dP/dp > 0$, then we can get more profit by raising the price. If $dP/dp < 0$, we can get more profit by lowering the price. At the price that maximizes profit, $dP/dp = 0$.

To see how this process works, let's revisit our widget factory. The price p and the production level x are related by

$$p = 10 - \frac{x}{10,000},$$
$$10,000p = 100,000 - x,$$
(5.67) $$x = 100,000 - 10,000p.$$

Since we can sell $100,000 - 10,000p$ items at price p, our demand function is

(5.68) $$D(p) = 100,000 - 10,000p.$$

Our revenue is then

(5.69) $$R = pD(p) = 100,000p - 10,000p^2,$$

our cost is

(5.70) $$C = 22,000 + 7D(p) = 722,000 - 70,000p,$$

and our profit is

(5.71) $$P = R - C = -10,000p^2 + 170,000p - 722,000.$$

We can then compute our derivatives, getting

$$dR/dp = 100,000 - 20,000p,$$
$$dC/dp = -70,000,$$
(5.72) $$dP/dp = 170,000 - 20,000p.$$

As long as $p < \$8.50$, we have $dP/dp > 0$, so we can make more profit by raising the price. If $p > \$8.50$, then $dP/dp < 0$, so we can make more profit by lowering the price. Profit is maximized when $dP/dp = 0$, namely at $p = \$8.50$. At this price, we can sell $100,000 - 10,000(8.5) = 15,000$ widgets per month for a revenue of $R(p) = 8.50 \times 15,000 = 127,500$ dollars, compared to our cost of $22,000 + 7 \times 15,000 = 127,000$ dollars, for a monthly profit of $500.

That's the same answer that we previously got by thinking of everything as a function of x. In this particular example, there are two ways to solve the problem, and you can use whichever you prefer. In most business applications, however, you don't have a choice. Sometimes you control the production level and you have to think in terms of the marginal cost, revenue, and profit $C'(x)$, $R'(x)$, and $P'(x)$, respectively. Sometimes you control the price and have to think in terms of dC/dp, dR/dp, and dP/dp. Either way, the overall strategy is the same: to maximize profit, take the derivative of profit with respect to the quantity you control, and set it equal to 0.

5.5. The Shape of a Graph

We have seen that the sign of the derivative $f'(x)$ of $f(x)$ tells us whether $f(x)$ is increasing, and that the places where $f'(x) = 0$ are candidates for being maxima or minima. In this section we'll elaborate on these ideas. We'll see how studying the sign of $f(x)$, the sign of $f'(x)$, and the sign of the derivative of $f'(x)$—written $f''(x)$ and called the **second derivative** of $f(x)$—tells us a lot about the graph of $f(x)$. We will use two polynomials for our case studies: $f(x) = x^3 - 3x$ and $g(x) = x^4 - 6x^2$.

The Graph of $f(x) = x^3 - 3x$. The first thing to ask about any function is

> *Where is it positive, and where is it negative?*

We don't actually have to check every point. Once we know that $f(-10) = -970$, we have a pretty good idea that $f(-10.01)$ and $f(-9.99)$ will also be negative. In fact, there are only two ways that a function can go from negative to positive or vice versa:

(1) it can cross through 0, or

(2) it can jump.

In this case, the function $f(x)$ is a continuous polynomial (actually, *all* polynomials are continuous), and so it can't jump. The only places where $f(x)$ can change sign are at the places where $f(x) = 0$. These are called the **roots** of $f(x)$. In this case, finding the roots is easy:

(5.73)
$$\begin{aligned} x^3 - 3x &= 0, \\ x(x^2 - 3) &= 0, \\ x = 0 \quad &\text{or} \quad x^2 = 3, \\ x = 0 \quad &\text{or} \quad x = \pm\sqrt{3}. \end{aligned}$$

We mark these on a number line as in Figure 5.3.

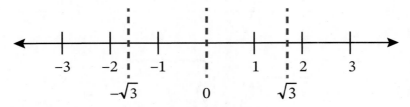

Figure 5.3. The roots of $f(x) = x^3 - 3x$

This divides the number line into four intervals: $(-\infty, -\sqrt{3})$, $(-\sqrt{3}, 0)$, $(0, \sqrt{3})$, and $(\sqrt{3}, \infty)$. The sign of $f(x)$ can't change within an interval; on each interval, the function is either completely positive or completely negative. To figure out which, we just pick a point in each interval and we test it:

(5.74)
$$\begin{aligned} f(-10) = -970 < 0 &\quad \text{so} \quad f(x) \text{ is negative on } (-\infty, -\sqrt{3}), \\ f(-1) = 2 > 0 &\quad \text{so} \quad f(x) \text{ is positive on } (-\sqrt{3}, 0), \\ f(1) = -2 < 0 &\quad \text{so} \quad f(x) \text{ is negative on } (0, \sqrt{3}), \\ f(10) = 970 > 0 &\quad \text{so} \quad f(x) \text{ is positive on } (\sqrt{3}, \infty). \end{aligned}$$

There isn't anything special about the numbers -10, -1, 1, and 10. We could just as well have picked -5π, -1.27, 0.3, and 1776. But it's easier to evaluate our function at ± 10 and ± 1.

We update our number line with this new information, producing Figure 5.4, and call it a **sign chart** for $f(x)$.

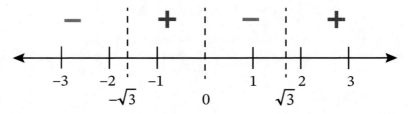

Figure 5.4. The sign chart of $f(x) = x^3 - 3x$

If we didn't know anything about the size of $f(x)$ and only had the information in the sign chart, we could take a first stab at graphing $f(x)$, and we'd get something that looks like Figure 5.5.

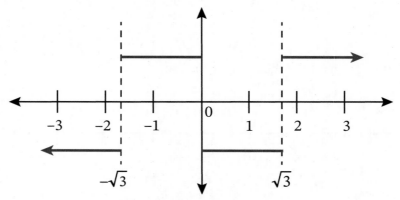

Figure 5.5. A schematic graph of $f(x) = x^3 - 3x$

5.5. The Shape of a Graph

Of course, this isn't a very *good* graph! On each interval we've drawn our function as a horizontal line because we don't know whether it's increasing or decreasing. To understand that, we need to look at $f'(x)$.

The next step is to make a sign chart for $f'(x) = 3x^2 - 3$. We repeat what we did before, only for $f'(x)$ instead of for $f(x)$. We first look for the places where $f'(x) = 0$ or $f'(x)$ doesn't exist. These are called the **critical points** of $f(x)$. Since $f'(x) = 3x^2 - 3 = 3(x^2 - 1)$, these occur at $x = \pm 1$. Proceeding as before, we break our number line into three intervals, $(-\infty, -1)$, $(-1, 1)$, and $(1, \infty)$. We evaluate $f'(x)$ at three test points, say -2, 0, and 2.

(5.75)
$$f'(-2) = 9 > 0 \quad \text{so} \quad f'(x) \text{ is positive on } (-\infty, -1),$$
$$\text{which means} \quad f(x) \text{ is increasing on } (-\infty, -1).$$

$$f'(0) = -3 < 0 \quad \text{so} \quad f'(x) \text{ is negative on } (-1, 1),$$
$$\text{which means} \quad f(x) \text{ is decreasing on } (-1, 1).$$

$$f'(2) = 9 > 0 \quad \text{so} \quad f'(x) \text{ is positive on } (1, \infty),$$
$$\text{which means} \quad f(x) \text{ is increasing on } (1, \infty).$$

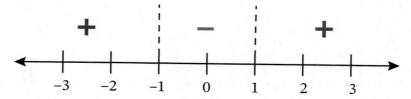

Figure 5.6. The sign chart of $f'(x) = 3x^2 - 3$

We can combine the sign chart for $f(x)$ (Figure 5.4) and the sign chart for $f'(x)$ (Figure 5.6) into one chart (Figure 5.7).

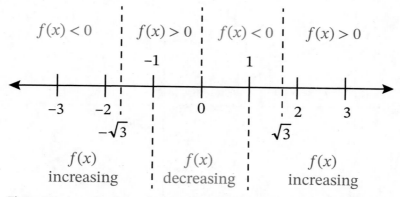

Figure 5.7. A combined sign chart for $f(x)$ and its derivative. The sign of the derivative tells us whether $f(x)$ is increasing or decreasing.

At $x = -1$, the function stops increasing and starts decreasing. In other words, $f(-1) = 2$ must be higher than the values of $f(x)$ at any nearby points. It isn't necessarily the highest point on the whole graph (in fact, $f(3) = 18$, which is much bigger than 2), but it's higher than all of its neighbors. We call a point like $(-1, 2)$ a **local maximum**. Similarly, when $x = 1$ the function stops decreasing and starts increasing, so that must be the bottom of a valley, and there is a **local minimum** at $(1, -2)$. The sign chart for $f'(x)$ tells us everything we need to know to find the local maxima and minima. We summarize these observations:

> **First Derivative Test:** The local maxima of a differentiable function occur where the derivative changes from positive to negative. The local minima occur where the derivative changes from negative to positive.

Now we're ready to make a better sketch, drawing an upward-sloping line wherever $f'(x) > 0$ and a downward sloping line wherever $f'(x) < 0$, making sure that the sign of the function agrees with our sign chart for $f(x)$. That is, there are four possible local pictures of our function, shown in Figure 5.8.

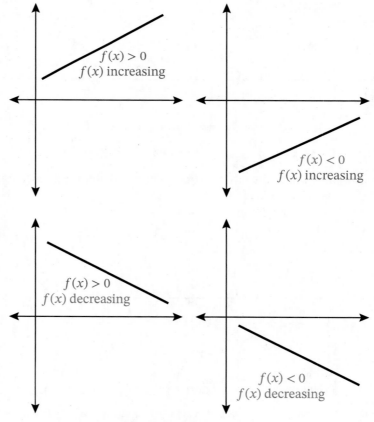

Figure 5.8. The four possible local pictures

5.5. The Shape of a Graph

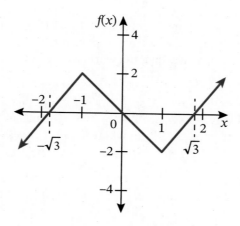

Figure 5.9. An improved sketch of the graph of $f(x) = x^3 - 3x$

Applying this technique to our number line, with the six intervals $(-\infty, -\sqrt{3})$, $(-\sqrt{3}, -1)$, $(-1, 0)$, $(0, 1)$, $(1, \sqrt{3})$, and $(\sqrt{3}, \infty)$, gives the improved sketch shown in Figure 5.9.

Finally, we want to consider which way the graph of $f(x)$ is curving. In a ∪-shaped graph like that of x^2, the derivative is always increasing, going from very negative to slightly negative to 0 to slightly positive to very positive. Curves with this shape are said to be **concave up**. In a ∩-shaped graph like that of $-x^2$, the derivative is always decreasing, going from very positive to slightly positive to 0 to slightly negative to very negative. Curves with this shape are said to be **concave down**. See Figure 5.10. The **concavity** of a graph depends on whether $f'(x)$ is increasing or decreasing, which depends on whether the derivative of $f'(x)$, written $f''(x)$, is positive or negative. That is:

The graph of $f(x)$ is concave up $\Leftrightarrow$ $f'(x)$ is increasing $\Leftrightarrow$ $f''(x) > 0$.
The graph of $f(x)$ is concave down $\Leftrightarrow$ $f'(x)$ is decreasing $\Leftrightarrow$ $f''(x) < 0$.

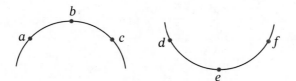

Figure 5.10. When the second derivative is negative, the first derivative can be positive (at a), 0 (at b), or negative (at c). When the second derivative is positive, the first derivative can be negative (at d), 0 (at e), or positive (at f).

When $f'(x) = 0$ and $f''(x) > 0$, the first derivative must be increasing, so it must be changing from negative to positive. Likewise, when $f'(x) = 0$ and $f''(x) < 0$, the first derivative must be decreasing, so it must be changing from positive to negative. Combined with the First Derivative Test, this gives:

> **Second Derivative Test**: Points where $f'(x) = 0$ and $f''(x) < 0$ are local maxima. Points where $f'(x) = 0$ and $f''(x) > 0$ are local minima. If $f'(x) = 0$ and $f''(x) = 0$, then the second derivative test does not work; we need to use the first derivative test instead.

In our example, $f'(x) = 3x^2 - 3$, so $f''(x) = (f'(x))' = (3x^2 - 3)' = 6x$, so our sign chart for $f''(x)$ is very simple; see Figure 5.11. The places where $f''(x) = 0$ are called **subcritical points**, and the places where $f''(x)$ changes sign are called **points of inflection**.

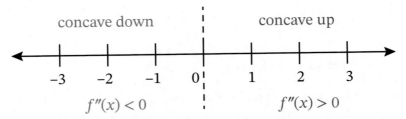

Figure 5.11. The sign chart for $f''(x) = 6x$

Since $f'(-1) = 0$ and $f''(-1) < 0$, there is a local maximum at $x = -1$, and since $f'(1) = 0$ and $f''(1) > 0$, there is a local minimum at $x = 1$. In this example, we already knew about the local maximum and local minimum from the sign chart of $f'(x)$, but sometimes it's easier to apply the second derivative test than the first derivative test.

Combining everything, we can create Table 5.5 showing what is happening to $f(x)$ in each interval and at the points where the intervals meet.

Table 5.5. Behavior of $f(x) = x^3 - 3x$

Interval	Sign of f	Direction	Concavity	Comment
$x < -\sqrt{3}$	negative	up	down	
$x = \sqrt{3}$	0	up	down	root
$-\sqrt{3} < x < -1$	positive	up	down	
$x = -1$	positive	flat	down	local max
$-1 < x < 0$	positive	down	down	
$x = 0$	0	down	0	inflection point
$0 < x < 1$	negative	down	up	
$x = 1$	negative	flat	up	local min
$1 < x < \sqrt{3}$	negative	up	up	
$x = \sqrt{3}$	0	up	up	root
$x > \sqrt{3}$	positive	up	up	

5.5. The Shape of a Graph

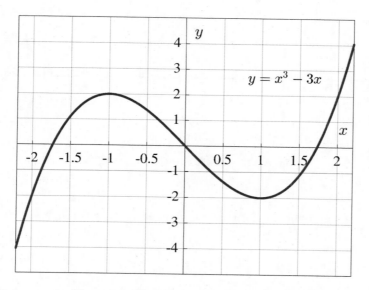

Figure 5.12. The actual graph of $f(x) = x^3 - 3x$

With this data we can adjust our sketch (Figure 5.9), smoothing the rough corners and making the graph concave down when $x < 0$ and concave up when $x > 0$. The resulting sketch probably wouldn't look *exactly* like the actual graph of $f(x)$, but it would have all the right qualitative features. The actual graph is shown in Figure 5.12.

The Graph of $x^4 - 6x^2$. As a second example, let's look at the function $g(x) = x^4 - 6x^2$. First we find the roots of $g(x)$:

$$x^4 - 6x^2 = 0,$$
$$x^2(x^2 - 6) = 0,$$
(5.76) $$x = 0 \quad \text{or} \quad x = \pm\sqrt{6} \approx \pm 2.45.$$

We pick test points at $-3, -1, 1,$ and 3, and we compute

$$f(-3) = 27 > 0 \quad \text{so} \quad f(x) \text{ is positive on } (-\infty, -\sqrt{6}),$$
$$f(-1) = -5 < 0 \quad \text{so} \quad f(x) \text{ is negative on } (-\sqrt{6}, 0),$$
$$f(1) = -5 < 0 \quad \text{so} \quad f(x) \text{ is negative on } (0, -\sqrt{6}),$$
(5.77) $$f(3) = 27 > 0 \quad \text{so} \quad f(x) \text{ is positive on } (\sqrt{6}, \infty),$$

as in the sign chart shown in Figure 5.13.

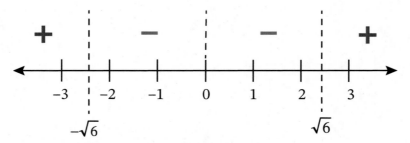

Figure 5.13. The sign chart for $g(x) = x^4 - 6x^2$

Using just the sign chart for $g(x)$, we can get the (very!) rough sketch shown in Figure 5.14:

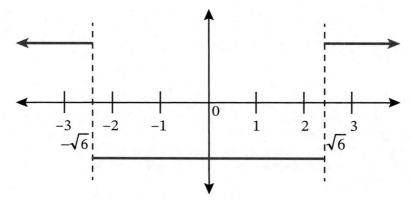

Figure 5.14. A schematic graph of $g(x) = x^4 - 6x^2$

Note that the signs do not simply alternate, as in the previous example! A continuous function can only change sign by crossing through 0, but it doesn't *have* to change sign every time it hits 0. (Think about the functions x^2 and $-x^2$, which hit 0 at $x = 0$ but never change sign.) Not every root is an axis crossing. Not every critical point is a local maximum or local minimum. Not every subcritical point is a point of inflection. Most are, but some are not.

Next we compute

(5.78) $$g'(x) = 4x^3 - 12x, \qquad g''(x) = 12x^2 - 12.$$

Note how $g(x)$ is related to our previous example. Since $g'(x) = 4f(x)$ and $g''(x) = 4f'(x)$, we already have our sign charts for $g'(x)$ (Figure 5.4) and $g''(x)$ (Figure 5.6), which we repeat in Figure 5.15 for convenience. $g'(x)$ is negative for $x < -\sqrt{3}$, positive for $-\sqrt{3} < x < 0$, negative for $0 < x < \sqrt{3}$, and positive for $x > \sqrt{3}$. This means that there must be local minima at $x = \pm\sqrt{3}$ and a local maximum at $x = 0$. At these extreme points, the values of g are $g(\pm\sqrt{3}) = -9$ and $g(0) = 0$.

5.5. The Shape of a Graph

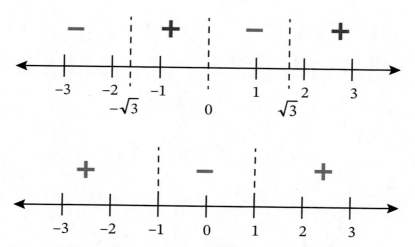

Figure 5.15. The sign charts of $g'(x)$ (top) and $g''(x)$ (bottom)

Using the information from the sign chart for $g'(x)$, we can make a better sketch (Figure 5.16) that shows the local maximum and local minima.

Meanwhile, $g''(x)$ is positive when $x < -1$ or $x > 1$ and is negative when $-1 < x < 1$. This means that the graph of $g(x)$ is concave up on $(-\infty, -1)$, concave down on $(-1, 1)$, and concave up on $(1, \infty)$. Using the information from the sign chart for $g''(x)$,

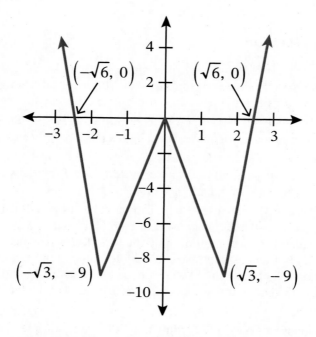

Figure 5.16. A better sketch of the graph of $g(x) = x^4 - 6x^2$

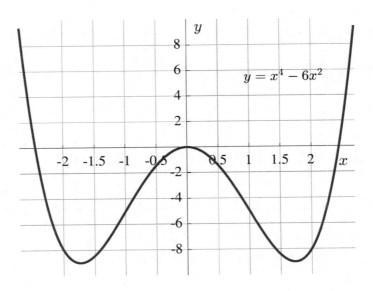

Figure 5.17. The actual graph of $g(x) = x^4 - 6x^2$

we can make a better sketch that resembles the true graph of g(x), shown in Figure 5.17.

5.6. Newton's Method

Graphing $f(x) = e^x + \frac{x^2}{2} - 4x + 1$. A key part of curve sketching was finding the roots, critical points, and subcritical points of $f(x)$. This involved solving the equations $f(x) = 0$, $f'(x) = 0$, and $f''(x) = 0$. In our examples, $f(x)$ and its derivatives were polynomials that were easily factored, which made producing our sign charts easy.

But what do we do when we have to deal with a more complicated function, such as $f(x) = e^x + \frac{x^2}{2} - 4x + 1$, with $f'(x) = e^x + x - 4$ and $f''(x) = e^x + 1$? In this example, $f''(x)$ is always positive, so $f'(x)$ is always increasing and $f(x)$ is always concave up. But what do the sign charts of f' and f look like?

Since $f'(x)$ is always increasing, there can be at most one critical point of $f(x)$. Since $f'(1) = e - 3 \approx -0.282$ is negative and $f'(2) = e^2 - 2 \approx 5.39$ is positive, that critical point (which is a local minimum by the First Derivative Test) is somewhere between 1 and 2. But where is it? We could try to home in on it with the bisection method of Section 3.7, but there is a better way, called Newton's method. We'll return to this example at the end of this section, but first let's develop the method in general.

Question 1: Suppose we have a linear function $L(x)$ with multiplier (a.k.a. slope) m, and that $L(a) = b$. Where does this function hit 0?

5.6. Newton's Method

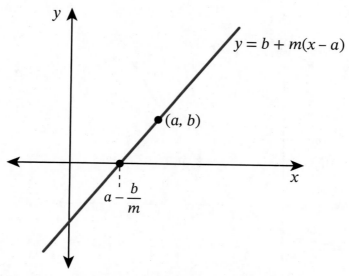

Figure 5.18. The x intercept of a line with slope m through (a, b) is $a - \frac{b}{m}$.

Answer: See Figure 5.18. The equation of the linear function in point-slope form is

(5.79) $$L(x) = b + m(x - a).$$

(Note that we are using point-slope form, not slope-intercept form, and that b is not the y-intercept!) Setting this equal to 0 gives

$$\begin{aligned} b + m(x - a) &= 0, \\ m(x - a) &= -b, \\ x - a &= -\frac{b}{m}, \end{aligned}$$

(5.80) $$x = a - \frac{b}{m}.$$

Question 2: Suppose that we have a *nonlinear* function $f(x)$ such that $f(a) = b$ and $f'(a) = m$. Use a linear approximation to estimate where $f(x) = 0$.

Answer: This is the same problem! See Figure 5.19. The best linear approximation to our function near $x = a$ is

(5.81) $$f(x) \approx b + m(x - a)$$

which is 0 at $x = a - \frac{b}{m}$.

Idea behind Newton's method: If $x_0 = a$ is close to a root of $f(x)$, then

(5.82) $$a - \frac{f(a)}{f'(a)}$$

is *very* close to a root of $f(x)$.

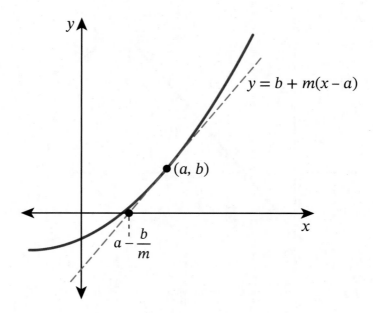

Figure 5.19. A root of a function with $f(a) = b$ and $f'(a) = m$ is approximately $a - \frac{b}{m}$.

The Actual Algorithm for Newton's Method. Newton's method uses this idea recursively to home in amazingly quickly on roots.

(1) Start with an initial point x_0 that is fairly close to a root. For solving $e^x + x - 4 = 0$, we will pick $x_0 = 1$. Unfortunately, there is no simple rule for how to pick x_0 or what "fairly close" means.

(2) Turn this into an improved guess, $x_1 = x_0 - \frac{f(x_0)}{f'(x_0)}$.

(3) Turn this into an improved guess, $x_2 = x_1 - \frac{f(x_1)}{f'(x_1)}$.

(4) Repeat as many times as necessary, always taking the new guess x_{n+1} to be

(5.83) $$x_{n+1} = x_n - \frac{f(x_n)}{f'(x_n)}.$$

(5) Stop when $f(x_n)$ is close enough to 0 (e.g., zero to fifteen decimal places).

(6) Convergence is typically extremely fast, with the number of decimal places doubling every step. For some functions, like x^2 (whose derivative is 0 at the root), convergence is slower, but the method still works.

(7) If the method didn't seem to work, either there isn't a root where we thought there was, or our initial guess x_0 wasn't good enough. We may need to use the bisection method a few times to get a better x_0, and then switch to Newton's method.

Returning to our example, we want to graph the function $f(x) = e^x + \frac{x^2}{2} - 4x + 1$, which requires finding the roots of the function $g(x) = f'(x) = e^x + x - 4$. Here $g'(x) = e^x + 1$. Since e is close to 3, $g(1) = e - 3$ is fairly close to 0, so it is reasonable to pick $x_0 = 1$ as our initial guess. Table 5.6 shows our calculations. It only takes three

5.6. Newton's Method

Table 5.6. Solving $g(x) = 0$ with Newton's method

n	x_n	$g(x_n)$	$g'(x_n)$	$x_n - \frac{g(x_n)}{g'(x_n)}$
0	1	-0.281718172	3.71828183	1.07576569
1	1.07576569	0.0008003	3.93223722592	1.0737304832
2	1.0737304832	6.06861×10^{-6}	3.92627558542	1.07372893756
3	1.07372893756	1.37×10^{-11}	3.92627106245	1.07372893756

steps to get from our initial guess $x_0 = 1$ to an estimate $x_3 = 1.07372893756$ that is accurate to eleven decimal places.

Now that we have found our root $x_3 = 1.07371893756$ of $g(x)$, which is a local minimum of $f(x)$, we evaluate $f(x_3) = 0.20780222789$. Since the value of $f(x)$ is positive at this minimum, $f(x)$ must be positive everywhere. More precisely, $f'(x)$ is negative for $x < x_3$ and is positive for $x > x_3$, and $f''(x)$ is positive everywhere. This means that the graph of $f(x)$ is roughly U-shaped, with a single minimum at $(1.07372893756, 0.20780222789)$, as shown in Figure 5.20.

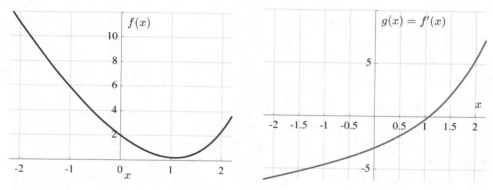

Figure 5.20. The graphs of $f(x) = e^x + \frac{x^2}{2} - 4x + 1$ and $g(x) = f'(x) = e^x + x - 4$

Return to Babylon. Next let's turn our attention to finding square roots. If a is a positive number, then the roots of $f(x) = x^2 - a$ are $\pm\sqrt{a}$. Here we have $f'(x) = 2x$, so the formula for Newton's method becomes

$$
\begin{aligned}
x_{n+1} &= x_n - \frac{f(x_n)}{f'(x_n)} \\
&= x_n - \frac{x_n^2 - a}{2x_n} \\
&= \frac{x_n}{2} + \frac{a}{2x_n} \\
&= \frac{1}{2}\left(x_n + \frac{a}{x_n}\right).
\end{aligned}
$$

(5.84)

For $f(x) = x^2 - a$, Newton's method is *exactly* the same as the Babylonian method. That's why our approximation of $\sqrt{5}$ in Chapter 3 converged so quickly.

As with the Babylonian method, it's easy to program a computer to do all the calculations in Newton's method. Here is a sample MATLAB program that implements Newton's method for $g(x) = e^x + x - 4$ and generates the data in Table 5.6.

```
M=zeros(4,5);                    % Initialize
x = 1;                           % Our initial guess
for j=1:4                        % Run the loop
    M(j,1) = j-1                 % n runs from 0 to 3
    M(j,2)=x                     % Write down x_n
    M(j,3) = exp(x)+x-4          % That's g(x_n)
    M(j,4) = exp(x)+1            % That's g'(x_n)
    M(j,5) = M(j,2)-M(j,3)/M(j,4) % x_n - g(x_n)/g'(x_n)
    x = M(j,5)                   % Set x to its new value
end;                             % Finish the loop
Answer=M                         % Print out the table.
```

When Newton's Method Fails. Newton's method can work brilliantly when three conditions are met:

(1) The function has a root.
(2) The function is differentiable near that root.
(3) Our initial guess is close to that root.

In such cases, our sequence of guesses converges quickly to that root, typically doubling the number of decimal places of accuracy at each step. However, things can fall apart when the conditions aren't met.

First, bad things can come from a bad initial guess, even if the function itself is well behaved. For instance, consider the function $f(x) = xe^{-x^2/2}$ shown in Figure 5.21. This function has a single root at $x = 0$ and critical points at $x = \pm 1$. If we choose x_0 close enough to the origin (anything between $-\sqrt{2}/2$ and $\sqrt{2}/2$ will do), the method works and we quickly converge to 0. This region is called the **basin of attraction** for the root $x = 0$. If $x_0 > 1$, however, the tangent line leads us in the wrong direction. For instance, if $x_0 = 2$, then $x_1 = 8/3 \approx 2.67$. If x is between $\sqrt{2}/2$ and 1 or between -1 and $-\sqrt{2}/2$, then the tangent line takes us in the right direction, but overshoots. For instance, if $x_0 = 0.8$, then $x_1 \approx -1.42$.

Next, Newton's method will often fail if $f(x)$ is not differentiable at the root. For instance, the graph of the function $f(x) = x^{1/3}$ (see Figure 5.22) has a vertical tangent at $x = 0$, so there is no good linear approximation to the function there. In this case, we wind up overshooting the mark no matter where we start, and

$$(5.85) \qquad x_{n+1} = x_n - \frac{x_n^{1/3}}{\frac{1}{3}x^{-2/3}} = x_n - 3x_n = -2x_n.$$

5.6. Newton's Method

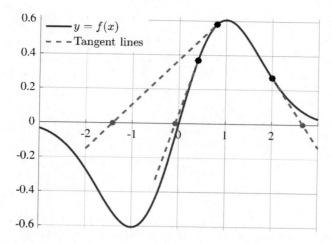

Figure 5.21. Newton's method for $f(x) = xe^{-x^2/2}$ works well starting from $x_0 = 0.4$ but not from $x_0 = 0.8$ or $x_0 = 2$.

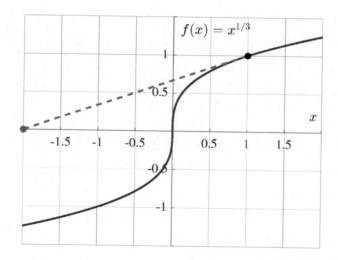

Figure 5.22. No matter where we start, Newton's method for $f(x) = x^{1/3}$ badly overshoots.

If our first guess is $x_0 = 0.01$, then we'll get $x_1 = -0.02$, $x_3 = 0.04$, $x_4 = -0.08$, and so on.

Finally, Newton's method can be deceived by points that are almost roots, but not quite. For instance, the graph of $f(x) = x^2 + 0.01$, shown in Figure 5.23, comes close to touching the x axis at $x = 0$. If we start at $x_0 = 1$, the first few estimates are $x_1 = 0.495$, $x_2 = 0.237399$, $x_3 = 0.97579$, and $x_4 = -0.0024510$, and it looks like we're approaching a root. However, at the fifth step we run away. The value of $f(x_4)$ is quite small, but the value of $f'(x_4)$ is even smaller, making $x_5 = x_4 - f(x_4)/f'(x_4) = 2.038758$ big.

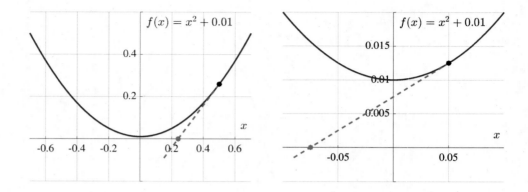

Figure 5.23. When doing one step of Newton's method on $f(x) = x^2 + 0.01$, the extra 0.01 doesn't really matter when x_0 is big, but can matter quite a bit when x_0 is small.

5.7. Chapter Summary

The Main Ideas.

- The derivatives of our basic "building block" functions are the following.

$f(x)$	x^n	$\sin(x)$	$\cos(x)$	$\tan(x)$	a^x	e^x
$f'(x)$	nx^{n-1}	$\cos(x)$	$-\sin(x)$	$\sec^2(x)$	$k_a a^x$	e^x

- Sums, differences, and scalar multiples of functions follow the obvious rule

(5.86) $$(af(x) + bg(x))' = af'(x) + bg'(x).$$

- The **product rule** is

(5.87) $$(f(x)g(x))' = f(x)g'(x) + g(x)f'(x).$$

The derivative is **not** $f'(x)g'(x)$.

- The **quotient rule** is

(5.88) $$\left(\frac{f(x)}{g(x)}\right)' = \frac{f'(x)g(x) - g'(x)f(x)}{(g(x))^2}.$$

- There are three forms of the chain rule. Learn all three, since there are places where each one is the most useful:

$$\frac{dz}{dx} = \frac{dz}{dy}\frac{dy}{dx} \qquad \text{Leibniz's chain rule;}$$

$$(f(g(x)))' = f'(g(x))\, g'(x) \qquad \text{Newton's chain rule;}$$

(5.89) $$\frac{d}{dx}f(u) = f'(u)\frac{du}{dx} \qquad \text{the hybrid chain rule.}$$

5.7. Chapter Summary

- *What goes up has to stop before it comes down.* Local maxima and minima can only occur at critical points. Almost all optimization problems involve taking the derivative of something and setting it equal to 0.
- Business optimization problems involve both the quantity x and the price p of an item being manufactured. These are related by the price function and the demand function. Sometimes you control x and have to write p as a function of x. Sometimes you control p and have to write x as a function of p. Either way, the revenue is x times p and the profit is the revenue minus the cost.
- First derivative test:
  1. Local maxima occur where $f'(x)$ switches from positive to negative.
  2. Local minima occur where $f'(x)$ switches from negative to positive.
- Second derivative test:
  1. If $f'(x) = 0$ and $f''(x) < 0$, then there is a local maximum at x.
  2. If $f'(x) = 0$ and $f''(x) > 0$, then there is a local minimum at x.
  3. If $f'(x) = 0$ and $f''(x) = 0$, then we can't tell.
- The key to graphing is making sign charts of $f(x)$, $f'(x)$, and $f''(x)$.
- The sign of $f'(x)$ tells whether $f(x)$ is increasing or decreasing. The sign of $f''(x)$ tells whether the graph of $f(x)$ is concave up or concave down.
- Newton's method is the iteration $x_{n+1} = x_n - f(x_n)/f'(x_n)$. It homes in extremely quickly to a root of $f(x)$ if $f(x)$ is differentiable and our initial guess is close enough to that root.

Expectations. You should be able to:

- State from memory the derivatives of the basic functions, the product and quotient rules, and all three versions of the chain rule.
- Use the rules to analyze the derivative of a complicated function, breaking it down step by step until all that is left are derivatives of building blocks.
- Solve related rates problems using the chain rule.
- Optimize a function by finding its critical points and applying the first and second derivative tests.
- Apply optimization to business problems involving cost, revenue, and profit, both in situations where you control the quantity (and price is a function of quantity) and in situations where you control the price (and demand is a function of price).
- Use sign charts for $f(x)$, $f'(x)$, and $f''(x)$ to sketch the graph of a function, correctly indicating
    - regions where the function is positive,
    - regions where the function is negative,
    - regions where the function is increasing,
    - regions where the function is decreasing,
    - regions where the graph is concave up,
    - regions where the graph is concave down, and
    - the locations of roots, local maxima, local minima, and points of inflection.

- Use Newton's method to find roots of a complicated functions.
- Write (or adapt) a computer program to automate that computation.

5.8. Supplemental Material: Small Angle Approximations

In Section 5.1, we computed the limits

(5.90) $$\lim_{h \to 0} \frac{\sin(h)}{h} \quad \text{and} \quad \lim_{h \to 0} \frac{\cos(h) - 1}{h}$$

numerically, finding that the first limit is 1 and the second limit is 0. Here we'll derive these limits rigorously, and in the process we'll develop two very useful approximations,

(5.91) $$\sin(x) \approx x \quad \text{and} \quad \cos(x) \approx 1 - \frac{x^2}{2},$$

that apply whenever the angle x is small (up to half a radian or so, or just under 30 degrees). These are called **small angle approximations**.

Suppose that h is an angle between 0 and $\pi/2$. If we take a wedge W of angle h from the unit circle, the area of W is a fraction $h/(2\pi)$ of the area of the entire circle and is therefore equal to $\frac{h}{2\pi}\pi = \frac{h}{2}$. We then consider a triangle T_1 inscribed in W and a triangle T_2 that contains W, as in Figure 5.24.

The triangle T_1 has a base of $\cos(h)$ and a height of $\sin(h)$, so it has an area of $\sin(h)\cos(h)/2$. The triangle T_2 has a base of 1 and a height of $\tan(h) = \frac{\sin(h)}{\cos(h)}$, so it has an area of $\frac{\sin(h)}{2\cos(h)}$. Since T_1 is contained in W, and W is contained in T_2, we have

$$\text{Area of } T_1 < \text{Area of } W < \text{Area of } T_2,$$

(5.92) $$\frac{\sin(h)\cos(h)}{2} < \frac{h}{2} < \frac{\sin(h)}{2\cos(h)}.$$

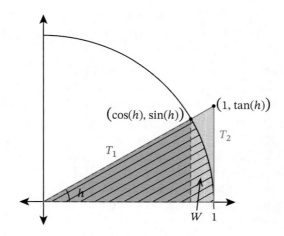

Figure 5.24. Two triangles and a slice of pie

5.8. Small Angle Approximations

Multiplying by $2/\sin(h)$ then gives

(5.93) $$\cos(h) < \frac{h}{\sin(h)} < \frac{1}{\cos(h)}.$$

Taking reciprocals switches the directions of the inequalities, giving us

(5.94) $$\frac{1}{\cos(h)} > \frac{\sin(h)}{h} > \cos(h).$$

As h gets closer and closer to 0, $\cos(h)$ and $\frac{1}{\cos(h)}$ both get closer and closer to 1. Since $\frac{\sin(h)}{h}$ is trapped (or **squeezed** or **sandwiched**) between two numbers, both of which are close to 1, $\frac{\sin(h)}{h}$ must itself be close to 1.

So far we've only considered positive angles. To get our limit, we have to also consider the possibility that $h < 0$. So suppose that $h = -\theta$, where θ is a small positive angle. Then

(5.95) $$\frac{\sin(h)}{h} = \frac{\sin(-\theta)}{-\theta} = \frac{-\sin(\theta)}{-\theta} = \frac{\sin(\theta)}{\theta} \approx 1.$$

Since $\frac{\sin(h)}{h}$ is close to 1 for *all* values of h that are close to 0, we have established that

(5.96) $$\lim_{h \to 0} \frac{\sin(h)}{h} = 1.$$

This also gives the first half of our small angle approximation. If x is small, $\frac{\sin(x)}{x}$ is close to 1, so $\sin(x) \approx x$.

To understand limits involving $\cos(h) - 1$, we first notice that

(5.97) $$\cos(h) - 1 = \frac{(\cos(h) - 1)(\cos(h) + 1)}{\cos(h) + 1} = \frac{\cos^2(h) - 1}{\cos(h) + 1} = \frac{-\sin^2(h)}{\cos(h) + 1}.$$

We then have

(5.98) $$\frac{\cos(h) - 1}{h} = \frac{-\sin^2(h)}{h(\cos(h) + 1)}$$
$$= -\left(\frac{\sin(h)}{h}\right)^2 \frac{h}{\cos(h) + 1}.$$

As $h \to 0$, $\frac{\sin(h)}{h}$ goes to 1 (as we just proved), while $\frac{h}{\cos(h)+1}$ goes to 0, so the product goes to 0.

In fact, we can say more. By the exact same argument,

(5.99) $$\frac{\cos(h) - 1}{h^2} = -\left(\frac{\sin(h)}{h}\right)^2 \frac{1}{\cos(h) + 1}.$$

As $h \to 0$, the right-hand side approaches $-(1)^2 \frac{1}{2} = -\frac{1}{2}$. Since $(\cos(h) - 1)/h^2 \approx -1/2$, we must have $\cos(h) - 1 \approx -h^2/2$, so $\cos(h) \approx 1 - \frac{h^2}{2}$. Replacing h with x, this is the second half of our small angle approximation.

5.9. Exercises

Derivative Rules

5.1. Find the derivative of each function.
(a) $g(x) = 3x^5 - 2x + 5$
(b) $f(x) = 3\sin(x) - 5e^x$
(c) $h(x) = \tan(x) - \pi x^3 + ex$
(d) $g(t) = 8\sqrt[3]{t^2} + \cos(t) - 6t + \pi^2$
(e) $f(z) = 5^z + \sin(z) - \cos(z) + \sqrt{3}$
(f) $h(r) = 5r^\pi - r^{4/3} + 5/2$

5.2. Use the product or quotient rule to find the derivative of each function.
(a) $f(x) = x^4 \sin(x)$
(b) $g(x) = 3x^3 e^x$
(c) $h(x) = 3^x \tan(x)$
(d) $f(x) = x^2 \cos(x) \sin(x)$
(e) $g(s) = \sqrt[4]{s^7} \tan(s)$
(f) $h(t) = \dfrac{t^4}{\tan(t)}$
(g) $f(x) = \dfrac{\sin(x)}{(2x^4 - 1)}$
(h) $g(t) = \dfrac{t^4 e^t}{5(t^4 - 2t)}$
(i) $j(r) = \dfrac{3^r}{2\cos(r)\tan(r)}$
(j) $f(x) = \dfrac{\tan(x)}{e^x(3x^{5/2} - 2x)}$

5.3. (a) Extend the product rule to express $(f(x)g(x)h(x))'$ in terms of $f(x)$, $g(x)$, $h(x)$, and their derivatives. (*Hint:* Think of $f(x)g(x)h(x)$ as $(f(x)g(x)) \times h(x)$, and apply the ordinary product rule twice.)
(b) Explain what each term in your formula represents.
(c) Suppose that $\ell(t)$, $w(t)$, and $h(t)$ are the length, width, and height of a box whose dimensions keep changing. At what rate would the volume of the box be changing if $\ell(t)$ changed but $w(t)$ and $h(t)$ were constant? If $w(t)$ changed but $\ell(t)$ and $h(t)$ were constant? If $h(t)$ changed but $\ell(t)$ and $w(t)$ were constant? If all three dimensions of the box changed?

5.4. Rewrite each of the following expressions in the form $f(u)$, where u is a function of x. (*Example:* $\sin^2(x) = u^2$, where $u = \sin(x)$.)
(a) e^{x^2+1}
(b) $\cos^3(x)$
(c) $\sin^2(x) + 3\sin(x) + 7$
(d) $\dfrac{e^x}{e^x + 1}$

5.5. Use the chain rule to find the derivative of each function with respect to x.
(a) $9(x^{5/2} - 3x)^{30}$
(b) $\cos^4(3x^2 - 5)$
(c) $\sqrt[3]{\tan(x) - e^x}$

5.9. Exercises

(d) $9e^{x^4+\pi x}$
(e) $\sin(5x^2+4)$
(f) $h(x) = (\sin^2(x) - 2.3x)^{7/4}$

5.6. Find the derivative of each function with respect to x or t. Indicate which rule(s) you are using.
   (a) $4x^4 \tan(x^3)$
   (b) $\dfrac{3x^2}{\cos(4x)}$
   (c) $\sqrt{e^{3x} + \pi x^2}$
   (d) $\dfrac{\sin^2(x)}{(5x^3 - 3x + 5)^3}$
   (e) $\cos^4(2\sqrt{t} - 5)$
   (f) $(3t^{5/3} - 6t^{2/3} + 8t^{1/3})^5$
   (g) $e^{x^2}(3x^5 - 2x)^8 \tan^2(x)$
   (h) $\dfrac{3t^2}{\cos(4t)}$

5.7. Let $h(x) = (2g(x))^3$, where $g(x)$ is a function with $g(5) = -6$ and $g'(5) = -10$. Find $h'(5)$.

5.8. Let $f(x) = g(2xh(x))$, where $g'(42) = -2$, $h'(3) = -1$, and $h(3) = 7$. Find $f'(3)$.

5.9. Let $h(t) = \tan(\cos(t))$. Find $h'(\pi/2)$ and $h'(3\pi/2)$.

5.10. Let $f(x)$ and $h(x)$ be functions satisfying
$$f(3) = 6, \quad h(3) = 5, \quad h(5) = 9,$$
$$f'(3) = 9, \quad f'(5) = 10, \quad h'(3) = h'(6) = 2.$$
Compute the derivative of each function with respect to x at $x = 3$.
   (a) $f(x) - h(x)$
   (b) $(f(x))^3$
   (c) $(h(x) - f(x))^2$
   (d) $f(h(x))$
   (e) $h(f(x))$
   (f) $f(x)h(x)$
   (g) $f\left(\dfrac{8}{9}x^2 - h(x)\right)$

Derivatives and Microscopes

5.11. Find the slope of the line tangent to the curve $y = 2x^2 - 4x$ at $x = 1$.

5.12. Determine if $h(x) = \cos^2(x)$ is increasing or decreasing at $x = 3\pi/4$. At what rate is it changing?

5.13. Consider the function $f(x) = \sin(2x) + 4$.
   (a) Find the equation of the line tangent to $y = f(x)$ at $x = 0$.
   (b) Write the microscope equation (linear approximation) for $f(x)$ near $x = 0$.
   (c) Use this linear approximation to estimate $g(0.05)$ and $g(0.1)$. Which of these approximations do you think is more accurate, and why?

5.14. Let $f(x) = \sqrt[3]{x}$.
   (a) Write the microscope equation (linear approximation) for $f(x)$ near $x = 8$.
   (b) Use the microscope equation to estimate $\sqrt[3]{8.1}$ and $\sqrt[3]{7.8}$.
   (c) Evaluate $\sqrt[3]{8.1}$ and $\sqrt[3]{7.8}$ with a calculator. Which cube root was approximated more accurately, and by (approximately) what factor?

5.15. Use the microscope equation to estimate $\sqrt{403}$ and $\sqrt[3]{1001}$.

Maximum Value

5.16. Let $f(x) = \dfrac{1}{2x^2 - 3x + 6}$.
   (a) Using MATLAB, graph $f(x)$ on the interval $[-3, 3]$. By zooming in on the graph, find the point where $f(x)$ is maximized to two decimal place accuracy.
   (b) Find the derivative of $f(x)$.
   (c) Using part (b), find the *exact* value of x where $f(x)$ reaches its maximum value.
   (d) At what point is the graph of $f(x)$ rising most steeply? Describe how you determined the location of this point. (*Hint*: Graphing $f'(x)$ might give valuable insight.)

Exercises 5.17 and 5.18 are about **per capita** quantities, meaning how much there is **per person**. For instance, if a country with a population of 50 million has a gross domestic product (GDP) of $1 trillion, then its per capita GDP is $1 trillion/50 million = $20,000.

5.17. The population of a U.S. state is currently 10,000,000 people and is shrinking by 10,000 people per year. In this state, the per capita yearly spending on energy is $2000 per person and is growing at a rate of $60 per year. Currently, how much is the whole state paying for energy? Is that total yearly energy expenditure growing or shrinking? How fast?

5.18. In the same state as in Exercise 5.17, the total yearly personal income is $500 billion and is falling by $250 million per year. What is the current per capita personal income? Is this per capita income rising or falling? By how much?

Position and Velocity

5.19. A ball is dropped at time $t = 0$ from the top of a tall building. As it falls, its height (in feet) from the ground is given by the formula $h(t) = 400 - 16t^2$, where t is measured in seconds.
   (a) Find a formula for the velocity $v(t) = h'(t)$ of the ball after t seconds.
   (b) Graph, on the same set of axes, the velocity and the height as functions of time. Your legend should include the different units for h and v.
   (c) At what time does the ball hit the ground? What is its velocity then?

5.20. A mass hanging from a spring is oscillating up and down. Its height in inches, measured from a particular reference point, is $z(t) = 5\cos(t) + 20$, where t is measured in seconds.
   (a) Find a formula for the velocity $v(t)$ of the ball.

(b) What is the mass doing when $v(t) < 0$? What is it doing when $v(t) > 0$?
(c) In what two situations is $v(t) = 0$?
(d) What is the maximum height of the mass?
(e) When and where does the mass reach its maximum velocity?

Circumference, Area, and Volume

5.21. The circumference of a circle of radius r is $C(r) = 2\pi r$.
   (a) If r is measured in meters, what are the units of C and C'?
   (b) Compute the derivative $C'(r)$. Does it have the correct units, according to part (a)?
   (c) How much does the circumference of a circle change if its radius increases from 100 meters to 101 meters?
   (d) Imagine that the Earth is a perfect sphere, and that a giant belt about 24,000 miles long is placed along the equator. How much fabric would have to be added to the belt to raise it one inch (everywhere) above the ground?

5.22. The area of a circle of radius r is $A(r) = \pi r^2$.
   (a) If r is measured in meters, what are the units of A and A'?
   (b) Compute the derivative $A'(r)$. Does it have the correct units, according to part (a)?
   (c) Imagine a circular public park of radius 100 meters. The city decides to put a sidewalk, 1 meter wide, around the outer edge of the park. Using a linear approximation for area near $r = 100$ meters, estimate the area of the sidewalk.
   (d) The area of the sidewalk is also approximately the same as the width of the sidewalk times the length of the sidewalk. (This is only approximate since the the outside and inside of the sidewalk have slightly different lengths, as we saw in Exercise 5.21.) What is the approximate length of the sidewalk?
   (e) Repeat parts (c) and (d) for a circle of radius r and a sidewalk of width Δr. What does this tell us about how the area and circumference of a circle are related?

5.23. The volume of a sphere of radius r is $V(r) = \frac{4}{3}\pi r^3$.
   (a) If r is measured in meters, what are the units of V and V'?
   (b) Imagine a sphere of radius 3 meters that we want to paint with a layer of paint that is 1 mm thick. The painted sphere will then have a radius of 3.001 meters. Using the microscope equation for V near $r = 3$ meters, estimate the volume of paint needed to do the job.
   (c) The amount of paint needed is approximately the surface area of the sphere times the thickness of the paint. What is the approximate surface area of the sphere?
   (d) Repeat parts (b) and (c) for a sphere of arbitrary radius r and paint of thickness Δr to derive a formula for the surface area of a sphere of radius r.

5.24. If the radius of a spherical balloon is r inches, its volume is $\frac{4}{3}\pi r^3$ cubic inches.
   (a) The radius of the balloon is changing. What is the change in the balloon's volume per change in its radius when the radius is 3 inches?

(b) When the radius is 3 inches, approximately how much does the radius have to increase to increase the volume by 25 cubic inches?

(c) Suppose that someone is inflating the balloon at the rate of 19 cubic inches of air per second. When the radius is 9 inches, at what rate is the radius increasing in inches per second?

Business Optimization

5.25. A company manufactures and sells x pounds of beef jerky per week. Suppose the cost and revenue functions are
$$C(x) = 4000 + 20x, \qquad R(x) = 40x - \frac{x^2}{100}, \qquad 0 \le x \le 600.$$

(a) At a production level $x = 400$, find the marginal cost $C'(400)$, the marginal revenue $R'(400)$, and the marginal profit $P'(400)$.

(b) Use the microscope equation to approximate the changes in revenue and profit if production is increased from 400 to 410 pounds per week.

(c) Suppose the company decides to expand and adds a second smoker oven. Now the cost and revenue functions are given by
$$C(x) = 5000 + 20x, \qquad R(x) = 40x - \frac{x^2}{100}, \qquad 0 \le x \le 1200.$$

(a) Find $C'(1100)$, $R'(1100)$, and $P'(1100)$.

(b) Use the microscope equation to approximate the changes in revenue and profit if production is increased from 1100 to 1120 pounds per week. Is increasing production above 1100 a good or bad idea?

(c) What is the most profitable level of production?

5.26. The average profit function, $\overline{P}(x)$, is given by
$$\overline{P}(x) = \frac{P(x)}{x},$$
where $P(x)$ is the profit when x units are produced.

(a) Describe in words what the average profit function means.

(b) If $P(x) < 0$, is it possible that $P'(x) > 0$? Explain.

(c) If $P(x) < 0$, is it possible that $\overline{P}'(x) > 0$? Explain.

(d) Find a condition involving $P(x)$ and $P'(x)$ that implies $\overline{P}'(x) > 0$. Interpret this condition.

5.27. The revenue function, $R(x)$, is given by $R(x) = xp(x)$, where $p(x)$ is the price function.

(a) Describe in words why this is the correct formula for the revenue function.

(b) Is it possible to have $R'(x) > 0$ if $p'(x) < 0$? Why or why not?

(c) Find a condition involving $p'(x)$ that implies $R'(x) > 0$. Interpret this condition.

Savings and Loan Associations (S&Ls). In the 1942 film "It's a Wonderful Life," the main character (George Bailey) operates Bailey's Savings and Loan. S&Ls specialize in collecting money from depositors and lending it out for home mortgages. The depositors earn interest at an annual interest rate of r_d, while the S&L charges a (higher!) annual interest rate of r_b on its mortgages. The S&L needs to make r_d big enough to entice people to deposit their money, but it doesn't want to pay any more interest than it has to.

5.28. From prior experience, the current operator of Bailey's Savings and Loan expects people to deposit 100 million times r_d dollars into the S&L. Due to government regulations, the S&L can only loan out 80% of the money deposited. Borrowers then pay 8% interest ($r_b = 0.08$) on their loans. For example, if Bailey's offers 3% interest on deposits ($r_d = 0.03$), then the deposits will total $0.03 \times 100{,}000{,}000 = 3{,}000{,}000$ dollars. They can then loan 80% of $3 million, or $2,400,000, at 8% interest, earning $0.08 \times \$2{,}400{,}000 = \$192{,}000$ per year in interest. Since they will also pay out $90,000/year in interest to depositors, their net yearly profit is $102,000.

If $r_b = 0.08$, what interest rate r_d on deposits will maximize Bailey's profit?

Optimizing Price. Exercises 5.29–5.32 involve picking the right price for a product to maximize profit. This depends on the demand function $D(p)$ and on the cost function $C(x)$. In many industries, the cost of producing x items is proportional to x. The proportionality constant is called the **unit manufacturing cost**. For instance, if the unit manufacturing cost is $1.83/item, then the cost function is $C(x) = 1.83x$.

5.29. Suppose that the demand function for a particular product is $D(p) = 3000 - 500p$ and the unit manufacturing cost is $1.00/item.
 (a) What price should the manufacturer charge to maximize its profit?
 (b) Suppose the unit manufacturing cost rises to $1.50/item. What is the new optimal price?
 (c) To maximize profit, what fraction of the rise in the unit manufacturing cost should be passed on to the consumer?

5.30. Suppose the demand function for a car is $D(p) = 100{,}000 - 2p$.
 (a) If the manufacturing cost is $20,000/car, how much should the auto company charge to maximize profit?
 (b) If the manufacturing cost increases to $24,000/car, how much should the auto company charge?

5.31. Suppose that a toy has a nonlinear demand function $D(p) = \dfrac{1{,}500{,}000}{100 + p^2}$. Find the price that will maximize profit when the unit manufacturing cost is $10, when the unit cost is $15, and when the unit cost is $20.

Food for Thought: Exercises 5.29–5.30 illustrate an interesting phenomenon. When the demand function is linear—but only when it is linear—exactly half of any *increase* in the unit manufacturing cost should be passed on to the consumer. We explore this in general in Exercise 5.32.

5.32. Consider a general linear demand function $D(p) = a - mp$, with unit cost c.
 (a) In terms of a, m, and c, what is the price that maximizes profit?
 (b) Is exactly half the cost of manufacturing passed on to the consumer? Explain.

Other Maxima and Minima

5.33. At what positive value of x does $f(x) = \frac{1}{4}x + \frac{1}{x^2}$ attain its minimum value? Explain how you found this value. (You should *not* use MATLAB or a graphing calculator to do this problem.)

5.34. What is the minimum value of the function $f(x) = \frac{1}{4}x + \frac{1}{x^2}$ on the interval $[3, 5]$? Where does this minimum occur? Explain how you found this value of x. (You should *not* use MATLAB or a graphing calculator to do this problem.)

Derivatives and Graphs

5.35. Let $f(x) = x^3 + 3x$.
   (a) Make a sign chart for $f(x)$.
   (b) Compute $f'(x)$ and make a sign chart for $f'(x)$.
   (c) Compute $f''(x)$ and make a sign chart for $f''(x)$.
   (d) Sketch the graph of $f(x)$ between $x = -2$ and $x = 2$.

5.36. Let $f(x) = x^4 - 4x^2 + 3$.
   (a) Make a sign chart for $f(x)$.
   (b) Compute $f'(x)$ and make a sign chart for $f'(x)$.
   (c) Compute $f''(x)$ and make a sign chart for $f''(x)$.
   (d) Sketch the graph of $f(x)$ between $x = -2.5$ and $x = 2.5$.

5.37. Let $f(x) = xe^{-x}$.
   (a) Make a sign chart for $f(x)$.
   (b) Compute $f'(x)$ and make a sign chart for $f'(x)$.
   (c) Compute $f''(x)$ and make a sign chart for $f''(x)$.
   (d) Sketch the graph of $f(x)$ between $x = -1$ and $x = 4$.

5.38. Let $f(x) = e^{-x^2}$. This function is always positive, so we don't have to bother with a sign chart for $f(x)$.
   (a) Compute $f'(x)$ and make a sign chart for $f'(x)$.
   (b) Compute $f''(x)$ and make a sign chart for $f''(x)$.
   (c) Sketch the graph of $f(x)$ between $x = -2$ and $x = 2$.

5.39. The graph below gives the annual profit $y = P(x)$ for a certain product, where x is the amount of capital expenditures.
   (a) Interpret $P(0)$.
   (b) Sketch a graph of $P'(x)$.
   (c) Describe in words how the profit changes as capital expenditures increase.

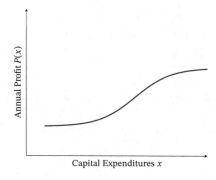

(d) Sketch a graph of $P''(x)$. What is the economic significance of the point of inflection of the original profit function $P(x)$? (*Hint*: Look up "diminishing returns.")

5.40. Answer the following questions for each graph.
- At the indicated point x^*, what is the sign $(+, -, \text{or } 0)$ of f'?
- Is f' (not f) increasing or decreasing at x^*?
- What is the sign of f'' (the derivative of f') at x^*?

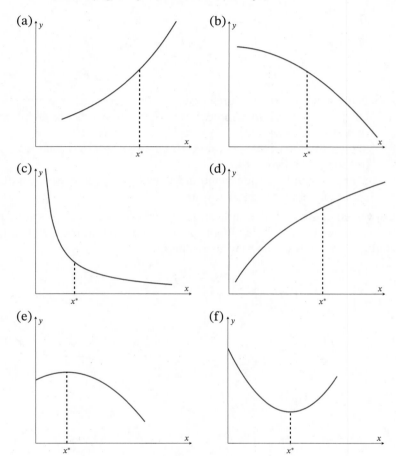

5.41. Sketch the graph of a function $f(x)$ that satisfies each set of conditions. (There are many such functions to choose from.) On each graph, mark any critical points or extrema.
(a) $f'(x) < 0$ for $x < 0$; $\quad f'(0) = 0$; $\quad f'(0) > 0$ for $0 < x < 3$; $f'(3) = 0$; $\quad f'(x) < 0$ for $x > 3$.
(b) $f'(x) < 0$ for $x < 1$; $\quad f'(1) = 0$; $\quad f'(x) > 0$ for $x > 1$.
(c) $f'(-2) = 0$; $\quad f''(-2) < 0$; $\quad f'(2) = 0$; $\quad f''(2) > 0$.
(d) $f(5/2) = 1$; $\quad f'(x) > 0$ everywhere; $\quad f''(x) > 0$ when $x < 5/2$; $f''(x) < 0$ when $x > 5/2$.

5.42. Consider the following graph of a function $f(x)$.

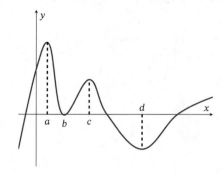

(a) Is the second derivative greater at point b or point d? Explain.
(b) On a copy of this curve, indicate where the curve is concave up and where it is concave down.
(c) Sketch a graph of the derivative $f'(x)$. Label the points corresponding to a through d on your sketch.
(d) Sketch a graph of the second derivative $f''(x)$. Label the points corresponding to a through d on your sketch.

5.43. Copy each graph and mark any critical points or extrema. Indicate which extrema are local and which are global. (Assume that at their ends the curves continue in the direction they are headed.)

(a) (b)

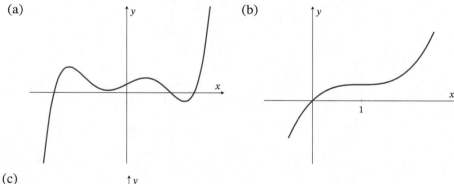

(c)

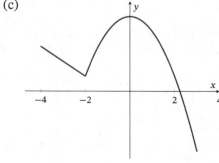

5.9. Exercises

Newton's Method

5.44. Newton used the example $x^3 - 2x - 5 = 0$ to introduce his method for finding roots.
   (a) Explain why there must be a root somewhere between $x = 2$ and $x = 3$.
   (b) Starting at $x_0 = 2$, find this root to an accuracy of at least four decimal places.

5.45. Use Newton's method to find a solution of $x^3 + 2x^2 = 6$ near the point $x = 1$, accurate to at least four decimal places.

5.46. Consider the equation $x^4 - 8x^2 + 5 = 0$.
   (a) Starting at $x_0 = \frac{1}{2}$, find a root to an accuracy of at least four decimal places.
   (b) Starting at $x_0 = -\frac{1}{2}$, find a different root to an accuracy of at least four decimal places.
   (c) Find the two other roots, again to at least four decimal places, by starting at $x_0 = 3$ and $x_0 = -3$.
   (d) What happens if you start at $x_0 = 0$, at $x_0 = 2$, or at $x_0 = -2$? Why are these really bad initial guesses?

Food for Thought: While there is no general formula for the roots of a polynomial of fifth order or higher, and the formulas for third and fourth-order polynomials are too complicated to be useful, we can sometimes solve higher-order polynomial equations by other means. For instance, the equation $x^4 - 8x^2 + 5$ can be solved exactly by defining $y = x^2$, which turns the equation into $y^2 - 8y + 5 = 0$. The quadratic formula then tells us that $y = 4 \pm \sqrt{11}$, so $x = \pm\sqrt{4 \pm \sqrt{11}}$. You should check that these are in fact the four roots that you approximated in Exercise 5.46.

Chapter 6

Models of Growth and Oscillation

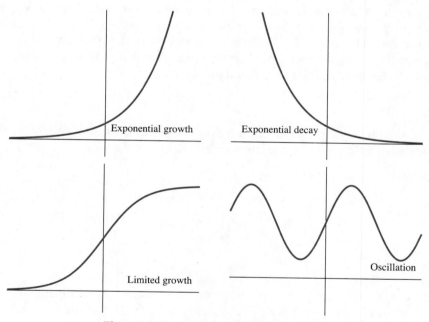

Figure 6.1. Four ways that a quantity can evolve

Figure 6.1 shows four kinds of graphs that you may see when studying real-world phenomena. The top left graph might represent the value of a bank account, the number of active users at the beginning of an marketing campaign, or the population of an invasive species. The top right graph might represent a quantity of radioactive material, the value of a failing investment, or the amount of pollution in a lake that is gradually flushing itself clean. The bottom left graph might represent the later stages of the top left picture. The number of people who have seen a viral video, or the population of an invasive species, can't keep growing forever. Eventually it levels out. Finally, the bottom right graph could represent any quantity that goes in a repeating cycle, from the motion of a wheel to a pendulum to the seasons to the business cycle. These four patterns appear over and over again in the real world. In this chapter we will study how and why that happens.

The world operates according to laws, but these laws typically don't tell us how *much* of something there will be at time t. Instead, these laws tell us how fast that quantity (or those quantities) are *changing*. Your bank's policies don't determine how much money is in your bank account, but they do determine how fast your balance is increasing. The law of gravity doesn't determine the velocity of a falling object, but it does say how fast that velocity is changing. Population biology doesn't determine how many rabbits are in a colony, but it does say how the population changes with time.

In other words, most of the laws that govern the natural world are expressed mathematically as **differential equations**. These are equations that relate the derivative (usually with respect to time) of one or more quantities to the current levels of those quantities and to the time.[1] For instance, we have already seen that if we invest a certain amount of money at 6% annual interest (compounded continuously), then our balance $y(t)$ follows the differential equation

$$\text{(6.1)} \qquad \frac{dy}{dt} = 0.06y.$$

We have also already seen that the numbers of Potential, Active, and Rejected users of a new product follow the SIR **system** of differential equations

$$\frac{dS}{dt} = -aSI,$$

$$\frac{dI}{dt} = aSI - bI,$$

$$\text{(6.2)} \qquad \frac{dR}{dt} = bI.$$

It turns out that very many phenomena are described, at least approximately, by a few very simple differential equations. The solutions to these differential equations then give the pictures in Figure 6.1.

In this chapter,

- We will see how to take the laws of the world (described in words) and express them as differential equations. This is the essence of modeling. Most of the time, as with modeling the spread of disease, the resulting differential equations will be too complicated for us to find a formula for the solution.

---

[1] Before we learned what derivatives were, we called such things *rate* equations, but almost everybody calls them *differential* equations, and we're going to follow suit.

- We will learn about the simplest differential equations, so simple that we *can* solve them exactly.

(6.3) $$\frac{dy}{dt} = ry \qquad \text{(exponential growth);}$$

(6.4) $$\frac{dy}{dt} = -ry \qquad \text{(exponential decay);}$$

(6.5) $$\frac{dy}{dt} = rx, \quad \frac{dx}{dt} = -ry \qquad \text{(circular motion).}$$

- We will learn about (or review) the kinds of functions that show up as solutions to these equations. These include exponentials, logs, and trigonometric (trig) functions. Exponentials and logs are explained in Section 6.2. The supplemental material in Section 6.7 is "A Crash Course in Trigonometry".
- We will learn about more realistic models that take more features into account. In particular, in Section 6.5 we will study the **logistic equation**

(6.6) $$\frac{dy}{dt} = ry\left(1 - \frac{y}{L}\right),$$

which describes the growth of a single population with limited resources, as well as the **Lotka–Volterra** equations that describe the interaction of predators with prey. Here the emphasis isn't on finding a formula for the solution, but on understanding what is happening qualitatively.

6.1. Modeling with Differential Equations

There are several ingredients to a model involving differential equations.

(1) The **differential equation** itself describes the rules of the game.
(2) The **initial conditions** describe our starting point. A differential equation together with an initial condition is called an **initial value problem**, or IVP.
(3) A **solution** to the IVP. This is a function whose derivative follows the rules established by the differential equation and whose starting value matches the initial condition.
(4) A **numerical** solution to the IVP. Most of the time we can't find a formula for the solution to the IVP, but that's OK. As long as we can approximate it on a computer, we can make useful predictions.
(5) **General** solutions to a differential equation. These are families of functions that solve a differential equation, involving parameters that can be adjusted to match whatever initial conditions we want. If we can find a general solution (and that's a big if!), then we can understand what happens without relying on a computer.

We'll go over these ingredients, one at a time.

The Rules of the Game.

Put your money to work. If we invest money in a bank, what are the rules of the game? At most banks, the main rule is that the bank gives us interest at a rate proportional to how much money we have invested. If we invest $100, we get a certain amount of interest. If we invest $100,000, we get 1000 times that much interest. This proportionality is expressed with an interest **rate** r, which might vary from bank to bank. Like the coefficients a and b in the SIR model, r is a **parameter** of the system. We can summarize the rule in two ways:

The bank pays us money at a rate r times our current balance.

(6.7) $$\frac{dy}{dt} = ry.$$

The differential equation and the sentence above it mean exactly the same thing! The only difference is that one is written in words and the other is written with a formula. dy/dt is a very efficient way of saying "the rate at which our bank balance is growing" and ry is an efficient way of saying "the interest rate r times our current balance".

Now suppose that the bank charged a fee of $F/year for having an account. Then the rules of the game would be different, and our differential equation would be different. The rules would then be

*Money comes in at a rate proportional to how much we have (ry)
and leaves at a constant rate (F).*

(6.8) $$\frac{dy}{dt} = ry - F.$$

Each term corresponds to a feature of the rule. Extra features mean extra terms.

Some banks have rules that are more complicated, in which case our differential equation will be more complicated. For instance, suppose that the fee is waived for accounts with balances of $10,000 or more. Our new rule, in both words and equations, is

*If $y \geq 10{,}000$, then money comes in at rate ry.
If $y < 10{,}000$, then money comes in at rate ry and goes out at rate F.*

(6.9) $$\frac{dy}{dt} = \begin{cases} ry & \text{if } y \geq 10{,}000, \\ ry - F & \text{if } y < 10{,}000. \end{cases}$$

Beware of falling rocks. Next let's consider the rules of a completely different game: a rock that is falling to the ground. We don't know the rock's velocity $v(t)$, but (ignoring air resistance) the law of gravity tells us how fast that velocity is changing:

(6.10) $$\frac{dv}{dt} = -32 \frac{\text{feet}}{\text{second}^2}.$$

Since it ignores air resistance, this differential equation does not describe reality exactly. It is only an approximation, but as we already know, *close is good enough*. As long as the rock is reasonably heavy and isn't moving too fast, equation (6.10) is accurate. If we are talking about a falling feather instead of a rock, or if the rock is a

6.1. Modeling with Differential Equations

meteorite moving at thousands of feet per second, then we have to add an additional term to the equation to account for air resistance.

You may have noticed that $v(t)$ is itself a rate of change. It is the derivative of the height $z(t)$ of our rock above the ground. The law of gravity tells us about the *second* derivative of z. That is,

$$(6.11) \qquad \frac{d^2z}{dt^2} = -32 \frac{\text{feet}}{\text{second}^2}.$$

Equation (6.11) is called a **second-order** differential equation, because it describes a second derivative, while (6.10) is called a **first-order** differential equation. Second-order differential equations come up a lot in physics, where Newton's Second Law of Motion, $F = ma$, relates the acceleration of an object (that is, the second derivative of the position) to the force applied to that object. However, every second-order system of equations can be converted to a first-order system of equations by defining additional variables, such as the velocity. In the case of a falling rock, the second-order differential equation (6.11) is equivalent to the system of two first-order equations,

$$(6.12) \qquad \begin{aligned} dz/dt &= v, \\ dv/dt &= -32 \text{ feet/second}^2. \end{aligned}$$

In this chapter we will mostly concentrate on first-order equations of the form

$$(6.13) \qquad \frac{dy}{dt} = f(y, t),$$

for a single function $y(t)$, and on sets of differential equations of the form

$$(6.14) \qquad \begin{aligned} \frac{dy_1}{dt} &= f_1(y_1, \ldots, y_n, t) \\ \frac{dy_2}{dt} &= f_2(y_1, \ldots, y_n, t) \\ &\vdots \\ \frac{dy_n}{dt} &= f_n(y_1, \ldots, y_n, t) \end{aligned}$$

for a larger set $\{y_1(t), \ldots, y_n(t)\}$ of functions. In our examples, $f(y, t)$ and $f_j(y_1, \ldots, y_n, t)$ will be fixed expressions, such as $0.05y - 100$ or $y_1^2 - y_2^2 + 3t$.

Initial Conditions and Initial Value Problems. By themselves, the rules of the game don't determine the outcome. Three investors who invest $1000, $2000, and $10,000 at the same bank are playing by the same rules, but the graphs of their bank balances over time will not be the same. If the bank gives 5% interest and charges a $100/year fee, then the differential equation is

$$(6.15) \qquad \frac{dy}{dt} = 0.05y - 100.$$

The $1000 investor loses money, since he pays more in fees ($100/year) than he gets in interest (initially $50/year). The investor who starts with $2000 breaks even, earning exactly as much interest as he pays in fees. Meanwhile, the investor who starts with $10,000 makes money. See Figure 6.2.

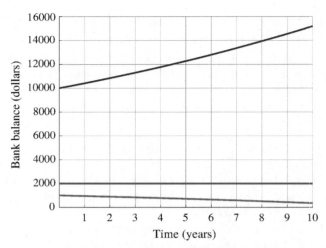

Figure 6.2. Three solutions to the differential equation (6.15)

In other words, the result of our model depends not just on the differential equation itself, but also on the starting point

(6.16) $\qquad y(0) = 1000 \quad$ or $\quad y(0) = 2000 \quad$ or $\quad y(0) = 10{,}000.$

We call the starting point an **initial condition**. For our bank problem, that means specifying the bank balance y at some initial time t_0. (We often start things at $t_0 = 0$, but not always.) Our falling rock problem involves two quantities, z and v, so our initial conditions have to specify both $v(t_0)$ and $z(t_0)$. For the SIR equations, we need to specify $S(t_0)$, $I(t_0)$, and $R(t_0)$. In general, we need to specify one piece of data for each variable in the problem.

> Do you think it's fair that a rich investor makes money from the bank while a poor investor loses money? Of course, this is a question about life in general, and not just about banks.
>
> It would be nice if life were fair, with everybody playing by the same rules and starting out with the same initial conditions, but that doesn't happen. Some people are born rich, others poor. Some are born with special abilities or with disabilities. Some have more educational opportunities than others. The list goes on.
>
> Almost everybody believes that things should be as fair as possible, but people disagree about what "fair" means. Some take the differences in initial conditions as given and want to make the rules of the game exactly the same for everybody. Others think that the rules should include handicaps to help those whose initial conditions are worse. Still others say that we should share resources so that everybody has the same initial conditions.
>
> So what do you think is fair? Math can't provide the answer, but at least it can frame the question. The better you understand how outcomes depend both on rules (differential equations) and on initial conditions, the better you'll be at pursuing whatever kind of fairness you believe in.

6.1. Modeling with Differential Equations

A differential equation, together with a set of initial conditions, is called an **initial value problem**, or IVP. Here are some examples of IVPs:

- The IVP

(6.17) $$\frac{dy}{dt} = -0.1(y - 70), \qquad y(0) = 170,$$

might describe the temperature $y(t)$ of a cup of coffee (in degrees Farenheit) t minutes after it was brewed. It loses heat at a rate proportional to how much hotter the coffee is than the room around it. The initial condition is the temperature of the coffee when it was brewed.

- The IVP

(6.18) $$\frac{dy}{dt} = 0.1(70 - y), \qquad y(0) = 20,$$

might describe the temperature of a container of ice cream t minutes after it was taken out of the freezer. It heats up at a rate proportional to how much hotter the room is than the ice cream, and the initial condition is its temperature when it came out of the freezer. Since $0.1(70 - y) = -0.1(y - 70)$, this is actually the same differential equation as the previous example, but the initial condition is different. The coffee started off hot, but the ice cream started off cold.

- The IVP

(6.19) $$\frac{dv}{dt} = -32 + .001v^2, \qquad v(0) = 0,$$

describes the vertical velocity of a skydiver, in feet/second, t seconds after jumping out of an airplane. The differential equation describes the forces involved: gravity pulling downward and air resistance pushing upward. The initial condition reflects the fact that the airplane is neither rising nor falling when the skydiver jumps.

Solutions. A function that satisfies the differential equation, and whose starting value matches the initial condition, is said to be a **solution** to the initial value problem. The IVP describes the rules of the game and the starting point, but the solution is what actually happens! *Building* a model involves figuring out what the IVP should be. *Running* the model means finding the solution.

As with all functions, solutions can be expressed via graphs, formulas, tables, or algorithms. For instance, the three graphs in Figure 6.2 show the solutions to three IVPs. The bottom graph is the solution to (6.15) with initial condition $y(0) = 1000$, the middle graph is the solution to (6.15) with initial condition $y(0) = 2000$, and the graph at the top is the solution to (6.15) with initial condition $y(0) = 10,000$.

Sometimes we describe our solutions with formulas. The functions $z(t) = 37 + 20t - 16t^2$ and $v(t) = 20 - 32t$ are solutions to (6.12) with initial conditions $z(1) = 41$ (feet) and $v(1) = -12$ (feet/second), as you can (and should!) check by taking the derivatives of $37 + 20t - 16t^2$ and $20 - 32t$ (to check the equations), and by plugging $t = 1$ into these expressions (to check the initial conditions).

It is theoretically possible to cook up IVPs that don't have solutions, or that have multiple solutions, but these require contrived (some would say *pathological*) choices of the expression $f(y, t)$. As long as $f(y, t)$ is remotely reasonable (and in particular, whenever $f(y, t)$ is a differentiable function of y and t), we have the following.

> **Existence and Uniqueness Principle**: *Under normal circumstances, there is one and only one solution to the initial value problem*
>
> (6.20) $$\frac{dy}{dt} = f(y, t), \qquad y(t_0) = y_0.$$

Numerical Solutions. Sometimes, when an IVP is particularly simple, we can find a formula for its solution. Some of these easy-to-solve problems are very important, and we will go over a few of them in the coming sections. The bad news, however, is that *most* real-world IVPs just can't be solved by finding a magic formula.

But there's good news, too. Even if we can't find a formula for the solution, we can solve the IVP on a computer anyway, using the ideas that we already used in Chapter 3 to tackle compound interest and the SIR equations. Here we repeat the algorithm, called **Euler's method**, to get a solution to any IVP of the form (6.20), with t running from an initial time t_0 to a final time t_f:

(1) Decide on a number N of steps we are going to take. This breaks the interval $[t_0, t_f]$ into subintervals of size $\Delta t = (t_f - t_0)/N$. For $j = 1, 2, \ldots, N$, let $t_j = t_0 + j\Delta t$. Set $y(t_0) = y_0$.

(2) For $j = 0, 1, \ldots, N-1$, use the known value of $y(t_j)$ to compute $y'(t_j) = f(y(t_j), t_j)$. Then use the linear approximation $y(t_{j+1}) \approx y(t_j) + y'(t_j)\Delta t$ to approximate $y(t_{j+1})$. These two steps can be combined into a single equation:

(6.21) $$y(t_{j+1}) \approx y(t_j) + f(y(t_j), t_j)\Delta t.$$

(3) Plot the array $[y(t_0), \ldots, y(t_f)]$ against the array $[t_0, \ldots, t_f]$ to get a graph of $y(t)$. If we prefer, we can print out a table of values.

(4) If we need more accuracy, we can repeat the process with a larger value of N and a smaller value of Δt. As N gets bigger and bigger, our approximate solution will get closer and closer to the true solution to (6.20).

Euler's method is a reliable crystal ball. With its help, and preferably with the help of a computer, we can solve even the ugliest-looking IVPs and get clean graphical results.[2] In many cases, solving an IVP numerically is the best way, or even the only practical way, to predict the future.

However, solving an IVP numerically has its drawbacks. If we want to understand how the final value $y(t_f)$ depends on the initial value $y(t_0) = y_0$, we have to run the program for a particular value of y_0, then run it again for a slightly different value, and so on, until we have enough data to see the pattern. If we want to understand how the solutions depend on the parameters of the differential equation (e.g., the interest rate, or the size of the bank fee), that's even more work. Computing $y(t_j)$ for (say) 100 different initial conditions and 100 different parameters requires $100 \times 100 = 10{,}000$ runs. That's a lot!

---

[2]There do exist much more efficient algorithms for solving IVPs numerically. However, those methods are outside the scope of this book so we're not going to study them here.

6.1. Modeling with Differential Equations

General Solutions. Sometimes there's a better way. Not always, but sometimes.

If a and b are any two numbers, then the functions

(6.22) $$z(t) = a + bt - 16t^2, \qquad v(t) = b - 32t,$$

are solutions to the differential equations (6.12) (namely, $dz/dt = v$ and $dv/dt = -32$). This is because the derivative of $z(t) = a + bt - 16t^2$ is $b - 32t$, which equals $v(t)$, and the derivative of $v(t) = b - 32t$ is -32, as required. If somebody gives us initial conditions, we can use them to determine what the parameters a and b should be. For instance, if $v(1) = -12$ and $z(1) = 41$, then

$$-12 = v(1) = b - 32 \quad \text{so} \quad b = 20,$$
(6.23) $$41 = z(1) = a + b - 16 = a + 4 \quad \text{so} \quad a = 37,$$

and our solution is $v(t) = 20 - 32t$, $z(t) = 37 + 20t - 16t^2$. Function (6.22) is called a **general solution** to the differential equations (6.12), while $v(t) = 20 - 32t$, $z(t) = 37 + 20t - 16t^2$ is a **particular solution** to the IVP with $v(1) = -12$, $z(1) = 41$.

You're probably wondering where the general solution (6.22) comes from. We'll see how to derive this solution in Chapter 7, but at this stage of the game it isn't important. All we need to know is that *it works*, and that's something that you can check for yourself by computing a few derivatives. Since the solutions to IVP problems are unique, any solution that works must be the one and only solution. "Guess and check" is a perfectly good strategy for solving IVPs!

As another example, the general solution to the equation $dy/dt = ry$ is $y(t) = Ae^{rt}$. It solves the equation because

(6.24) $$\frac{dy}{dt} = \frac{d(Ae^{rt})}{dt} = Are^{rt} = ry(t),$$

and because we can use the parameter A to fit any initial condition. If $y(0) = y_0$, then

(6.25) $$y_0 = Ae^{r0} = Ae^0 = A, \quad \text{so } A = y_0 \quad \text{and } y(t) = y_0 e^{rt}.$$

If we want to understand the effect of changing y_0, we just adjust A. What's more, our solution has the interest rate built into it as the parameter r. If we want to understand the effect of changing the interest rate from 5% to 6%, we just change $y_0 e^{0.05t}$ to $y_0 e^{0.06t}$.

All of the simple differential equations that we consider in this chapter have general solutions that you should learn to recognize. Understanding the functions (like e^{rt}) that appear in these solutions goes a long ways toward understanding the real-world systems that these differential equations model.

However, we must be careful not to overstate the importance of general solutions. They are often very hard to obtain. Whole textbooks have been written on techniques for finding them, and even those only touch on a small subset of the relevant differential equations out there. Most complicated differential equations don't even have general solutions that can be easily written down! To understand solutions to differential equations, we should be ready to

(1) Use general solutions when we can find them, both to solve specific IVPs and to get insight into how the solutions depend on initial conditions and on parameters, and

(2) Use other techniques, such as Euler's method, when we can't find a general solution or can't even find a formula for the solution to a specific IVP.

6.2. Exponential Functions and Logarithms

Exponential functions, logarithms, and trig functions are needed to understand all of the behaviors seen in Figure 6.1, but their properties take some getting used to. Most calculus students have already seen their definitions in previous courses, but fairly few are comfortable with equations such as

$$(6.26) \qquad \ln(5e^{x^2}) - e^{2\ln(x)} = \ln(3)\log_3(5).$$

If you are already comfortable with equation (6.26), feel free to skip ahead to Section 6.3 and to use this section as a reference. If not, read on for an overview of exponentials and logs. Trigonometric functions are covered in Section 6.7, Supplemental Materials.

What Does a^x Mean? You were probably introduced to exponentials back in middle school, where you learned that 2^3 means $2 \times 2 \times 2$, and that 2^5 means $2 \times 2 \times 2 \times 2 \times 2$. In general, a^x means "a multiplied by itself x times". Based on that, we have

$$(6.27) \qquad 2^3 \times 2^5 = (2 \times 2 \times 2) \times (2 \times 2 \times 2 \times 2 \times 2).$$

The product of 2^3 and 2^5 has $3 + 5 = 8$ factors of 2, three from 2^3 and five from 2^5, so $2^3 \times 2^5 = 2^{3+5}$. Similarly, if a is any real number and b and c are positive integers, then

$$(6.28) \qquad a^b \times a^c = a^{b+c}.$$

We can also look at $(2^3)^5 = (2^3) \times (2^3) \times (2^3) \times (2^3) \times (2^3)$. This has three factors of 2, repeated five times, for a total of 3×5 factors of 2. In general,

$$(6.29) \qquad (a^b)^c = a^{bc}.$$

Our naive definition that a^x means "a multiplied by itself x times" works fine for 2^3 or 2^{17}, but what do 2^0, 2^{-3}, and $2^{1.363}$ mean? You can't multiply 2 by itself 0 times, or -3 times, or 1.363 times! Our challenge is to expand our definition while preserving the rules (6.28) and (6.29).

We begin with a^0. Since $0 + b = b$, we must have

$$(6.30) \qquad a^b = a^{0+b} = a^0 a^b$$

by rule (6.28). The only way for that to work is if $a^0 = 1$. So we define

$$(6.31) \qquad a^0 = 1 \quad \text{for all } a \neq 0.$$

(The expression 0^0, just like the ratio $0/0$, is not defined.)

Next we look at negative powers. Since $b + (-b) = 0$, we must have

$$(6.32) \qquad a^b a^{-b} = a^{b+(-b)} = a^0 = 1.$$

For this to work, a^{-b} must equal $1/a^b$. Put another way, "multiply a by itself -3 times" essentially means "start at 1 and divide by a 3 times". This makes sense for all $a \neq 0$.

Next we look at fractional powers. If p and q are integers, with $q \neq 0$, we have

$$(6.33) \qquad \left(a^{p/q}\right)^q = a^p,$$

6.2. Exponential Functions and Logarithms

so $a^{p/q}$ is the qth root of a^p, so $2^{1.363}$ means the 1000th root of $2^{1,363}$. It's also $(a^{1/q})^p$, which is the pth power of the qth root of a, so $2^{1.363}$ is the 1363rd power of the 1000th root of 2.

Finally, to make sense of an irrational power like 2^π, we have to use approximations. *Close is good enough!* Since $3.14159 < \pi < 1.314160$, 2^π is somewhere between $2^{3.14159}$ and $2^{3.14160}$. If $r_1, r_2, \ldots$ is a sequence of rational numbers getting closer and closer to π, then $2^{r_1}, 2^{r_2}, \ldots$ is a sequence of real numbers getting closer and closer to 2^π.

In this way, we've defined what a^x means for every positive real number a and for every real number x. Specifically:

- If x is a positive integer, a^x is the product of a with itself x times.
- $a^0 = 1$.
- If x is a positive integer, then $a^{-x} = 1/a^x$.
- If $x = p/q$ is a rational number, then a^x is the qth root of a^p, and it is also the pth power of the qth root of a.
- If x is an irrational number, then we can approximate x arbitrarily well by rational numbers r. a^x is the limit of a^r as r approaches x.

These last two definitions generally require $a > 0$, since if $a < 0$, then the qth root of a may not exist. Exponentials follow the

Three Laws of Exponents: If $a > 0$ and b and c are any real numbers, then

(6.34)
$$a^b \times a^c = a^{b+c},$$
$$a^b/a^c = a^{b-c},$$
$$\left(a^b\right)^c = a^{bc}.$$

The Graph of a^x. When $a > 1$, the function a^x grows very quickly with x. For instance, 2^{20} is a little over a million, but 2^{19} is only about half a million. Half of the growth of the function 2^x between $x = 0$ and $x = 20$ occurs between $x = 19$ and $x = 20$. And then 2^{21} is over *two* million. In the other direction, 2^x shrinks rapidly as x goes negative. 2^{-20} is a little less than a millionth. 2^{-21} is less than a two-millionth.

Figure 6.3 shows the graphs of three exponential functions, namely $f(x) = 2^x$, $g(x) = (1/2)^x$, and $h(x) = 4^x$. Notice that all three graphs go through $(0,1)$, since $a^0 = 1$ no matter what a is (as long as $a \neq 0$). All three graphs lie above the x axis, since a^x is always positive. All three graphs have similar shapes. In fact, the graphs of $f(x)$ and $g(x)$ are mirror images of each other, since

(6.35)
$$g(x) = (1/2)^x = (2^{-1})^x = 2^{-x} = f(-x).$$

The graph of $f(x)$ rises quickly to the right and dies off quickly to the left. The graph of $g(x)$ rises quickly to the left and dies off quickly to the right. We say that $f(x)$ shows **exponential growth**[3] while $g(x)$ shows **exponential decay**, but they're two sides of the same coin. Exponential decay is just growth run backward.

---

[3]When things grow quickly, people often describe the growth as "exponential", but true exponential growth is much faster than what people usually mean.

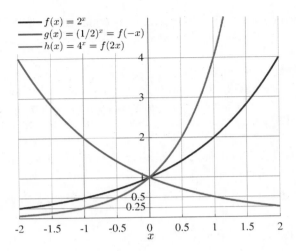

Figure 6.3. The graph of a^x for $a = 2, \frac{1}{2},$ and 4

The function $h(x)$ behaves like $f(x)$, except that its growth (to the right) and decay (to the left) is twice as fast.

(6.36) $$h(x) = 4^x = (2^2)^x = 2^{2x} = f(2x).$$

In fact, we'll soon see that *all* of the functions a^x (with different values of $a \neq 1$) are essentially the same, up to reflection and horizontal scale.

Logarithms. The function $\log_a(x)$, usually pronounced "logarithm (or log) base a of x", is the inverse of the function a^x. That is

(6.37) $$\begin{aligned} y = \log_a(x) \quad &\text{means} \quad x = a^y, \\ x = \log_a(y) \quad &\text{means} \quad y = a^x. \end{aligned}$$

In particular, $a^{\log_a(x)} = x$ (that's the first line) and $\log_a(a^x) = x$ (that's the second).

If that doesn't make sense to you right away, think about the inverse functions "square" and "square root". If x and y are nonnegative, $y = \sqrt{x}$ means $x = y^2$ and $x = \sqrt{y}$ means $y = x^2$. Both $(\sqrt{x})^2$ and $\sqrt{x^2}$ are equal to x. If you apply a function and then apply its inverse, or if you first apply the inverse function and then apply the original function, you get back to where you started.

The graph of $\log_a(x)$, shown in Figure 6.4, is just like the graph of a^x, only with the roles of x and y reversed. Instead of going through $(0, 1)$, it goes through $(1, 0)$. Instead of lying above the x axis, it lies to the right of the y axis. This means that $\log_a(x)$ *is only defined when $x > 0$*. Instead of shooting up extremely quickly, with $2^{21} \approx 2{,}000{,}000$ being twice as big as $2^{20} \approx 1{,}000{,}000$, it grows extremely slowly, with $\log_2(2{,}000{,}0000) \approx 21$ being only one greater than $\log_2(1{,}000{,}000) \approx 20$. Instead of a^x being greater than 1 for $x > 0$ and between 0 and 1 for $x < 0$, $\log_a(x)$ is positive for $x > 1$ and negative for $0 < x < 1$.

6.2. Exponential Functions and Logarithms

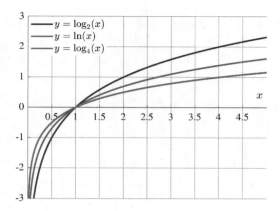

Figure 6.4. The graph of $\log_a(x)$ for $a = 2, e,$ and 4

Let's compute some examples:

$$
\begin{aligned}
64 &= 2^6 & \text{so} & \quad \log_2(64) = 6, \\
1/16 &= 2^{-4} & \text{so} & \quad \log_2(1/16) = -4, \\
64 &= 4^3 & \text{so} & \quad \log_4(64) = 3, \\
1/16 &= 4^{-2} & \text{so} & \quad \log_4(1/16) = -2.
\end{aligned}
\tag{6.38}
$$

Note that for $x = 64$ and for $x = 1/16$, $\log_4(x)$ is half as big as $\log_2(x)$. This is because if $y = \log_4(x)$, then $x = 4^y = 2^{2y}$, so $\log_2(x) = 2y$. That is, the functions $\log_2(x)$ and $\log_4(x)$ are the same, up to a constant multiple:

$$\log_2(x) = 2\log_4(x). \tag{6.39}$$

We will soon see that all of the functions $\log_a(x)$ are the same, up to a constant factor, so we only really need to learn about one log function.

A particularly important logarithm is the log base e. This is called the **natural log** and is usually written $\ln(x)$ rather than $\log_e(x)$. Some people pronounce $\ln(x)$ as "lin x", while others just call it "log x", with the "natural" part understood.

Closely related to the laws of exponents are the following.

Laws of Logarithms: If x, y, and a are positive numbers with $a \neq 1$, and if r is any real number, then

$$
\begin{aligned}
\log_a(xy) &= \log_a(x) + \log_a(y), \\
\log_a(x/y) &= \log_a(x) - \log_a(y), \\
\log_a(x^r) &= r\log_a(x), \\
\log_a(x) &= \log_a(b)\log_b(x), \quad \text{and} \\
\log_a(x) &= \frac{\ln(x)}{\ln(a)}.
\end{aligned}
\tag{6.40}
$$

The first law of logs comes from the first law of exponents (6.34). Let $b = \log_a(x)$, so $x = a^b$, and let $c = \log_a(y)$, so $y = a^c$. Then
$$xy = a^b a^c = a^{b+c}, \quad \text{so}$$
(6.41)
$$\log_a(xy) = b + c = \log_a(x) + \log_a(y).$$

Similarly, the second law of logs comes from the second law of exponents:
$$\frac{x}{y} = \frac{a^b}{a^c} = a^{b-c}, \quad \text{so}$$
(6.42)
$$\log_a(x/y) = b - c = \log_a(x) - \log_a(y).$$

The third law of logs comes from the third law of exponents. If $x = a^b$, then $x^r = a^{rb}$, so $\log_a(x^r) = rb = r \log_a(x)$.

The fourth and fifth laws, sometimes called the *change of base rules*, are a little trickier. By definition, $x = b^{\log_b(x)}$ and $b = a^{\log_a(b)}$. But then
$$\begin{aligned} x &= b^{\log_b(x)} \\ &= \left(a^{\log_a(b)}\right)^{\log_b(x)} \end{aligned}$$
(6.43)
$$= a^{\log_a(b)\log_b(x)},$$

so $\log_a(x) = \log_a(b) \log_b(x)$. Plugging in $x = a$ also gives the useful fact that $\log_a(b)$ and $\log_b(a)$ are reciprocals. In particular, $\log_a(e) = 1/\ln(a)$. Finally,

(6.44)
$$\log_a(x) = \log_a(e) \log_e(x) = \ln(x)/\ln(a).$$

Look back at equation (6.26) at the beginning of this section. With our rules in hand, we can now make sense of it:
$$\begin{aligned} \ln(5e^{x^2}) &= \ln(5) + \ln(e^{x^2}) \\ &= \ln(5) + x^2 \\ e^{2\ln(x)} &= e^{\ln(x^2)} \\ &= x^2 \\ \ln\left(5e^{x^2}\right) - e^{2\ln(x)} &= \ln(5) + x^2 - x^2 \\ &= \ln(5) \\ &= \ln(3)\frac{\ln(5)}{\ln(3)} \end{aligned}$$
(6.45)
$$= \ln(3) \log_3(5).$$

Derivatives of Exponentials and Logarithms. Next we turn to the derivative of a^x. We already saw that
$$\begin{aligned} (a^x)' &= \lim_{h \to 0} \frac{a^{x+h} - a^x}{h} \\ &= \lim_{h \to 0} \frac{a^h - 1}{h} a^x \end{aligned}$$
(6.46)
$$= k_a a^x.$$

6.2. Exponential Functions and Logarithms

That is,

> The rate at which an exponential function grows is proportional to how much is already there.

Another way of putting this is that a^x is the (unique!) solution to the IVP

(6.47) $$\frac{dy}{dx} = k_a y, \qquad y(0) = 1.$$

At this point, we don't understand much about the constant k_a, except that $k_2 \approx 0.69$, $k_3 \approx 1.1$, and that there is a number called e for which $k_e = 1$.

However, the function $e^{k_a x}$ also solves the IVP (6.47), since

(6.48) $$(e^{k_a x})' = e^{k_a x}(k_a x)' = k_a e^{k_a x} \quad \text{and} \quad e^{k_a 0} = e^0 = 1.$$

This means that a^x and $e^{k_a x}$ must be the same function! In particular,

(6.49) $$a = a^1 = e^{k_a(1)} = e^{k_a} \quad \text{so} \quad k_a = \log_e(a) = \ln(a).$$

There are two lessons here. First, our mystery constants k_a are really natural logs. Second, all exponential functions are the same, up to a horizontal scaling:

(6.50) $$a^x = (e^{\ln(a)})^x = e^{\ln(a)x}.$$

If we understand the functions e^{kx}, where k is an arbitrary constant, then we understand *all* exponentials.

Finally, we consider the derivatives of our log functions. Since $x = e^{\ln(x)}$, the chain rule tells us that

$$\begin{aligned} (x)' &= \left(e^{\ln(x)}\right)', \\ 1 &= e^{\ln(x)}(\ln(x))', \\ 1 &= x(\ln(x))', \end{aligned}$$

(6.51) $$(\ln(x))' = \frac{1}{x}.$$

Since $\log_a(x)$ is just $\frac{\ln(x)}{\ln(a)}$, the derivative of $\log_a(x)$ is just $\frac{1}{x \ln(a)}$. Putting everything together, we have the following.

> **Derivatives of Exponentials and Logs:**
>
> $$\frac{de^x}{dx} = e^x,$$
>
> $$\frac{da^x}{dx} = \ln(a) a^x,$$
>
> $$\frac{d \ln(x)}{dx} = \frac{1}{x},$$
>
> (6.52) $$\frac{d \log_a(x)}{dx} = \frac{1}{x \ln(a)}.$$

6.3. Simple Models of Growth and Decay

There are many situations where the rate at which a real-word quantity changes is a linear function of that quantity. These include the following.

(1) *Banks.* If we have $y(t)$ dollars in a bank that offers an interest rate of r, then $dy/dt = ry$. If the bank also charges a fee of F/year, then $dy/dt = ry - F$, which is still a linear function of y. If we invest an additional I/year in the account (e.g., saving for retirement), then $dy/dt = ry + I - F$, which is still a linear function.

(2) *Populations.* The same ideas apply to the number of humans in a city, the number of rabbits on an island, or the number of bacteria in a Petri dish. Both the number of births and the number of deaths per unit of time are roughly proportional to how many individuals are already there, so $dy/dt = ry$. If additional individuals are joining or leaving the population at a constant rate (e.g., people migrating to the city or rabbits being hunted), then $dy/dt = ry + M$, where M (for "migration") is the influx minus outflux.

(3) *Economies.* Most of the things that drive an economy, such as investment capital and demand for products, are proportional to the existing size of the economy. As a result, economic growth often follows the rule $dy/dt = ry$.

(4) *Early stage marketing.* When we first introduce a product, almost everybody is a Potential, and $I'(t) = aSI - bI = (aS - b)I$ is well approximated by $(aT - b)I$, where T is the total population. The same considerations apply to the early stages of an epidemic, since epidemics are governed by the same SIR equations.

(5) *Late stage marketing.* Toward the end of a successful product's life cycle, it will run out of Potentials, and the number of Actives will decrease by the rule $I'(t) = -bI$.

(6) *Radioactive decay.* The rate at which radioactive nuclei disintegrate is proportional to how many are still left. If $y(t)$ is the mass of radioactive material remaining at time t, then $dy/dt = -ry$.

(7) *Heating and cooling.* If the temperature of a cup of coffee is $y(t)$, then the cup will lose heat at a rate proportional to the difference between the coffee's temperature and room temperature (say, 70 degrees Fahrenheit). The temperature will drop at a rate proportional to the temperature difference, so $dy/dt = -r(y - 70)$. If we have something colder than room temperature, like a cold drink or a dish of ice cream, it will heat up at rate $r(70 - y) = -r(y - 70)$. The same differential equation describes both heating and cooling.

Once we figure out how to solve the initial value problems

(6.53) $$dy/dt = ry, \qquad y(0) = y_0$$

and

(6.54) $$dy/dt = ry + p, \qquad y(0) = y_0,$$

where r and p are constants, we will be able to understand *all* of these real-world systems.

6.3. Simple Models of Growth and Decay

Exponential Growth and Decay. The solution to (6.53) is easy. We have already seen that the derivative of $y(t) = Ae^{rt}$ is $r(Ae^{rt})$, and that $y(0) = Ae^0 = A$, so $A = y_0$. In other words, the solution to (6.53) is

$$y(t) = y_0 e^{rt}. \tag{6.55}$$

The IVP (6.53) and the solution (6.55) are *incredibly* common. Not only should you memorize them, but getting from one to the other should become a spinal reflex for you.

If our initial condition occurs at a time $t_0 \neq 0$, we just adjust our formula to count the time since we started. If $dy/dt = ry$ and $y(t_0) = y_0$, then

$$y(t) = y_0 e^{r(t-t_0)}. \tag{6.56}$$

For instance, suppose that a city's population is growing at 2.5%/year, and that its population in the year 2020 was 1,000,000 people. How many people will be in the city in the year 2030? When will the population hit 2,000,000?

To figure out $y(2030)$, we just plug $y_0 = 1{,}000{,}000$, $r = 0.025$, and $t = 2030$ into equation (6.56):

$$y(2030) = 1{,}000{,}000 e^{0.025(2030-2020)} = 1{,}000{,}000 e^{0.25} \approx 1{,}284{,}000. \tag{6.57}$$

To figure out when the population will hit 2,000,000, we solve

$$\begin{aligned}
y(t) &= 2{,}000{,}000, \\
1{,}000{,}000 e^{r(t-t_0)} &= 2{,}000{,}000, \\
e^{r(t-t_0)} &= 2, \\
r(t - t_0) &= \ln(2), \\
t &= t_0 + \frac{\ln(2)}{r} \\
&\approx 2020 + \frac{0.693}{0.025} \\
&= 2047.7.
\end{aligned} \tag{6.58}$$

When a quantity is growing exponentially, we call the time it takes for the quantity to double the **doubling time**. When a quantity is shrinking exponentially, we call the time it takes to drop to half its initial value the **half-life**. The solutions to $e^{rt} = 2$ and $e^{-rt} = 1/2$ are both $t = \ln(2)/r$. Since $\ln(2)$ is approximately 0.70, this leads to a very simple rule of thumb.

> **The Law of 70**: The doubling time (or half-life) is approximately 70 divided by the growth (or decay) rate in percent.

If a city (or an investment, or an economy) grows at 2%/year, it will double in about 35 years. If it grows at 7%/year, it will double in 10 years. If it decays at 10%/year, it will be down to half of its initial value in 7 years, a quarter of its initial value in 14 years, and an eighth of its initial value in 21 years.

The same methods work for all problems involving exponential growth. To figure out how much we'll have at time t, just plug t into (6.56). To figure out when we'll reach a certain level, set up equation (6.56), divide by y_0, take the natural log of both sides, and do a little algebra.

We can also use logs to compute growth (or decay) rates from data. For instance, suppose that the population of a city is growing exponentially. In 2020 there are 1.3 million people. In 2030 there are 1.65 million people. What is the growth rate r?

We set this up almost exactly like the previous problem, except that we're solving for r instead of t:

$$
\begin{aligned}
y(2030) &= 1{,}650{,}000, \\
y(2020)e^{r(2030-2020)} &= 1{,}650{,}000, \\
1{,}300{,}000 e^{10r} &= 1{,}650{,}000, \\
e^{10r} &= 1.65/1.3 = 1.269, \\
10r &= \ln(1.269) = 0.2384, \\
r &= 0.02384.
\end{aligned}
$$

(6.59)

Our city is growing at between 2.3% and 2.4% per year.

To recap:

- The IVP (6.53) comes up in many applications. The differential equation describes the growth process, and the initial condition sets a starting point.
- The *solution* to this IVP is the exponential function $y(t) = y_0 e^{rt}$. If the initial condition is $y(t_0) = y_0$ instead of $y(0) = y_0$, then the solution is $y(t) = y_0 e^{r(t-t_0)}$. These solutions are worth memorizing.
- If we know y_0, r, and t, then we use exponentials to find $y(t)$. If we know y_0, $y(t)$, and t, then we use logs to find r. If we know y_0, $y(t)$, and r, we use logs to find t.

Growth or Decay with an Offset. We finish this section with the slightly more complicated differential equation (6.54). The expression for dy/dt, namely $ry + p$, can be rewritten as $r(y + \frac{p}{r})$. If we define a new variable $z = y + \frac{p}{r}$, then z and y only differ by a constant, so their derivatives are the same.

(6.60) $$\frac{dz}{dt} = \frac{dy}{dt} = r\left(y + \frac{p}{r}\right) = rz.$$

But that's something we already know how to solve! Since the derivative of z is proportional to z, we have

$$
\begin{aligned}
z(t) &= z(t_0)e^{r(t-t_0)}, \\
y(t) + \frac{p}{r} &= \left(y_0 + \frac{p}{r}\right)e^{r(t-t_0)}, \\
y(t) &= \left(y_0 + \frac{p}{r}\right)e^{r(t-t_0)} - \frac{p}{r}.
\end{aligned}
$$

(6.61)

Although it comes up fairly often, the solution (6.61) is *not* worth memorizing. Instead, you should remember the method we used to get it. Define a new variable z that turns our IVP into the spinal reflex problem (6.53). Then use a little algebra.

6.4. Two Models of Oscillation

Returning to our hot coffee example, if the temperature $y(t)$ of a cup of coffee satisfies $dy/dt = -0.1(y - 70)$ and $y(0) = 170$, where we're measuring time in minutes, then we define $z = y - 70$ to get

$$dz/dt = -0.1z \text{ and } z(0) = 100,$$
$$z(t) = 100e^{-0.1t},$$
(6.62) $$y(t) = 70 + 100e^{-0.1t}.$$

The coffee's temperature $y(t)$ doesn't decay exponentially, but the *difference* $z(t)$ between the coffee's temperature and room temperature does. Every $70/10 = 7$ minutes, the coffee gets halfway toward room temperature.

A final note: Don't get hung up on the sign of r or the sign of p. If the quantity of radioactive material follows the rule $dy/dt = -0.02y$, then we have described the situation as $dy/dt = rt$ with $r = -0.02$, whose solution is $y(t) = y_0 e^{r(t-t_0)} = y_0 e^{-0.02(t-t_0)}$. Others describe it as $dy/dt = -rt$ with $r = +0.02$, whose solution is $y(t) = y_0 e^{-r(t-t_0)} = y_0 e^{-0.02(t-t_0)}$. In the end, the solutions are the same! It's purely a matter of taste whether we talk about the material growing at a negative rate or shrinking at a positive rate.

6.4. Two Models of Oscillation

The models of the last section only went in one direction. Either the quantity in question grew forever or it shrank forever. However, many things in the real world go in cycles. These include everything from sunrise and sunset to seasons to the proverbial business cycle. In this section we explore two very simple models that give rise to sine and cosine functions, and in Section 6.5 we consider some more complicated models of oscillation.

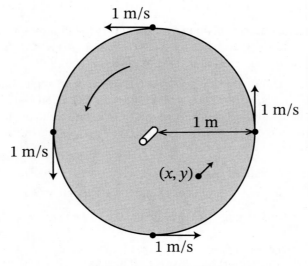

Figure 6.5. A rotating wheel

Wheels. Imagine that a wheel of radius 1 meter is rotating counterclockwise, as in Figure 6.5, and that the rim of the wheel is moving at a speed of 1 meter/second. The right-hand side of the wheel is moving up, the top of the wheel is moving to the left, the left-hand side of the wheel is moving down, and the bottom of the wheel is moving to the right. The rate at which a point on the wheel is moving up is proportional to its x-coordinate, and the rate at which it is moving to the left is proportional to its y coordinate. Since a point at position $(1,0)$ is moving up at speed 1 and a point at position $(0,1)$ is moving left at speed 1, those proportionality constants must both be 1. In other words, the coordinates $x(t)$ and $y(t)$ of a point on the wheel satisfy the differential equations

$$(6.63) \qquad \frac{dx}{dt} = -y, \qquad \frac{dy}{dt} = x.$$

The derivative of $\cos(t)$ is $-\sin(t)$, and the derivative of $\sin(t)$ is $\cos(t)$, so one solution to these equations is

$$(6.64) \qquad x(t) = \cos(t), \qquad y(t) = \sin(t).$$

This describes the trajectory of a spot that starts at $(1,0)$. After time t it has rotated t radians counterclockwise, so its new coordinates are $(\cos(t), \sin(t))$.

Since the derivative of $-\sin(t)$ is $-\cos(t)$ and the derivative of $\cos(t)$ is $-\sin(t)$, another solution is

$$(6.65) \qquad x(t) = -\sin(t), \qquad y(t) = \cos(t).$$

This describes the trajectory of a spot that starts at $(0,1)$. (Remember to "guess and check". We don't need to know where the solutions (6.64) and (6.65) come from to see that they work.) Combining this with our other solution, we get the *general* solution to our equations:

$$(6.66) \qquad x(t) = a\cos(t) - b\sin(t), \qquad y(t) = a\sin(t) + b\cos(t).$$

You should check that these functions really do satisfy the differential equations (6.63). Since $\cos(0) = 1$ and $\sin(0) = 0$, we have $x(0) = a$ and $y(0) = b$, so this describes the trajectory of a spot on the wheel that starts at position (a, b).

Now suppose that our starting point is on the rim, at an angle A above the x axis, as in Figure 6.6. That is, $x(0) = \cos(A)$ and $y(0) = \sin(A)$. After time t, our wheel has rotated $B = t$ radians counterclockwise, so our new angle should be $A + B = A + t$, and our new coordinates should be $(\cos(A+B), \sin(A+B))$. Comparing this to our general solution with $a = \cos(A)$ and $b = \sin(A)$ gives

$$\begin{aligned}\cos(A+B) &= \cos(A)\cos(B) - \sin(A)\sin(B), \\ \sin(A+B) &= \cos(A)\sin(B) + \sin(A)\cos(B).\end{aligned}$$
(6.67)

These are the **addition of angle formulas** for sine and cosine.[4]

---

[4]This approach to addition of angles is something of a "cheat", because we previously *used* the addition of angle formulas to compute the derivatives of $\sin(x)$ and $\cos(x)$. To avoid circular reasoning (pun intended), we will derive the formulas (6.67) from scratch in the "Crash Course in Trigonometry" in Section 6.7.

6.4. Two Models of Oscillation

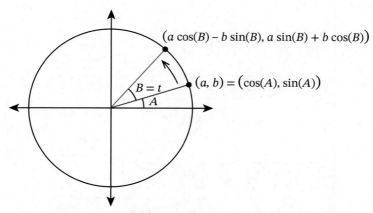

Figure 6.6. Understanding addition of angles through differential equations

We have just described a wheel that is rotating at 1 radian/second. More generally, a wheel that is rotating at angular speed ω (measured in radians/second) is governed by the differential equations

$$\text{(6.68)} \qquad \frac{dx}{dt} = -\omega y, \qquad \frac{dy}{dt} = \omega x,$$

whose general solution is

$$\text{(6.69)} \qquad x(t) = a\cos(\omega t) - b\sin(\omega t), \qquad y(t) = a\sin(\omega t) + b\cos(\omega t).$$

The example of the rotating wheel shouldn't come as a surprise. The sine and cosine functions were defined in terms of positions on a circle, so we should expect to see them whenever we see circles. The next example shows how we get sines and cosines from oscillations that don't seem to have anything to do with circles.

Springs. Figure 6.7 shows a block of mass m connected to a spring. Hooke's law says that if the spring is stretched a distance x, then it will pull the block back toward equilibrium with a force proportional to x. If the spring is compressed, it will push the block. In general, the force F on the spring is

$$\text{(6.70)} \qquad F = -kx,$$

where x is the displacement from equilibrium. The proportionality constant k is called the **spring constant**. Meanwhile, Newton's second law, $F = ma$, says that the acceleration of the block is proportional to the force applied to it. Combining these, we get that

$$\text{(6.71)} \qquad \frac{d^2x}{dt^2} = a = \frac{F}{m} = -\frac{k}{m}x = -\omega^2 x,$$

where $\omega = \sqrt{k/m}$. That is, the oscillations of the block and spring are governed by the second-order differential equation (6.71). This is similar to (6.53), except that it involves a second derivative instead of a first derivative.

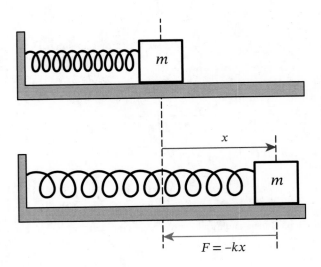

Figure 6.7. A mass on a spring

The functions e^{kt} have the property that their first derivatives are proportional to themselves, and their second derivatives are positive multiples of themselves: $\left(e^{kt}\right)'' = k^2 e^{kt}$. The functions $\sin(\omega t)$ and $\cos(\omega t)$ have the property that their second derivatives are *negative* multiples of themselves:

$$\frac{d^2}{dt^2}\sin(\omega t) = \frac{d}{dt}\omega\cos(\omega t) = -\omega^2 \sin(\omega t),$$

(6.72) $$\frac{d^2}{dt^2}\cos(\omega t) = \frac{d}{dt}(-\omega\sin(\omega t)) = -\omega^2 \cos(\omega t).$$

In fact, the general solution to (6.71) is

(6.73) $$x(t) = a\cos(\omega t) + b\sin(\omega t).$$

At time $t = 0$, we have $x = a$ and $\frac{dx}{dt} = \omega b$, so we can relate a and b to the initial position x_0 and the initial velocity v_0, and we write

(6.74) $$x(t) = x_0 \cos(\omega t) + \frac{v_0}{\omega}\sin(\omega t).$$

The position of the block as a function of time is exactly the same as the x-coordinate of a spot on a wheel that is rotating at ω radians/second. What's more, it takes $2\pi/\omega$ seconds to complete a cycle, regardless of the size of the oscillation. This allows us to use oscillations to count time. Until electronics took over in the the mid-20th century, most clocks and wrist watches were based on this principle.

Hooke's law is specific to springs, but most physical oscillations follow similar rules. For instance, the force pulling a pendulum back toward the center is (approximately) proportional to the angle of displacement, so the period of a pendulum doesn't depend on the size of the oscillation. Even in the electronics age, clocks still count oscillations, but the thing that is oscillating is more complicated than a mass on a spring. In fact, small fluctuations around almost *any* stable equilibrium are governed by a rule

6.5. More Sophisticated Models

similar to Hooke's law. Each system exhibits a characteristic frequency ω, and fluctuations are described by (6.74). This pattern is called **simple harmonic motion**.

6.5. More Sophisticated Models

In this section we will consider three problems, two realistic and one a fantasy, where the behavior of the system is more complicated than simple exponential growth, exponential decay, or sines and cosines. The first two lead to **limited growth**, as in the third graph in Figure 6.1 (on page 135). The last leads to oscillations, as in the last graph in Figure 6.1.

The Australian Rabbit Problem. Rabbits are known for three things. They are very cute. They deliver Easter baskets. And they breed like rabbits.

In 1859, Thomas Austin released 24 rabbits in Australia, in the hope that they would breed and make for good hunting. As he put it,

The introduction of a few rabbits could do little harm and might provide a touch of home, in addition to a spot of hunting.

He was wrong. The rabbit population increased at a rate of about 10%/month, and after 10 years (120 months) there were roughly

$$(6.75) \qquad 24e^{0.1(120)} = 24e^{12} = 3.9 \text{ million}$$

rabbits in that part of Australia.[5] So how many rabbits were there after 20 years? After 30 years?

If the rabbit population continued to grow exponentially, there would have been $24e^{24} = 636$ billion rabbits in Australia in 1879, or about 214,000 rabbits per square mile. By 1889 there would have been about 100 quadrillion rabbits, or about 1250 rabbits per square *foot*.

Those numbers are absurd. Exponential growth is so fast that it simply *can't go on forever*. At some point, the presence of so many rabbits changes the rules of the game, and the simple differential equation $dy/dt = ry$ no longer applies. To model situations like the Australian Rabbit Problem, we need a more sophisticated model.

When there is a shortage of resources, such as food and nesting sites, rabbits are forced to compete with one another. When two rabbits want the same piece of food, at least one of them will go away hungry (or perhaps injured or dead, if they fight over it). This diminishes the population in proportion to how often two rabbits meet, which is proportional to the *square* of the population. That is,

$$(6.76) \qquad \frac{dy}{dt} = ry - \alpha y^2,$$

where we have introduced a second parameter α that gives the strength of the competition. Equation (6.76) is called the **logistic equation**, and it is usually written in a

---

[5]That figure is just a rough estimate. All that is known for sure is that millions of rabbits were being hunted each year, yet their population still grew.

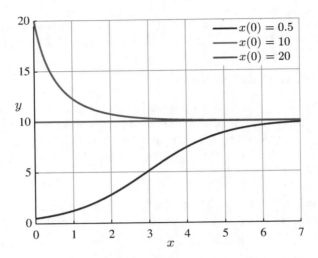

Figure 6.8. Three solutions to the logistic equation with $r = 1$ and $L = 10$.

slightly different form,

$$\begin{aligned}
\frac{dy}{dt} &= ry - \alpha y^2 \\
&= ry\left(1 - \frac{\alpha y}{r}\right) \\
&= ry\left(1 - \frac{y}{L}\right),
\end{aligned}$$

(6.77)

where $L = r/\alpha$. The parameter L is called the **carrying capacity** of the system. For Australian rabbits, it is about 600 million.

Notice that dy/dt is positive if $0 < y < L$ and negative if $y > L$. In the logistic model, also called the **limited growth** model, the population grows until it approaches the carrying capacity, but it never goes above the carrying capacity. If for some reason we start with $y > L$, as in the top solution in Figure 6.8, then the population shrinks until it approaches the carrying capacity.

The formula for the general solution to (6.77) is known, but isn't very important. (It happens to be $y(t) = LAe^{rt}/(1 + Ae^{rt})$.) Much more important is understanding its qualitative behavior.

When $y \ll L$, then $(1 - \frac{y}{L}) \approx 1$ and $\frac{dy}{dt} \approx ry$. In other words, the early stages look just like exponential growth, proportional to e^{rt}. After y reaches $L/2$, the system slows down, and growth gets slower and slower. When y is close to L, we can approximate $ry(1 - \frac{y}{L})$ as $rL(1 - \frac{y}{L}) = r(L - y)$. We have already seen how to solve $dy/dt = r(L - y)$. Since the derivative of $L - y$ is approximately $-r(L - y)$, the quantity $L - y$ decays exponentially, like a constant times e^{-rt}. In this way, the limited growth model combines the features of exponential growth (at the beginning) and exponential decay (at the end).

6.5. More Sophisticated Models

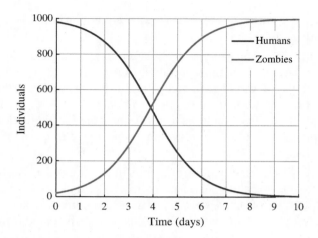

Figure 6.9. The zombie apocalypse. Starting on day 0 with 20 zombies and 980 humans, the human population is reduced to just two on day 10.

Humans versus Zombies. Suppose that a town of 1000 people is attacked by zombies. At time t, there are $Z(t)$ zombies and $H(t) = 1000 - Z(t)$ humans. On any given day, each human has a 1% chance of running into each of the zombies. If they meet, there is a 10% chance of the zombie turning the human into a zombie, and a 90% chance of the human getting away. If zombies never die, and if we start with 20 zombies and 980 humans, how many humans and zombies will there be after t days?

This is exactly the SIR model, only with no attrition, since zombies stay zombies forever, and with no Rejecteds/Recovereds. The humans are all Potential zombies, and the zombies are the Actives. (If a new app is so addictive that the Actives spend all day staring at their phones, you might well call them zombies.) The transmission coefficient is $a = (0.01)(0.1) = 0.001$. We then have

$$\begin{aligned} \frac{dH}{dt} &= -0.001HZ, \\ \frac{dZ}{dt} &= 0.001HZ \\ &= 0.001Z(1000 - Z) \\ &= Z\left(1 - \frac{Z}{1000}\right). \end{aligned}$$
(6.78)

The equation for Z in the SIR model with no attrition is *exactly* the logistic equation with $r = 1$ and $L = 1000$. In the early stages of the zombie apocalypse, the number of zombies grows exponentially as e^t. In the late stages, the number of humans shrinks exponentially as e^{-t}, as seen in Figure 6.9.

Predators and Prey. We are mostly interested in differential equations where the rules of the game don't change with time, so the right-hand side $f(y, t)$ is actually only a function of y. First-order differential equations of this sort, with a single variable, never lead to oscillations. To get oscillations, we either need a second-order equation, as with the mass on a spring (6.71), or first-order equations in two or more variables, as with the rotating wheel (6.63).

The dynamics of predators and prey provides a good example of the second kind of system. Suppose that a predator that eats rabbits, such as foxes, takes hold in Austrialia. At first the fox population would grow quickly, as the foxes would have lots of food. The large fox population would then cause the rabbit population to drop. The low rabbit population would then cause the foxes to starve, and the fox population would drop. The low fox population would then allow the rabbits to recover. And the cycle would repeat.

Reality check: Using foxes to control rabbits in Australia is a *very* bad idea, since foxes eat more than just rabbits. Foxes are themselves considered an invasive species and threaten a number of native species, such as the bilby. (Easter celebrations in Australia sometimes feature an Easter bilby instead of the hated rabbit.)

Just as the logistic model applies to all sorts of limited growth problems, there is a model, called the Lotka-Volterra model, that applies to all sorts of predator-prey problems. We can talk about foxes and rabbits, or owls and mice, or wolves and elk, or deer and the plants that they eat. Mathematically, these are all essentially the same.

Let $x(t)$ be the population of a prey (e.g., rabbits), and let $y(t)$ be the population of a predator (e.g., foxes). The simplest version of the Lotka-Volterra model says that

(6.79)
$$\frac{dx}{dt} = \alpha x - \beta xy,$$
$$\frac{dy}{dt} = \gamma xy - \delta y,$$

where α, β, γ, and δ are positive parameters. In the absence of predators, the prey population would grow exponentially as $e^{\alpha t}$. In the absence of prey, the predator population would decay as $e^{-\delta t}$. However, every time a predator meets a prey animal (something that happens with a frequency proportional to xy), it's good news for the predator (add γxy to dy/dt) and bad news for the prey (subtract βxy from dx/dt).

A typical trajectory for this system, with $(\alpha, \beta, \gamma, \delta) = \left(\frac{2}{3}, \frac{4}{3}, 1, 1\right)$ and initial conditions $x(0) = 0.5$ and $y(0) = 0.25$ is shown in Figure 6.10.

The functions aren't exactly sines and cosines, but they have the same qualitative behavior. The solutions are periodic, with a cycle of growth of the prey, followed by growth of the predator, followed by decline of the prey, followed by decline of the predator. Since ecosystems always have many more prey than predators, the specific numbers for the populations should not be taken literally and should perhaps be thought of as tens of predators and thousands of prey.

After reading about Australian Rabbits, you might object that we didn't take competition among prey, or among predators, into account. In fact, a more sophisticated version of the model adds a small competition term $-\mu x^2$ to dx/dt and sometimes a competition term $-\nu y^2$ to dy/dt. With those terms in place, the populations of predators and prey gradually spiral inward toward a steady state, with x settling close to δ/γ and y settling close to α/β.

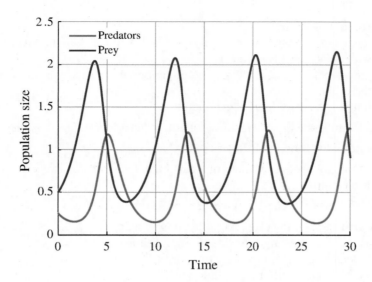

Figure 6.10. A solution to the Lotka–Volterra equations with $(\alpha, \beta, \gamma, \delta) = \left(\frac{2}{3}, \frac{4}{3}, 1, 1\right)$

In the real world, populations of predators and prey do tend to stabilize over time. However, random events like forest fires or a hard winter sometimes knock things out of equilibrium, after which the oscillations start all over again.

6.6. Chapter Summary

The Main Ideas.
- Many things in the real world are described by differential equations. Each term in a differential equation quantifies something that makes a quantity grow or shrink. Modeling a system with a differential equation amounts to writing these factors down.
- Most differential equations are too complicated to be solved with a formula. Instead, we solve them numerically with Euler's method and similar techniques.
- A few very simple differential equations occur repeatedly in the real world, and their solutions *are* given by simple formulas. These include
  - $dy/dt = ry$, which gives exponential growth (if $r > 0$) or exponential decay (if $r < 0$).
  - $dx/dt = -\omega y$, $dy/dt = \omega x$, which gives circular motion (sines and cosines of ωt).
  - $d^2x/dt^2 = -\omega^2 x$, simple harmonic motion, which also gives sines and cosines of ωt.
- These simple models can be used to approximate more complicated equations. For instance, the logistic model looks like exponential growth of x at the beginning, and exponential decay of $L - x$ at the end.

- The exponential function $a^x = e^{x \ln(a)}$
    - means more than just "multiply a by itself x times", which only makes sense when x is a positive integer.
    - grows extremely quickly in one direction and decays extremely quickly in the other.
    - follows the three laws of exponents.
- A logarithm is the inverse of an exponential. That is, $y = \log_a(x)$ means $x = a^y$.
- The natural log function $\ln(x)$ is the log base e.
- Logarithms follow the five laws of logarithms. In particular, every log function is a multiple of $\ln(x)$.
- The derivatives of e^x and $\ln(x)$ are e^x and $1/x$, respectively.
- Sines, cosines, and the other trig functions appear
    - as ratios of the length of sides of right triangles.
    - as locations of points on the unit circle.
    - as solutions to differential equations.
- The solution to $dy/dt = ry$ with $y(t_0) = y_0$ is $y(t) = y_0 e^{r(t-t_0)}$. That is, whenever something changes in proportion to how much is already there, you get exponential growth or decay.
- The law of 70: In exponential growth/decay, the doubling time or half-life is approximately 70 divided by the growth rate in percent.
- The logistic model describes population growth with limited resources, and also describes the interaction of warring populations (e.g., humans versus zombies).
- The Lotka-Volterra model describes interactions between predators and prey, and it gives rise to oscillations.

Expectations. You should be able to:

- Turn a verbal description of a situation, such as a bank giving such-and-such interest and charging such-and-such fees, into a differential equation.
- Modify a differential equation to incorporate a new feature of a model, such as competition for resources.
- Solve differential equations numerically with Euler's method, both by hand and with the help of a computer.
- Manipulate exponential functions, logs, and trig functions, using
    - the three laws of exponents.
    - the laws of logs.
    - the Pythagorean identities for trig functions.
- Sketch the graphs of e^x, $\ln(x)$, $\sin(x)$, and $\cos(x)$ accurately.
- Recite from memory the values of the sine, cosine, and tangent of 0, $\pi/6$, $\pi/4$, $\pi/3$, and $\pi/2$.
- Solve the differential equations $dy/dt = ry$ and $dy/dt = ry + p$.

- Use the solutions of these differential equations in three ways:
  - Use exponentials to find $y(t)$ at some later time, given y_0 and r.
  - Use logs to find when $y(t)$ will reach a certain value.
  - Use logs to deduce r from data.
- Use sines and cosines to describe simple harmonic motion.
- Analyze a differential equation qualitatively by approximating it with something whose solution you already understand.

6.7. Supplemental Material: A Crash Course in Trigonometry

Sines, Cosines, and Triangles. The trigonometric functions $\sin(x)$ and $\cos(x)$ show up in three different settings:

- Describing the shape of a right triangle.
- Describing the location of a point on a circle.
- As solutions to some very common differential equations, as we saw in Section 6.4.

Right Triangles. How can we parametrize the shape of a right triangle? We might give the lengths of the three sides, which certainly *determine* the shape, but those aren't properties *of* the shape, because two triangles can have the same shape without having the same size, as in Figure 6.11. To describe the shape—and only the shape—we either need to specify one of the angles (such as θ in Figure 6.11) or the ratio of two of the sides. Those are quantities that are the same for all triangles of the same shape.

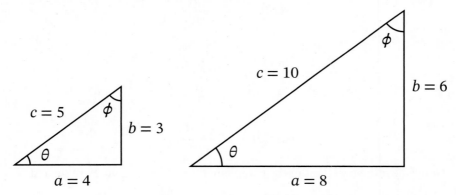

Figure 6.11. A 3-4-5 right triangle and a 6-8-10 right triangle have the same shape.

From the point of view of the angle θ, the side labeled a is called **adjacent** and the side labeled b is called **opposite**. The side labeled c is called the **hypotenuse**. Their lengths are related by the Pythagorean theorem,

$$a^2 + b^2 = c^2. \tag{6.80}$$

The angles θ and ϕ are said to be **complementary**, which means that they add up to $\pi/2$ radians. From ϕ's point of view, b is adjacent, a is opposite, and c is still the hypotenuse.

There are six possible ratios of side lengths, and they are called the sine, cosine, tangent, cotangent, secant, and cosecant of the angle:

$$\sin(\theta) = \frac{\text{Opposite}}{\text{Hypotenuse}} = \frac{b}{c},$$

$$\cos(\theta) = \frac{\text{Adjacent}}{\text{Hypotenuse}} = \frac{a}{c},$$

$$\tan(\theta) = \frac{\text{Opposite}}{\text{Adjacent}} = \frac{b}{a},$$

$$\cot(\theta) = \frac{\text{Adjacent}}{\text{Opposite}} = \frac{a}{b},$$

$$\sec(\theta) = \frac{\text{Hypotenuse}}{\text{Adjacent}} = \frac{c}{a},$$

(6.81) $$\csc(\theta) = \frac{\text{Hypotenuse}}{\text{Opposite}} = \frac{c}{b}.$$

The mnemonic SOH-CAH-TOA summarizes the first three. **S**ine equals **O**pposite over **H**ypotenuse. **C**osine equals **A**djacent over **H**ypotenuse. **T**angent equals **O**pposite over **A**djacent.

The functions tangent, cotangent, secant, and cosecant can all be expressed in terms of sine and cosine:

$$\tan(\theta) = \frac{\sin(\theta)}{\cos(\theta)},$$

$$\cot(\theta) = \frac{\cos(\theta)}{\sin(\theta)},$$

$$\sec(\theta) = \frac{1}{\cos(\theta)},$$

(6.82) $$\csc(\theta) = \frac{1}{\sin(\theta)}.$$

Notice how the functions of the complementary angles θ and ϕ are related:

$$\cos(\phi) = \frac{b}{c} = \sin(\theta),$$

$$\cot(\phi) = \frac{b}{a} = \tan(\theta),$$

(6.83) $$\csc(\phi) = \frac{c}{a} = \sec(\theta),$$

and similarly $\sin(\phi) = \cos(\theta)$, $\tan(\phi) = \cot(\theta)$, and $\sec(\phi) = \csc(\theta)$. The prefix "co" in cosine, cotangent and cosecant stands for "complementary". The cofunctions of the angle θ are the regular functions of its complementary angle ϕ, and vice versa.

There are three special triangles associated with each angle θ, as shown in Figure 6.12. The first has $c = 1$ and is the key to understanding the sine and cosine functions. The second has $a = 1$ and is the key to understanding tangent, and secant, while

6.7. Supplemental Material: A Crash Course in Trigonometry

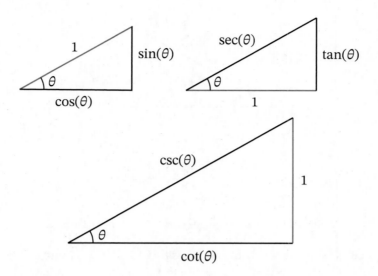

Figure 6.12. The three fundamental triangles for an angle θ

the third has $b = 1$ and is the key to understanding cotangent and cosecant. In the first triangle, with $c = 1$, we have $a = c\cos(\theta) = \cos(\theta)$ and $b = c\sin(\theta) = \sin(\theta)$. The Pythagorean theorem then tells us that

$$\sin^2(\theta) + \cos^2(\theta) = 1, \tag{6.84}$$

where the notation $\sin^2(\theta)$ means $(\sin(\theta))^2$. The second triangle has $a = 1$, $b = \tan(\theta)$, and $c = \sec(\theta)$, and the Pythagorean theorem tells us that

$$\tan^2(\theta) + 1 = \sec^2(\theta). \tag{6.85}$$

Similarly, the third triangle shows that

$$\cot^2(\theta) + 1 = \csc^2(\theta). \tag{6.86}$$

Equations (6.84), (6.85), (6.86) are called the **Pythagorean identities**.

The angles $\pi/6$ (also known as 30 degrees), $\pi/4$ (also known as 45 degrees), and $\pi/3$ (also known as 60 degrees) come up often. Let's compute the six functions of these angles.

In an equilateral triangle, say with side length 1, all three angles are $\pi/3$. If we cut an equilateral triangle in half, as in Figure 6.13, we get a right triangle with a base angle of $(\pi/3)/2 = \pi/6$ and with sides $c = 1$, $b = 1/2$, and $a = \sqrt{c^2 - b^2} = \sqrt{3}/2$. This tells us that

$$\sin(\pi/6) = \cos(\pi/3) = \tfrac{1}{2}, \qquad \cos(\pi/6) = \sin(\pi/3) = \tfrac{\sqrt{3}}{2},$$

$$\tan(\pi/6) = \cot(\pi/3) = \tfrac{\sqrt{3}}{3}, \qquad \cot(\pi/6) = \tan(\pi/3) = \sqrt{3},$$

$$\sec(\pi/6) = \csc(\pi/3) = \tfrac{2\sqrt{3}}{3}, \qquad \csc(\pi/6) = \sec(\pi/3) = 2. \tag{6.87}$$

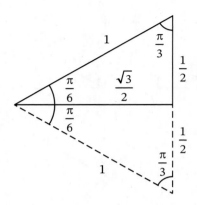

Figure 6.13. A 30-60-90 triangle is half of an equilateral triangle.

An isosceles right triangle with $a = 1$ has $b = 1$ and $c = \sqrt{2}$ and $\theta = \phi = \pi/4$, as in Figure 6.14. This tells us that

(6.88)
$$\begin{aligned} \sin(\pi/4) = \cos(\pi/4) &= \sqrt{2}/2, \\ \tan(\pi/4) = \cot(\pi/4) &= 1, \\ \sec(\pi/4) = \csc(\pi/4) &= \sqrt{2}. \end{aligned}$$

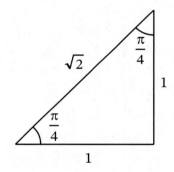

Figure 6.14. An isosceles right triangle

If A and B are angles, then the sines and cosines of $A \pm B$ and $2A$ are given by the following formulas.

(6.89)
$$\begin{aligned} \sin(A + B) &= \sin(A)\cos(B) + \cos(A)\sin(B), \\ \sin(A - B) &= \sin(A)\cos(B) - \cos(A)\sin(B), \\ \cos(A + B) &= \cos(A)\cos(B) - \sin(A)\sin(B), \\ \cos(A - B) &= \cos(A)\cos(B) + \sin(A)\sin(B), \\ \sin(2A) &= 2\sin(A)\cos(A), \\ \cos(2A) &= \cos^2(A) - \sin^2(A) = 2\cos^2(A) - 1 = 1 - 2\sin^2(A). \end{aligned}$$

These are called the **addition of angle** formulas and the **double angle** formulas.

6.7. Supplemental Material: A Crash Course in Trigonometry

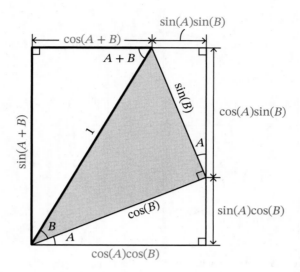

Figure 6.15. The sine and cosine of $A + B$

The addition of angle formulas are explained in Figure 6.15.[6] The height of the triangle on the left is the sum of the heights of the triangles on the right, which gives the formula for $\sin(A + B)$. The width of the triangle on the left is the width of the bottom right triangle minus the width of the top right triangle, which gives the formula for $\cos(A+B)$. To get the formulas for $\sin(A-B)$ and $\cos(A-B)$, just replace B with $-B$ and use the facts that $\sin(-B) = -\sin(B)$ and $\cos(-B) = \cos(B)$. To get the formulas for $\sin(2A)$ and $\cos(2A)$, just set $B = A$.

The Unit Circle. Next we look at the coordinates of points on the unit circle. If the line from the origin to a point on the unit circle makes an angle θ with the x-axis, then we have a triangle of hypotenuse 1 inscribed in the circle, as in Figure 6.16. The base and height of this triangle are $\cos(\theta)$ and $\sin(\theta)$, respectively, so the coordinates of our point must be $(\cos(\theta), \sin(\theta))$.

When dealing with right triangles, we only considered angles θ between 0 and $\pi/2$. However, the unit circle allows us to work with arbitrary angles. We simply *define* $\sin(\theta)$ and $\cos(\theta)$ to be the y and x coordinates (in that order!) of the corresponding point on the unit circle. The other trig functions are defined as ratios of sines and cosines, as in (6.82), as long as this doesn't involve dividing by 0. For instance, $\sin(0) = 0$ and $\cos(0) = 1$, so

(6.90) $\qquad \tan(0) = 0/1 = 0 \qquad$ and $\qquad \sec(0) = 1/1 = 1.$

Similarly, $\sin(\pi/2) = 1$ and $\cos(\pi/2) = 0$, so

(6.91) $\qquad \cot(\pi/2) = 0 \qquad$ and $\qquad \csc(\pi/2) = 1.$

---

[6] This way of seeing the proof of the formulas in a single figure was popularized by Blue the Trigonographer. You can find many more trigonometric proofs by picture, or "trigonographs", at Blue's blog: trigonography.com/toc.

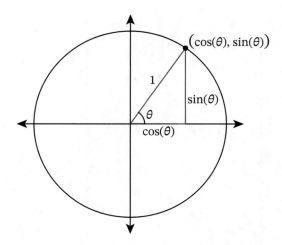

Figure 6.16. The coordinates of a point on the unit circle are $(\cos(\theta), \sin(\theta))$.

However, $\cot(0)$, $\csc(0)$, $\tan(\pi/2)$, and $\sec(\pi/2)$ are undefined, since they involve division by 0. The graphs of $\sin(\theta)$ and $\cos(\theta)$ are shown in Figure 6.17.

Since points on the unit circle have $x = \cos(\theta)$ and $y = \sin(\theta)$, symmetries of the unit circle give us relations among the sine and cosine functions. Rotating a circle by 2π sends (x, y) to itself, so the sine and cosine functions repeat every 2π.

(6.92) $\qquad \sin(\theta + 2\pi) = \sin(\theta) \qquad \text{and} \qquad \cos(\theta + 2\pi) = \cos(\theta).$

Four other symmetries are shown in Figure 6.18.

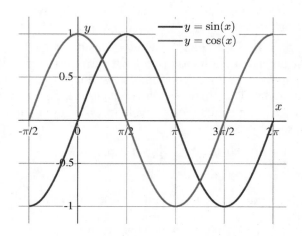

Figure 6.17. The graphs of $\sin(x)$ and $\cos(x)$

6.7. Supplemental Material: A Crash Course in Trigonometry

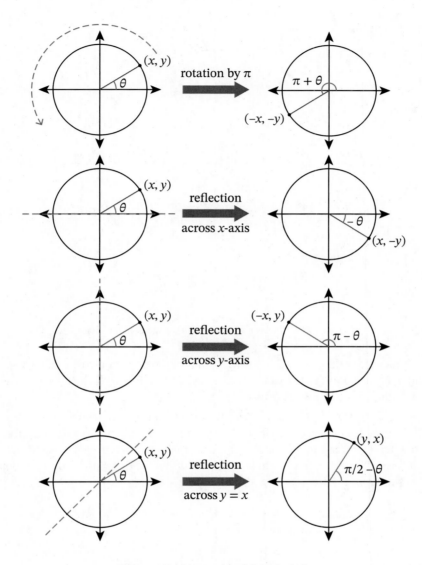

Figure 6.18. Symmetries of the unit circle

Rotating by π sends (x, y) to $(-x, -y)$ and θ to $\theta + \pi$:
(6.93) $\qquad \sin(\theta + \pi) = -\sin(\theta) \qquad$ and $\qquad \cos(\theta + \pi) = -\cos(\theta)$.

Reflecting about the x axis sends θ to $-\theta$ and sends (x, y) to $(x, -y)$:
(6.94) $\qquad \sin(-\theta) = -\sin(\theta) \qquad$ and $\qquad \cos(-\theta) = \cos(\theta)$.

Reflecting about the y axis sends θ to $\pi - \theta$ and sends (x, y) to $(-x, y)$:
(6.95) $\qquad \sin(\pi - \theta) = \sin(\theta) \qquad$ and $\qquad \cos(\pi - \theta) = -\cos(\theta)$.

Finally, reflecting across the line $y = x$ sends θ to $\pi/2 - \theta$ and sends (x, y) to (y, x):
(6.96) $\qquad \sin\left(\frac{\pi}{2} - \theta\right) = \cos(\theta) \qquad$ and $\qquad \cos\left(\frac{\pi}{2} - \theta\right) = \sin(\theta)$.

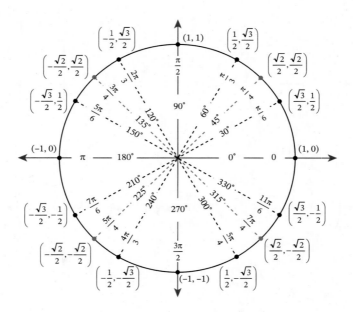

Figure 6.19. Angles and coordinates around the unit circle

With these identities, plus the known values of the sines and cosines of 0, $\pi/6$, $\pi/4$, $\pi/3$, and $\pi/2$, we can fill in the values of the sine and cosine of all the important angles between 0 and 2π, as in Figure 6.19.

If $-\pi/2 < \theta < \pi/2$, then $\tan(\theta)$ is the slope of the line that makes an angle of θ with the x axis. This is negative when $\theta < 0$ and is positive when $\theta > 0$. It is large and positive when θ is close to $\pi/2$, and is large and negative when θ is close to $-\pi/2$. It is not defined when $\theta = \pm\pi/2$, since that would involve dividing by 0. The graph of $\tan(\theta)$ is shown in Figure 6.20.

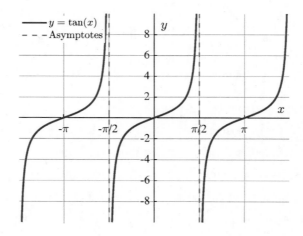

Figure 6.20. The graph of $\tan(x)$. Note the discontinuities at $x = \pm\pi/2$.

6.7. Supplemental Material: A Crash Course in Trigonometry

Note that

(6.97) $$\tan(\theta + \pi) = \frac{\sin(\theta + \pi)}{\cos(\theta + \pi)} = \frac{-\sin(\theta)}{-\cos(\theta)} = \frac{\sin(\theta)}{\cos(\theta)} = \tan(\theta),$$

so the graph repeats every π.

Derivatives and Inverse Trig Functions. We previously computed the derivatives of the six basic trig functions in Sections 5.1 and 5.2:

$$\frac{d}{dx}\sin(x) = \cos(x), \qquad \frac{d}{dx}\cos(x) = -\sin(x),$$

$$\frac{d}{dx}\tan(x) = \sec^2(x), \qquad \frac{d}{dx}\cot(x) = -\csc^2(x),$$

(6.98) $$\frac{d}{dx}\sec(x) = \sec(x)\tan(x), \qquad \frac{d}{dx}\csc(x) = -\csc(x)\cot(x).$$

The derivatives of the cofunctions all have minus signs, but otherwise they look like "co" of the derivatives of the original functions. This is because

$$\text{cofunction}(x) = \text{function}\left(\frac{\pi}{2} - x\right),$$

$$(\text{cofunction})'(x) = \text{function}'\left(\frac{\pi}{2} - x\right) \times \left(\frac{\pi}{2} - x\right)'$$

$$= -\text{function}'\left(\frac{\pi}{2} - x\right)$$

(6.99) $$= -\text{co(function}')(x).$$

Also note that the derivative of $\tan(x)$ is always positive. This fits the fact that $\tan(x)$ increases on the entire interval from $x = -\pi/2$ to $x = \pi/2$, and then jumps down and starts all over again.

The **inverse sine** of x, also called the **arcsine** of x (written $\arcsin(x)$ or $\sin^{-1}(x)$), is the angle between $-\pi/2$ and $\pi/2$ whose sine is x. That is, $\theta = \sin^{-1}(x)$ means the same thing as $x = \sin(\theta)$. The **inverse tangent** or **arctangent** of x is the angle between $-\pi/2$ and $\pi/2$ whose tangent is x. The graphs of $\sin^{-1}(x)$ and $\tan^{-1}(x)$ are obtained by taking the graphs of $\sin(x)$ and $\tan(x)$, restricting the input to be between $-\pi/2$ and $\pi/2$, and then swapping the input and output, as in Figure 6.21.

Warning: The notations $\tan^{-1}(x)$ and $\sin^{-1}(x)$ are standard but unfortunate, because they do not follow the same pattern as $\tan^2(x)$ or $\sin^2(x)$. $\tan^2(x)$ means $(\tan(x))^2$, but $\tan^{-1}(x)$ does **not** mean $(\tan(x))^{-1}$. It means the *inverse* tangent of x. The notation $\sin^n(x)$ is only used when n is a positive integer, in which case it means $(\sin(x))^n$, or when $n = -1$, in which case it means an inverse function. Expressions like $\sin^{-2}(x)$ don't mean anything at all.

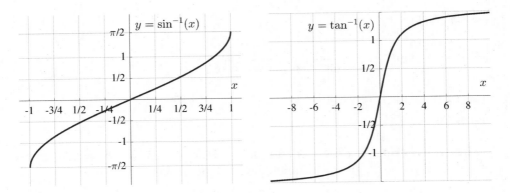

Figure 6.21. The graphs $y = \sin^{-1}(x)$ and $y = \tan^{-1}(x)$ are the same as $x = \sin(y)$ and $x = \tan(y)$, respectively, with y between $-\pi/2$ and $\pi/2$.

The other inverse trig functions exist but are much less interesting. They are related to $\sin^{-1}(x)$ and $\tan^{-1}(x)$ by the following identities:

$$\cos^{-1}(x) = \frac{\pi}{2} - \sin^{-1}(x),$$
$$\cot^{-1}(x) = \tan^{-1}(1/x),$$
$$\sec^{-1}(x) = \frac{\pi}{2} - \sin^{-1}(1/x),$$
(6.100) $$\csc^{-1}(x) = \sin^{-1}(1/x).$$

In practice, $\tan^{-1}(x)$ comes up frequently, $\sin^{-1}(x)$ comes up occasionally, $\cos^{-1}(x)$ and $\sec^{-1}(x)$ come up rarely, and $\csc^{-1}(x)$ and $\cot^{-1}(x)$ almost never come up.

We get the derivatives of the inverse trig functions from the chain rule:

$$x = \tan(\tan^{-1}(x)),$$
$$x' = \tan'(\tan^{-1}(x))(\tan^{-1}(x))',$$
$$1 = \sec^2(\tan^{-1}(x))(\tan^{-1}(x))',$$
$$1 = (1 + \tan^2(\tan^{-1}(x)))(\tan^{-1}(x))',$$
$$1 = (1 + (\tan(\tan^{-1}(x)))^2)(\tan^{-1}(x))',$$
$$1 = (1 + x^2)(\tan^{-1}(x))',$$
(6.101) $$\frac{d}{dx}\tan^{-1}(x) = \frac{1}{1 + x^2};$$

$$x = \sin(\sin^{-1}(x)),$$
$$x' = \sin'(\sin^{-1}(x))(\sin^{-1}(x))',$$
$$1 = \cos(\sin^{-1}(x))(\sin^{-1}(x))',$$
$$1 = \sqrt{1 - \sin^2(\sin^{-1}(x))}\,(\sin^{-1}(x))',$$
$$1 = \sqrt{1 - x^2}\,(\sin^{-1}(x))',$$
(6.102) $$\frac{d}{dx}\sin^{-1}(x) = \frac{1}{\sqrt{1 - x^2}}.$$

6.7. Supplemental Material: A Crash Course in Trigonometry

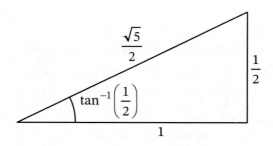

Figure 6.22. A fundamental triangle featuring $\tan^{-1}(1/2)$

We can visualize inverse trig functions using the fundamental triangles of Figure 6.12. For instance, if $\theta = \tan^{-1}(1/2)$, then $\tan(\theta) = 1/2$. Replacing $\tan(\theta)$ with $1/2$ in the second fundamental triangle, and using the Pythagorean theorem to compute the hypotenuse, we get the triangle of Figure 6.22. If you prefer to avoid fractions, you can rescale this triangle to have sides 1 and 2 and hypotenuse $\sqrt{5}$, but that doesn't really change anything. Similar triangles have the same angles and the same ratios of side lengths.

From either of these two triangles, we can compute all the trig functions of $\tan^{-1}(1/2)$ using SOH-CAH-TOA:

$$
\begin{aligned}
\sin(\tan^{-1}(1/2)) &= 1/\sqrt{5}, & \cos(\tan^{-1}(1/2)) &= 2/\sqrt{5}, \\
\tan(\tan^{-1}(1/2)) &= 1/2, & \cot(\tan^{-1}(1/2)) &= 2, \\
\sec(\tan^{-1}(1/2)) &= \sqrt{5}/2, & \csc(\tan^{-1}(1/2)) &= \sqrt{5}.
\end{aligned}
$$
(6.103)

The same method works for the inverse tangent of any positive number x, not just $\tan^{-1}(1/2)$. Just draw a triangle with opposite side x, adjacent side 1, and hypotenuse $\sqrt{1+x^2}$. SOH-CAH-TOA does the rest.

Figure 6.23 shows two triangles featuring the angles $\sin^{-1}(x)$ and $\sec^{-1}(x)$. The remaining triangles are left to the exercises.

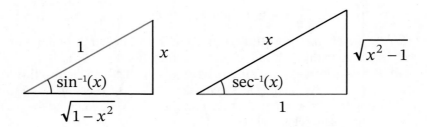

Figure 6.23. The fundamental triangles featuring $\sin^{-1}(x)$ and $\sec^{-1}(x)$

6.8. Exercises

Differential Equations, Initial Conditions, and Solutions

6.1. For parts (a), (c), (e), and (g), answer in complete sentences.
 (a) Explain what a differential equation represents.
 (b) Write down an example of a differential equation.
 (c) Explain what an initial condition is.
 (d) Write down an example of an initial condition.
 (e) Explain what an Initial Value Problem (IVP) is.
 (f) Write down an example of an IVP.
 (g) What does it mean to say "$y = f(t)$ is a solution to an IVP"?

6.2. Verify that $y(t) = \tan\left(t + \frac{\pi}{4}\right)$ is a solution to the IVP $y' = y^2 + 1$, $y(0) = 1$.

6.3. Verify that $y(t) = \frac{1}{3}(4e^{3t} - 1)$ is a solution to the IVP $y' = 3y + 1$, $y(0) = 1$.

6.4. (a) Show that for any constant C and for $t > -2C$, $y(t) = \left(\frac{t}{2} + C\right)^2$ is a solution to the differential equation $y' = \sqrt{y}$.
 (b) Why isn't this a solution when $t < -2C$?
 (c) By finding the correct C, solve the IVP $y' = \sqrt{y}$, $y(2) = 9$.

Food for Thought: The existence and uniqueness principle gives us a solution to every (reasonable) IVP for a short while. However, the solutions can blow up or have other problems, so the solutions may not work every time. For instance, the solution to Exercise 6.2 only works for $\frac{-3\pi}{4} < t < \frac{\pi}{4}$, and the solution to Exercise 6.4 only works for $t > -2C$. Fortunately, the differential equations introduced at the beginning of this chapter and discussed at length in Sections 6.3–6.5, have solutions that work for all values of t.

6.5. Suppose that you take out a $200,000 mortgage at 2.5% annual interest (compounded continuously). You make payments of $2500/month. Write a differential equation that describes how the amount $P(t)$ that you owe the bank changes with time. What is the initial condition? (You are not expected to solve this initial value problem, just to write it down.)

6.6. The vertical velocity of a skydiver after jumping out of an airplane satisfies the IVP (6.19):

$$\frac{dv}{dt} = -32 + 0.001v^2, \qquad v(0) = 0.$$

 (a) By modifying the SIREulers program from Chapter 3, find a numerical solution to this IVP and graph the solution for $0 \leq t \leq 15$.
 (b) The velocity is approaching a certain value, called the **terminal velocity**. As the skydiver approaches that velocity, v' gets closer and closer to 0. By using the differential equation and setting $v' = 0$, compute the terminal velocity. How close is $v(15)$ to that value?
 (c) The shape of the graph of $v(t)$ should remind you of one of the growth models in this chapter. Which one is it?

6.8. Exercises

Exponentials and Logs

6.7. Determine the *exact* value of each expression. You may leave your answers in terms of $\ln(2)$ and $\ln(10)$. For instance, if the question asked for $\ln(10e)$, you might write $\ln(10e) = \ln(10) + \ln(e) = 1 + \ln(10)$.

(a) $\ln(e/4)$ (b) $\ln(\sqrt{2e})$

(c) $\ln(10)e^{4\ln(2)}$ (d) $(e^{\ln(3)})^3$

(e) $e^{-\ln(5)}$ (f) $\ln(4/e)$

(g) $\ln(320)$ (h) $\ln(20^3)$

(i) $\ln(e^{500})$ (j) $e^{\ln(500)}$

6.8. Find the derivative of each function with respect to x or t.

(a) $\ln(3x)$ (b) $\ln(2^t)$

(c) $\ln(2x^2 - 3x)$ (d) $\log_5(3t)$

(e) $5^{(x^2)}$ (f) $e^{\cos(x)+x^2}$

(g) $\sin^{-1}(3x)$ (h) $\tan^{-1}(t^2)$

(i) $e^{3x}\tan^{-1}(2x)$ (j) $2^x \sin^{-1}(x^2)$

Exponential Growth and Decay

6.9. The rate of growth of an investment is proportional to its value. The investment was worth $40,000 in 2010 and $45,000 in 2015. What will it be worth in 2050?

6.10. Suppose that a bacterial population grows so that its mass is
$$P(t) = 0.2e^{1.2t} \text{ grams}$$
after t hours. Its initial mass is $P(0) = 0.2$ grams.
(a) When will its mass double, to 0.4 grams?
(b) How much longer will it take to double again, to 0.8 grams?
(c) After the population reaches a mass of 0.8 grams, how long will it take for yet another doubling to happen?
(d) What is the doubling time of this population?
(e) At what time will there be 1 kg of bacteria? (Yuck!)

6.11. Since the Civil War, the population of the Austin, Texas, metropolitan area has grown at a rate proportional to its population. In fact, it has grown at an amazingly steady rate of 3.5% per year. In the year 2000, the population was (approximately) 1,000,000. Assuming this growth pattern continues:
(a) What will the population of Austin be t years after the year 2000? (That is, let $t = 0$ correspond to the year 2000.)
(b) What will the population be in 2060?
(c) When will the population hit 5,000,000?

6.12. Poaching of black rhinos in east Africa, and later in the rest of Africa, was a severe problem from the 1960s until the 1990s, nearly driving the rhinos to extinction. The population of black rhinos dropped from an estimated 100,000 individuals in 1960 to approximately 2400 in 1990. The good news is that the population has slowly rebounded since then, with the rhino population $R(t)$ increasing at a rate proportional to the number of rhinos present. In 1990 the population was increasing at approximately 100 rhinos per year.
  (a) Write an initial value problem that governs the population of rhinos.
  (b) Write the solution to this initial value problem.
  (c) Use your model to predict the number of black rhinos in 2010. The actual number in the wild in 2010 was estimated to be 4900. Did your model provide a fairly close estimate?
  (d) Use your model to predict the number of black rhinos in 2020. The actual number in the wild in 2020 was estimated to be 5630. Did your model provide a close estimate?
  (e) Discuss the implications of your answer to part (d).

6.13. Suppose that sugar dissolves in water at a rate proportional to the amount left undissolved. Write a differential equation to describe this. If 50 lbs of sugar is added to a large quantity of water and 8 lbs are still undissolved after 6 hours, how long will it take until 98% of the sugar is dissolved?

6.14. In the early stages of an epidemic, almost everybody is Susceptible, and we can treat S as a constant. This means that I' is nearly proportional to I, and I grows exponentially. In March 2020, during the first wave of the Covid-19 pandemic, the number of confirmed cases in the United States rose from 541 on March 8 to 55,500 on March 24.
  (a) If we assume that the number of Infecteds was growing exponentially, on what date did the number of cases hit 5000? On what date did the number of cases hit 20,000?
  (b) How long did it take the number of cases to double?
  (c) If the number of cases continued to grow exponentially, how long would it have taken to reach 50 million cases?

Food for Thought: In fact, things weren't quite so bad. The entire world took drastic steps to limit the spread of the virus, reducing the transmission coefficient enough that infections started decreasing after reaching a peak of about 1 million in late May. Sadly, there were additional waves of infection after those emergency measures lapsed.

Present Value, explored in Exercises 6.15–6.18, is an important concept in economic analysis. Present value is used to compare the values of different possible payments made at different times.

For instance, suppose you own land and are offered $5000 for the mineral rights to that land. If you wait ten years for the minerals to increase in value, you will be able to sell the mineral rights for $8000. To compare these options, you need to convert the prospect of future money into an equivalent amount of money now—its present value. The **present value** of $8000 ten years from now is defined to be the amount of money

6.8. Exercises

you would need to invest now to have $8000 in ten years. If you invested $5000 now at an annual interest rate of 4% compounded continuously, then in ten years you would have $5000e^{0.4} = \$7459.12$. That is, $7459.12 in ten years has a present value of $5000.

6.15. At 4% interest, $8000 in ten years must have a present value slightly greater than $5000 (since $8000 > \$7459$). What is that present value?

6.16. If you can get a higher interest rate than 4%, the present value of a fixed future payment of $8,000 will be less than what you calculated in Exercise 6.15, since it will take less money now to grow to $8000 later. If the annual interest rate is 8%, what is the present value of a payment of $8000 ten years from now?

6.17. At what interest rate does a $8000 payment in ten years have the a present value of $5000?

6.18. In general, what is the formula for the present value of a payment of size P_0, t years in the future, at interest rate r?

Food for Thought: Exercises 6.16–6.18 explain why bond prices go up when interest rates go down, and vice versa. A bond is a promise, by a government or a corporation, to pay you a certain amount of money at a later date. The higher the interest rate is, the lower the present value of that promised payment, and the less the bond is worth.

Present value is also used to analyze mortgages and pensions. For a mortgage, the amount that the bank gives the homeowner up front equals the total present value of all of the payments that the homeowner will make later on. We will take this up again in Chapter 7, when we learn how to add up, or **integrate**, all of those contributions.

Growth and Decay with Offsets

6.19. Danielle just graduated from college. However, to help pay for her education, she took out loans totaling $50,000. Now that she has graduated, the loans accrue interest at a rate of 6% per year (compounded continuously). Danielle has set up an online payment plan from her checking account which continuously pays off the loan at a rate of $6000 per year.
   (a) Write down an initial value problem that describes the balance of Danielle's loan at time t.
   (b) We saw how to solve initial value problems of the form $y' = ry + p$, $y(0) = y_0$ in equation (6.61). Using this information, solve the IVP that you found in part (a).
   (c) By setting $y(t) = 0$, compute how long it takes Danielle to pay off her loan.

6.20. Suppose you open a saving account with an initial deposit of $5000, you continue to deposit money into the account at a rate of $2000 per year, and the account earns interest of 3% per year, compounded continuously. Write down an IVP that describes your savings account balance $b(t)$. (*Hint*: "change in money = interest + savings".) In this exercise, you do *not* need to find the solution to this IVP.

6.21. Solve the IVP of Exercise 6.20 as follows.
   (a) Define a variable z such that z' is proportional to z.
   (b) Write down the IVP that governs $z(t)$.

(c) Solve for $z(t)$.

(d) Since you know how $b(t)$ is related to $z(t)$, write down the formula for $b(t)$.

6.22. A food-truck business makes a profit of $10,000 the first year and $25,000 every subsequent year. At the end of the first year, the operators use the first-year profits to open a small business bank account that earns 3% interest per year (compounded continuously). From then on, they deposit their profits daily (which is fast enough that you can treat it as continuously). Write down an IVP that describes the amount of money in the bank account t years after the business started.

6.23. Solve the IVP of Exercise 6.22 using the method outlined in Exercise 6.21.

6.24. Bill is starting to save money for retirement. He sets aside $10,000/year to put into an investment account that grows at 7%/year. (*Note*: 7% is the historical average return for the stock market, adjusted for inflation.)
(a) Write down an IVP that describes this situation.
(b) Find the solution to this IVP. How much money will Bill have in t years?
(c) Bill plans to retire when he has $1,000,000 in the account. When will that be?

6.25. A company is using a radio advertising campaign to tell as many people as possible about a new product. It is thought that the rate at which new people hear the advertisement is proportional to the number of listeners who have not already heard it. No one is aware of the product at the start of the campaign. After 10 days, 30% of the listeners are aware of the product.
(a) Explain why the above can be mathematically modeled by

$$\frac{dN}{dt} = -k(N-L), \quad N(0) = 0, \quad N(10) = 0.3L,$$

where $N(t)$ is the number of people who have heard the ad, and L is the total population of radio listeners.
(b) Solve the differential equation given in part (a).
(c) After how many days will 70% of the listeners have heard the ad?
(d) Use MATLAB to graph the solution in part (b) for $0 \leq t \leq 100$. The curve looks the same for all values of L; $L = 1{,}000{,}000$ is a reasonable choice.

6.26. The interior temperature $T(t)$ of a Thanksgiving turkey, t minutes after being removed from the oven, satisfies the IVP

$$\frac{dT}{dt} = -k(T-70), \qquad T(0) = 165,$$

for some unknown constant k, where we are measuring temperature in degrees Fahrenheit.
(a) If $T(10) = 140$, what is k?
(b) What will the temperature of the turkey be 30 minutes after it comes out of the oven?
(c) How long do the dinner guests need to wait before the turkey reaches 100 degrees?

6.8. Exercises

Supply and Demand Curves. When the price of a product goes up, the demand goes down but the supply goes up. The amount $x = s(p)$ of supply at price p is called the **supply function**. But economists also talk about price as a function of supply. The function $p(x)$ is the price required to get manufacturers to produce a quantity x of the product. This is called the **supply curve**. Mathematically, the supply *curve* is the inverse of the supply *function*.

The same (confusing!) terminology applies to demand. The demand *function* gives demand as a function of price. The demand *curve* gives price as a function of demand. Exercises 6.27–6.29 explore supply curves and demand curves.

6.27. A convenience store depends on wholesalers to supply gasoline. The supply function (in gallons/week) is $x(p) = 2{,}000 + 400(p-2)$, where p is the wholesale price (in dollars per gallon) that the store is willing to pay for gas.
  (a) Graph the supply function $x(p)$. Label the axes appropriately.
  (b) On your graph, indicate the point corresponding to a wholesale price of $2.25. Include a sentence that provides a clear interpretation of this point.
  (c) Find the supply curve $p(x)$ corresponding to the given supply function. Provide a clear and concise interpretation for the supply curve.
  (d) Graph the supply curve $p(x)$. Label the axes appropriately.
  (e) On your graph, indicate the point corresponding to a supply of 2100 gallons. Include a sentence that provides a clear interpretation of this point.
  (f) Use your results from parts (a)–(e) to explain the inverse relationship between the supply function and the supply curve.

6.28. A convenience store has determined that the demand function for cheese popcorn (in bags/week) is $x(p) = 144/p^2$, where $\$2.00 \leq p \leq \$4.00$.
  (a) Graph the demand function $x(p)$. Label the axes appropriately.
  (b) On your graph, indicate the point corresponding to a price of $3. Include a sentence that provides a clear interpretation of this point.
  (c) Find the demand curve $p(x)$ corresponding to the given demand function. Provide a clear and concise interpretation for the demand curve.
  (d) Graph the demand curve $p(x)$. Label the axes appropriately.
  (e) On your graph, indicate the point corresponding to a demand of 16 bags per week of cheese popcorn. Include a sentence that provides a clear interpretation of this point.
  (f) Use your results from parts (a)–(e) to explain the inverse relationship between the demand function and the demand curve.

6.29. Suppose that, for a particular product, the marginal price dp/dx at x units of supply per day is proportional to the price p. At a price of $10 per unit, there is no supply at all. At a price of $12/unit, there is a daily supply of 40 units.
  (a) Write down an IVP for this situation.
  (b) What is the solution to this IVP?
  (c) At what price is there a supply of 80 units per day?
  (d) Use MATLAB to graph the supply curve for $0 \leq x \leq 200$.

Seasonal Supply and Demand. The **law of supply and demand** says that the price of a product goes up when the demand (how many consumers want to buy at the current price) is greater than the supply (how much producers are willing to sell at the current price). Likewise, the price drops when supply exceeds demand. However, the market doesn't respond instantly. It takes time for consumers to change their buying habits and for producers to change their production. Exercises 6.30 and 6.31 explore what happens in the meantime.

6.30. Many products have fluctuating supply and demand. For example, the demand for swimsuits is high in the spring and summer but low during the fall and winter, while the demand for sweaters is high in the fall and winter but low in the spring and summer. Supply may also be seasonal, especially for food items. When demand and supply depend on time, we write $d = d(t)$ and $s = s(t)$. As the supply and demand change, so does the price $p(t)$.
   (a) The rate of change in the price of a product is often proportional to the difference between the demand and the supply. Explain how this is expressed by the differential equation
   $$\frac{dp}{dt} = k(d - s),$$
   where k is a positive constant.
   (b) What does the differential equation in part (a) predict will happen to the price if demand exceeds supply?
   (c) What does the differential equation in part (a) predict will happen to the price if supply exceeds demand?

6.31. Consider seasonal supply and demand functions, as described in Exercise 6.30. (There are many correct answers to all three parts of this exercise.)
   (a) Pick a product whose *supply* is seasonal, and neatly sketch by hand a graph that shows how the supply varies over the course of a year. Explain why it is going up and down.
   (b) Pick a product whose *demand* is seasonal, and neatly sketch by hand a graph that shows how the demand varies over the course of a year. Explain why it is going up and down.
   (c) Suppose that the peak demand for a product occurs half-way through the year (say on day 182) and demand hits a minimum on the first day of the year (day 1). Using the functions discussed in this chapter, write down a mathematical formula for a quantity that behaves this way.

Rotations. In Section 6.4 we saw that $(x(t), y(t)) = (\cos(t), \sin(t))$ and $(x(t), y(t)) = (-\sin(t), \cos(t))$ are solutions to the differential equations

$$y' = x, \qquad x' = -y,$$

with initial conditions $(x(0), y(0)) = (1, 0)$ and $(x(0), y(0)) = (0, 1)$, respectively. These are said to be **basic** or **fundamental** solutions to the differential equations. All other solutions are obtained by taking a multiple of one basic solution plus a multiple of the other, also called a **linear combination** of the two solutions.

6.8. Exercises

6.32. Find solutions the differential equations $y' = x$, $x' = -y$ with the following initial conditions.
  (a) $x(0) = \sqrt{2}$ and $y(0) = 0$.
  (b) $x(0) = 0$ and $y(0) = \pi^2$.
  (c) $x(0) = \sqrt{2}$ and $y(0) = \pi^2$.

6.33. For the differential equations $y' = -x$, $x' = y$:
  (a) Find a solution with $x(0) = 1$ and $y(0) = 0$.
  (b) Find a solution with $x(0) = 0$ and $y(0) = 1$.
  (c) Find a solution with $x(0) = \sqrt{2}$ and $y(0) = \pi^2$.

6.34. Let ω be an arbitrary number. For the differential equations $y' = \omega x$, $x' = -\omega y$:
  (a) Find a solution with $x(0) = 1$ and $y(0) = 0$.
  (b) Find a solution with $x(0) = 0$ and $y(0) = 1$.
  (c) Find a solution with $x(0) = \sqrt{2}$ and $y(0) = \pi^2$.

Vibrations. In Exercises 6.35–6.37 we investigate solutions to the second-order differential equation
$$\frac{d^2x}{dt^2} = -\omega^2 x$$
that governs the vibration of a block of mass m on a spring with spring constant k, where $\omega = \sqrt{\frac{k}{m}}$.

6.35. Suppose that $\omega = 1$.
  (a) Find a solution with $x(0) = 1$ and $\frac{dx}{dt}(0) = 0$.
  (b) Find a solution with $x(0) = 0$ and $\frac{dx}{dt}(0) = 1$.
  (c) Find a solution with $x(0) = \sqrt{2}$ and $\frac{dx}{dt}(0) = \pi^2$.

6.36. Now suppose that $\omega = 2$, so $\frac{d^2x}{dt^2} = -4x$.
  (a) Find the solution with initial conditions $x(0) = 1$ and $\frac{dx}{dt}(0) = 0$.
  (b) Find the solution with initial conditions $x(0) = 0$ and $\frac{dx}{dt}(0) = 1$.
  (c) Find the solution with initial conditions $x(0) = 3$ and $\frac{dx}{dt}(0) = 0$.
  (d) Find the solution with initial conditions $x(0) = 0$ and $\frac{dx}{dt}(0) = 2$.
  (e) Find the solution with initial conditions $x(0) = 3$, $\frac{dx}{dt}(0) = 2$.
  (f) In terms of the parameters x_0 and v_0, find the solution with initial conditions $x(0) = x_0$ and $\frac{dx}{dt}(0) = v_0$.

6.37. Now consider an arbitrary value of ω and arbitrary parameters x_0 and v_0.
  (a) Find the solution with initial conditions $x(0) = 1$ and $\frac{dx}{dt}(0) = 0$.
  (b) Find the solution with initial conditions $x(0) = 0$ and $\frac{dx}{dt}(0) = 1$.
  (c) Find the solution with initial conditions $x(0) = x_0$ and $\frac{dx}{dt}(0) = v_0$.

Pendulums. Imagine a pendulum consisting of a heavy object (a *plumb bob*) at the end of a string of length ℓ. The angle θ that the pendulum makes with the vertical axis is governed by the differential equation
$$\frac{d^2\theta}{dt^2} = -\frac{g}{\ell}\sin(\theta),$$

where $g \approx 32 \text{ ft/s}^2 \approx 9.8 \text{ m/s}^2$ is the gravitational constant. Since $\sin(\theta) \approx \theta$ whenever θ is small,
$$\frac{d^2\theta}{dt^2} \approx -\frac{g}{\ell}\theta.$$
This is essentially the same as the equation that governs the position of a mass on a spring, only involving θ instead of x and g/ℓ instead of k/m. Note that the equation involves the length ℓ of the pendulum, but does not depend on the mass of the plumb bob.

6.38.
 (a) Compute the time it takes a pendulum with $\ell = 10$ cm to undergo a complete cycle, swinging from one side to the other and back again. This is called the **period** of oscillation.
 (b) Compute the period of a pendulum with $\ell = 40$ cm.
 (c) Compute the period of a pendulum with $\ell = 90$ cm.
 (d) Suppose that a pendulum has a period of 10 seconds. What is ℓ?

6.39. Make a pendulum by tying a small plumb bob to the end of a 90-cm (about 35.4-inch) piece of string.
 (a) Measure the period of oscillation. To avoid round-off errors, you might want to count how long it takes to make ten oscillations and divide by 10. If feasible, try the experiment with several different plumb bobs to check that the period really only depends on ℓ, and not on the mass of the bob. Compare your result to the answer you computed in part (c) of Exercise 6.38.
 (b) Now hold the string 40 cm from the middle of the bob and repeat the experiment. Compare your measured period to part (b) of Exercise 6.38.
 (c) Finally, hold the string 10 cm from the middle of the bob, repeat the experiment, and compare your measured period to part (a) of Exercise 6.38. (Be sure to use a bob that is much less than 10 cm across. If the mass is spread out, with one part of the bob being much closer to your fingers than another part, then the calculation becomes more complicated and involves the **moment of inertia** of the bob, a quantity we will learn about in Chapter 7.)

6.40. **Solving the Logistics Equation with MATLAB**. Modify the SIREulers program (see Section 3.6) to create a MATLAB program, called Logistics, to solve the logistic equation $y' = ry\left(1 - \frac{y}{L}\right)$. Your input parameters should be the time t, the number of steps N, the growth rate r, the carrying capacity L, and the initial value y_0. Your output should be arrays for t and y and a graph of y as a function of t.

Dimensionless Models. When formulating a mathematical model for a process, we often recast the original model, which involves variables with units of dollars or concentration or time, using dimensionless variables. There are several benefits to doing this, including reducing the number of parameters used in the model. Exercises 6.41–6.44 lead us through several examples of that approach.

6.8. Exercises

6.41. (a) Consider the exponential population growth model
$$\frac{dy}{dt} = ry, \qquad y(0) = y_0.$$
The dimension for the independent variable t is time, and the dimension for the dependent variable y is individuals. What are the dimensions of the parameter r?
(b) Consider a new independent variable $\tau = rt$. What are the dimensions for τ?
(c) Consider a new dependent variable $Y = y/y_0$. What are the dimensions for Y?
(d) We can rewrite our original dimensionful differential equation in terms of our new dimensionless variables. Show that $\frac{dy}{dt} = y_0 r \frac{dY}{d\tau}$.
(e) Rewrite the right-hand side of the differential equation in part (a) in terms of Y.
(f) Combining the results of parts (d) and (e), show that $\frac{dY}{d\tau} = Y$. What happened to the parameter r?
(g) Restate the initial condition in terms of Y. What happened to the parameter y_0?

6.42. In Exercise 6.41, we started with a dimensionful IVP
$$\frac{dy}{dt} = ry, \qquad y(0) = y_0,$$
and created an equivalent dimensionless IVP
$$\frac{dY}{d\tau} = Y, \qquad Y(0) = 1.$$
(a) If $r = 0.15$ and $y_0 = 20$, find the solution $y(t)$ to the dimensionful IVP.
(b) Find the solution $Y(\tau)$ to the dimensionless IVP.
(c) Using MATLAB, graph both $y(t)$ and $Y(\tau)$ on the same set of axes. What do you notice about the qualitative behavior of the two curves?

6.43. Consider the logistic model
$$\frac{dy}{dt} = ry\left(1 - \frac{y}{L}\right), \qquad y(0) = y_0.$$
(a) List the three parameters in the model, and give the dimensions of each one.
(b) What are the dimensions of $\tau = rt$ and $Y = y/L$?
(c) Show that when you rewrite the differential equation in terms of τ and Y, you get
$$\frac{dY}{d\tau} = Y(1 - Y).$$
(d) Let $\alpha = y_0/L$. The initial condition $y(0) = y_0$ then becomes $Y(0) = \alpha$. How many parameters does your new dimensionless differential equation contain? How many parameters does your new dimensionless IVP contain?

6.44. In Exercise 6.43, we used the dimensionless variable $Y = y/L$, as well as the dimensionless variable $\tau = rt$, to get rid of any dependence on L, but we wound up getting stuck with a dependence on y_0. Instead, consider the dimensionless variables $\tilde{Y} = Y/y_0$ and $\tau = rt$. Derive an IVP that governs $\tilde{Y}$ as a function of τ. When you compare the two dimensionless models, what differences do you see?

Trigonometry

6.45. Draw a right triangle where one of the angles is $\tan^{-1}(3/4)$. Use this triangle to compute $\sin(\tan^{-1}(3/4))$ and $\sec(\tan^{-1}(3/4))$.

6.46. Draw a right triangle where one of the angles is $\sin^{-1}(5/13)$. Use this triangle to compute $\cos(\sin^{-1}(5/13))$ and $\tan(\sin^{-1}(5/13))$.

6.47. Draw a right triangle featuring $\cos^{-1}(x)$, where x is an unspecified number between 0 and 1. In terms of x, what are all six trig functions of $\cos^{-1}(x)$?

6.48. Draw a right triangle featuring $\sec^{-1}(x)$, where x is an unspecified number greater than 1. In terms of x, what are all six trig functions of $\sec^{-1}(x)$?

Chapter 7

The Whole Is the Sum of the Parts

Chapters 2 through 6 have been about **differential calculus**. We developed the idea of a derivative, we developed rules for computing derivatives, and we used derivatives in a number of applications. Finally, we acknowledged the fact that most of the laws that govern the natural world are best described as differential equations, and we saw how the exponential, log, and trig functions that come from solving the simplest of these differential equations appear over and over in the natural world.

In this chapter we change gears completely and introduce **integration**. The root meaning of "integral" is *whole*. An *integer* is a whole number. An airplane has *structural integrity* if it stays in one piece. *Personal integrity* means not speaking with a forked tongue. *Social integration* means bringing different social or racial groups into a unified community. *E Pluribus Unum*: Out of many, one.

Mathematically, an **integral** represents a bulk quantity, accumulated over a region in space or time. In all cases, we compute the whole by breaking it up into little pieces, estimating the contribution of each piece, and adding up the pieces. This is the 4th Pillar of Calculus:

> *The whole is the sum of the parts.*

If you have taken calculus before, you may be thinking that integration is the opposite of differentiation. *What in blazes do sums have to do with it?!* The truth is that anti-derivatives have nothing to do with what integrals *are*, and very little to do with that they are *good for*, but they have a tremendous amount to do with how integrals are *calculated*. Meanwhile, sums have everything to do with what integrals are and what they are good for.

In this chapter we're going to talk about what integrals are (sums!) and what they're good for. We'll start in Section 7.1 with a variety of problems that can be addressed with a slice-and-dice strategy. In Section 7.2 we'll develop notation for estimating all the parts of the whole and computing their sum. In Section 7.3 we will define the integral of an arbitrary function over an arbitrary interval. Finally, in Section 7.4 we will consider accumulation functions, which are running totals.

Once we've learned what integrals are and how they can be used, we can turn our attention to computing them. In Chapter 8, we'll learn how to do that using antiderivatives.

7.1. Slicing and Dicing

We start with three problems. Their real-world meanings are very different, but mathematically they are almost identical.

(1) A start-up company lost money in its first four years, bleeding money at a rate of $\$100,000\sqrt{16-t^2}$/year, where t is measured in years. How much money did it lose in all?

(2) A projectile was shot into a vat of viscous fluid (e.g., molasses) that slowed it down and stopped it after 4 seconds. The velocity of the projectile t seconds after hitting the fluid was $10\sqrt{16-t^2}$ cm/second. How far did the projectile go before it stopped?

(3) What is the area under the curve $y = \sqrt{16-t^2}$ between $t = 0$ and $t = 4$?

If the company lost money at a constant rate, or if the projectile moved at a constant speed, or if the graph was a horizontal line, these problems would be easy:

$$\begin{aligned} \text{loss} &= \text{burn rate} \times \text{time}, \\ \text{distance} &= \text{velocity} \times \text{time}, \\ \text{area} &= \text{height} \times \text{width}. \end{aligned}$$

Unfortunately, the function $\sqrt{16-t^2}$ is *not* constant. However, it is *continuous*, which means that it only changes a little over a short period of time. We can get a good approximation for each total by breaking the interval $[0, 4]$ into smaller pieces, and then approximating $\sqrt{16-t^2}$ by a constant on each piece.

The Startup. We can treat the losses at the start-up company one year at a time. Let $f(t) = \$100,000\sqrt{16-t^2}$/year be the burn rate, as shown in Table 7.1.

Table 7.1. Burn rate as a function of time

Time (in years)	Burn rate (in $/year)
0	400,000
1	387,000
2	346,000
3	265,000
4	0

7.1. Slicing and Dicing

In the first year, the company lost money at rate of $f(0) = \$400,000$/year at the beginning of the year, and lost money at rate $f(1) \approx \$387,000$/year at the end of the year. All year long, its burn rate was somewhere between \$387,000/year and \$400,000/year, so its loss for the year was somewhere between \$387,000/year × 1 year = \$387,000 and \$400,000. Similarly, the loss in the second year was somewhere between $f(2) \times 1$ year and $f(1) \times 1$ year, the loss in the third year was somewhere between $f(3) \times 1$ year and $f(2) \times 1$ year, and the loss in the fourth year was somewhere between $f(4) \times 1$ year and $f(3) \times 1$ year. Adding these up, we get

\$387,000	<	first year loss	<	\$400,000
\$346,000	<	second year loss	<	\$387,000
\$265,000	<	third year loss	<	\$346,000
\$0	<	fourth year loss	<	\$265,000
\$998,000	<	total loss	<	\$1,398,000

This calculation is very rough. All we know is that the company lost between \$998,000 and \$1,398,000. A reasonable guess is the average between the **lower bound** of \$998,000 and the **upper bound** of \$1,398,000, namely \$1,198,000. But that's still just a rough guess.

The trouble with this calculation is that the burn rate can change by a lot in the course of a single year. In the fourth year, it went all the way from \$265,000/year to 0! To do better, we can redo the calculation half a year at a time, or one quarter at a time, or one month at a time, or one day at a time. The downside is that doing things by months requires 48 estimates and doing things by days requires $4 \times 365 = 1460$ estimates. As with Euler's method, smaller time steps bring us greater accuracy, but at the cost of more work.

The Projectile. We can estimate the distance traveled by our projectile the same way that we estimated the start-up company's losses. We make a table of the velocity $v(t) = 10\sqrt{16 - t^2}$ as a function of time.

Time (in seconds)	Velocity (in cm/sec)
0	40
1	38.7
2	34.6
3	26.5
4	0

In the first second, the projectile was going between 38.7 cm/sec and 40 cm/sec, in the second second it was going between 34.6 cm/sec and 38.7 cm/sec, and so on.

38.7 cm	<	distance traveled in first second	<	40 cm
34.6 cm	<	distance traveled in second second	<	38.7 cm
26.5 cm	<	distance traveled in third second	<	34.6 cm
0 cm	<	distance traveled in fourth second	<	26.5 cm
99.8 cm	<	total distance traveled	<	139.8 cm

Averaging the upper and lower bounds, we get an estimated distance of 119.8 cm.

The Area Under an Arc. The area under the curve is similar. We break the interval $[0, 4]$ into four pieces of width $\Delta t = 1$, which breaks the area under the curve into four vertical strips. See Figure 7.1. The area of each strip is greater than the area of the darkly shaded rectangle that sits inside it, and is less than the area of the lightly shaded rectangle. That is,

3.87	<	area of first strip	<	4
3.46	<	area of second strip	<	3.87
2.65	<	area of third strip	<	3.46
0	<	area of fourth strip	<	2.65
9.98	<	total area	<	13.98

Averaging the upper and lower bounds gives an estimated area of 11.98.

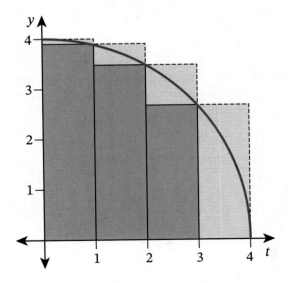

Figure 7.1. The area under $y = \sqrt{16 - t^2}$

7.1. Slicing and Dicing

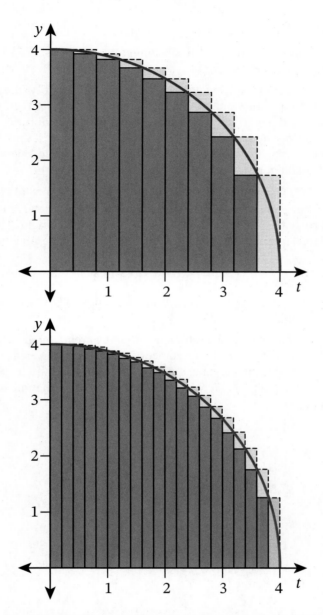

Figure 7.2. With ten strips we see that the area is between 11.618 and 13.218; with twenty, between 12.114 and 12.914.

As with the start-up and projectile problems, we can get greater accuracy by using more and more strips, as seen in Figure 7.2.

Geometry to the Rescue. The calculations in the three problems are *exactly* the same, except for factors of 10 cm and $100,000. The amount of money lost by the startup is equal to the area under the curve $y = \$100{,}000\sqrt{16 - t^2}$ between $t = 0$ and

$t = 4$, which is \$100,000 times the area under the curve $y = \sqrt{16 - t^2}$. The distance traveled by the projectile is equal to the area under the curve $y = 10\sqrt{16 - t^2}$, which is 10 (cm) times the area under $y = \sqrt{16 - t^2}$.

This is a general pattern. The total accumulation of a quantity (the **integral**) can often be visualized as the area under the graph of that quantity, just as the rate of change of that quantity (the **derivative**) can be visualized as the slope of the graph. In calculus, we do a lot of slope and area-under-the-curve problems, but it's not because we're obsessed with geometry. It's because problems that don't seem to have anything to do with geometry can often be solved with a little geometrical insight.

In this case, we can solve the start-up and projectile problems exactly with the help of some high school geometry. If $y = \sqrt{16 - t^2}$, then $y^2 = 16 - t^2$ and $t^2 + y^2 = 16 = 4^2$, so our curve is part of a circle of radius 4 with its center at the origin. The area we are looking at is the region in the first quadrant enclosed by this circle, and therefore equals

$$\tag{7.1} \frac{1}{4}\pi r^2 = \frac{1}{4}16\pi = 4\pi \approx 12.57.$$

This means that the projectile traveled a distance of 40π cm ≈ 125.7 cm, and that the startup lost \$400,000$\pi \approx$ \$1,257,000.

Even with only four pieces, the estimates we got by averaging the upper and lower bounds (namely a loss of \$1,198,000, a distance of 119.8 cm, and an area of 11.98) turned out to be accurate to within 5%. This approximation is called the **trapezoidal rule**, and it is pictured in Figure 7.3. We will revisit this method, with more detail about its accuracy, in Chapter 9.

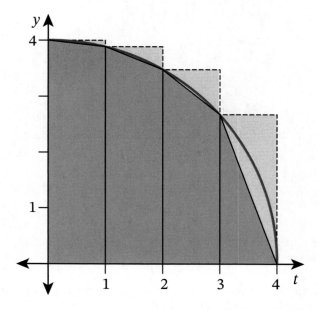

Figure 7.3. The area of each trapezoid is the average of the areas of the outer and inner rectangles.

7.1. Slicing and Dicing

And now for something completely different. Or maybe not. At first glance, the next two problems look like they have nothing to do with startups losing money or bullets moving through molasses. However, they are solved with the exact same slice-and-dice algorithm. That's the heart of integration.

(1) The **moment of inertia** of a particle of mass m at distance r from the origin is mr^2. This quantity measures how much effort (torque) it takes to rotate the particle around the origin. The moment of inertia of a larger object is the sum of the moments of inertia of its pieces. What is the moment of inertia of a thin uniform rod of mass 1 kg and length 1 m, sitting on the x-axis between $x = 0$ and $x = 1$ (meter)?

(2) What is the volume of a pyramid with a 1 m $\times$ 1 m square base and a height of 1 m?

(3) What is the area under the parabola $y = x^2$ between $x = 0$ and $x = 1$?

We claim that all three problems have the same answer, so solving any one of them is tantamount to solving all three. To see why, let's break each of the three quantities—namely the moment of inertia of the rod, the volume of the pyramid, and the area under the parabola—into ten pieces and estimate the size of each piece.

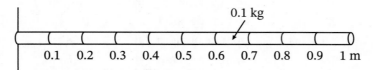

Figure 7.4. A rod broken into 10 pieces

The Rod. If we break the rod into ten pieces, as shown in Figure 7.4, then each piece has a mass of $1 \text{ kg}/10 = 0.1$ kg. The points in the first piece have a distance to the origin that ranges from 0 to 0.1 m, so the moment of inertia of the first piece is somewhere between $0.1 \times (0)^2$ kg-m^2 and $0.1(0.1)^2$ kg-m^2. The second piece has mass 0.1 kg, and every point in the second piece is at distance at least 0.1 m from the origin and at most 0.2 m from the origin, so the moment of inertia of the second piece is between $0.1(0.1)^2$ kg-m^2 and $0.1(0.2)^2$ kg-m^2. Continuing in this way we get

0 kg-m^2	<	moment of inertia of first piece	<	0.001 kg-m^2
0.001 kg-m^2	<	moment of inertia of second piece	<	0.004 kg-m^2
0.004 kg-m^2	<	moment of inertia of third piece	<	0.009 kg-m^2
0.009 kg-m^2	<	moment of inertia of fourth piece	<	0.016 kg-m^2
0.016 kg-m^2	<	moment of inertia of fifth piece	<	0.025 kg-m^2
0.025 kg-m^2	<	moment of inertia of sixth piece	<	0.036 kg-m^2
0.036 kg-m^2	<	moment of inertia of seventh piece	<	0.049 kg-m^2
0.049 kg-m^2	<	moment of inertia of eigth piece	<	0.064 kg-m^2
0.064 kg-m^2	<	moment of inertia of ninth piece	<	0.081 kg-m^2
0.081 kg-m^2	<	moment of inertia of tenth piece	<	0.100 kg-m^2
0.285 kg-m^2	<	total moment of inertia	<	0.385 kg-m^2

We can also use the trapezoidal rule, averaging the upper and lower bounds, to get an approximate moment of inertia of 0.335 kg-m^2.

The Pyramid. To find the volume of the pyramid, we lay it on its side and slice it into 10 slabs; see Figure 7.5. Each of these slabs is a square tile with beveled edges. The ith tile, shown in Figure 7.6, has thickness 0.1 m. On the right its cross-section is a square of side length $\frac{i}{10}$ meters and area $\frac{i^2}{100}$ square meters, and on the left its cross-section is a square of side length $\frac{i-1}{10}$ meters and area $\frac{(i-1)^2}{100}$ square meters. This means that its volume is somewhere between $\frac{(i-1)^2}{1000}$ and $\frac{i^2}{1000}$ cubic meters. That is,

0 m^3	<	volume of first slab	<	0.001 m^3
0.001 m^3	<	volume of second slab	<	0.004 m^3
0.004 m^3	<	volume of third slab	<	0.009 m^3
0.009 m^3	<	volume of fourth slab	<	0.016 m^3
0.016 m^3	<	volume of fifth slab	<	0.025 m^3
0.025 m^3	<	volume of sixth slab	<	0.036 m^3
0.036 m^3	<	volume of seventh slab	<	0.049 m^3
0.049 m^3	<	volume of eighth slab	<	0.064 m^3
0.064 m^3	<	volume of ninth slab	<	0.081 m^3
0.081 m^3	<	volume of tenth slab	<	0.100 m^3
0.285 m^3	<	total volume	<	0.385 m^3

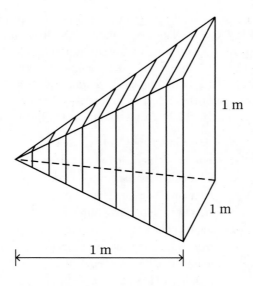

Figure 7.5. A pyramid sliced into 10 pieces

7.1. Slicing and Dicing

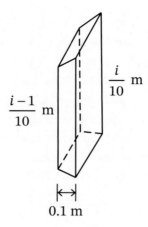

Figure 7.6. The ith slice of the pyramid

The Parabola. Finally, we look at the area under the parabola $y = x^2$ and cut the region into ten strips; see Figure 7.7. The area of each strip is at least the area of biggest rectangle that fits inside it and at most the area of the smallest rectangle that surrounds it.

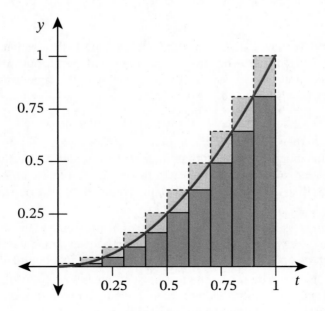

Figure 7.7. The area under a parabola sliced into ten pieces

0	<	area of first strip	<	0.001
0.001	<	area of second strip	<	0.004
0.004	<	area of third strip	<	0.009
0.009	<	area of fourth strip	<	0.016
0.016	<	area of fifth strip	<	0.025
0.025	<	area of sixth strip	<	0.036
0.036	<	area of seventh strip	<	0.049
0.049	<	area of eighth strip	<	0.064
0.064	<	area of ninth strip	<	0.081
0.081	<	area of tenth strip	<	0.100
0.285	<	total area	<	0.385

The upper and lower bounds for the volume of the pyramid (in cubic meters) are exactly the same as the upper and lower bounds for the moment of inertia of the rod (in kg-m^2), which are exactly the same as the upper and lower bounds for the area under the parabola. If we want more accuracy, we can cut our rod and our pyramid and the region under the parabola into more pieces. The more pieces we use, the closer our upper and lower bounds will be to the true answers. But no matter how many pieces we use, the volume (in m^3) will still equal the moment of inertia (in kg-m^2) and the area. To a mathematician, the three problems are the same.

In Section 8.3 we will see that the exact answer to all three problems is $1/3 \approx 0.333$. The trapezoidal rule estimate of $(0.285 + 0.385)/2 = 0.335$ is only off by half a percent.

Summary. We have looked at six different problems in this section. One involved money, one involved distance, one involved a moment of inertia, one involved volume, and two involved areas. However, all six were solved with the same strategy, following the slogan *the whole is the sum of the parts*.

> (1) Break the quantity we are trying to study into little pieces.
>
> (2) Estimate the size of each piece. In our examples were were able to say "at least this much and at most that much", which is ideal, but estimates of the form "approximately this much" are also useful. If we have upper and lower bounds, then the average of the two (the **trapezoidal rule**) is a good approximate value.
>
> (3) Add up the estimated sizes of the pieces to get an estimated total. We can also add up the lower bounds for the pieces to get a lower bound for the total, and add up the upper bounds for the pieces to get an upper bound for the total.
>
> (4) For more accuracy, repeat the process with more pieces. As the number of pieces goes to infinity, the lower bound, the upper bound, and any estimate in between will all home in on the exact answer.

7.2. Riemann Sums

In each case, the estimated contribution of each piece was of the form $f(x)\Delta x$ or $f(t)\Delta t$, where Δx (or Δt) was the width of each piece. This is the same thing that we would get if we studied the area under the curve $y = f(x)$ (or $y = f(t)$). In other words, *most bulk quantities can be visualized as the area under some curve.* For the start-up problem, $f(t)$ was the burn rate. For the projectile problem, it was the velocity. For the moment of inertia of the rod, it was the square of the distance from the origin, and for the volume of the pyramid, it was the cross-sectional area.

There no single magic formula that tells us what $f(x)$ or $f(t)$ is supposed to be. Depending on the application, we have to reason that out using our understanding of economics or physics or three-dimensional geometry. Calculus doesn't tell us what functions to study! It just tells us what we can do with those functions once we've found them.

Finally, the phrase "area under the curve" needs to be taken with a grain of salt. When $f(x)$ is negative, the negative contribution $f(x)\Delta x$ isn't the approximate area of a strip lying *below* the curve $y = f(x)$. Instead, it is *minus* the area of a strip lying *above* the curve. For instance, suppose that $f(x) = \cos(x)$ and that we are working on the interval $[0, 2\pi/3]$, as in Figure 7.8. When we add up terms of the form $f(x)\Delta x$, we are actually estimating the area of the blue region *minus* the area of the red region. But we still refer to the result, somewhat inaccurately, as "the area under $y = \cos(x)$ as x goes from 0 to $2\pi/3$".

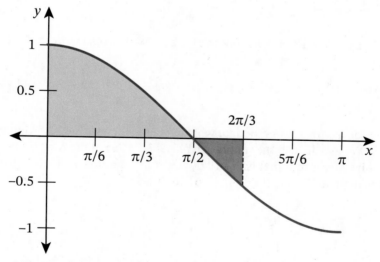

Figure 7.8. The area under $y = \cos(x)$ is actually a difference of two areas.

7.2. Riemann Sums

In the last section we saw that all sorts of bulk quantities can be represented as areas under (or minus areas over) a curve $y = f(x)$. In this section we develop the formalism for expressing that area as the limit of a sum.

Since we are always adding many terms, we need a compact notation for sums. We use a oversized capital Greek letter Sigma, written $\sum$. The formula

$$(7.2) \qquad z = \sum_{j=1}^{5} j^2$$

is a shorthand way of saying "add up five numbers, namely j^2 when j equals 1, 2, 3, 4, and 5, and call the result z". Underneath the $\sum$ we write the name of the counter and its starting value, and on top we write its ending value. You might think of it as a loop in a computer program:

```
sum = 0;                    % Start your sum at zero.
for j=1:5                   % Do a loop with j running from 1 to 5.
    sum = sum + j*j;        % Add j^2 to our running total.
end;                        % Finish the loop.
z = sum                     % Set z equal to the total.
```

We can call the counter in the sum i, j, m, n, or any other letter that we like, e.g.,

$$(7.3) \qquad \sum_{i=1}^{5} i^2 = \sum_{j=1}^{5} j^2 = \sum_{m=1}^{5} m^2 = \sum_{n=1}^{5} n^2 = 1^2 + 2^2 + 3^2 + 4^2 + 5^2 = 55.$$

The counter usually starts at 1 but doesn't have to. We can also write

$$(7.4) \qquad \sum_{k=9}^{13} (k-10)^3 = (9-10)^3 + (10-10)^3 + (11-10)^3 + (12-10)^3 + (13-10)^3.$$

Next we develop some notation for what happens when we break the interval $[a, b]$ into pieces. We usually call the number of pieces N. Most of the time we break $[a, b]$ into equal-size pieces, but that isn't strictly necessary. For instance, we *might* choose to break the interval $[2, 12]$ into $[2, 5]$, $[5, 6]$, $[6, 10]$, and $[10, 12]$, as in Figure 7.9.

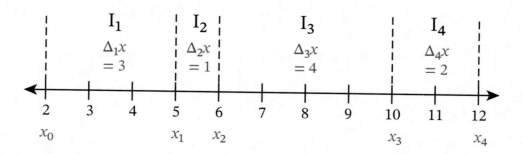

Figure 7.9. The interval $[2,12]$ broken into four unequal pieces

7.2. Riemann Sums

If our input variable is called x, then the endpoints of these intervals are called x_0, $x_1, \ldots, x_N$, so in this case $a = x_0 = 2$, $x_1 = 5$, $x_2 = 6$, $x_3 = 10$, and $b = x_4 = 12$. (If the input variable is called t, then we call the endpoints t_0, t_1, etc.) We call the ith interval I_i (or call the jth interval I_j or call the nth interval I_n). I_i runs from x_{i-1} to x_i, and its length is $\Delta_i x = x_i - x_{i-1}$. In this case $\Delta_1 x = 3$, $\Delta_2 x = 1$, $\Delta_3 x = 4$, and $\Delta_4 x = 2$. Breaking the interval $I = [a, b]$ into subintervals I_i is equivalent to cutting the region under the curve $y = f(x)$ between $x = a$ and $x = b$ into vertical strips, each one sitting over an interval I_i.

We need to estimate the area of the strip sitting above I_i. This strip is approximately a rectangle of width $\Delta_i x$ and height $f(x)$, so its area is approximately $f(x)\Delta_i x$. But which value of x do we plug into the function $f(x)$?

There are many ways to pick a point x_i^* to represent the interval I_i. We could use the **left endpoint** x_{i-1}, the **right endpoint** x_i, or any point in between. Sometimes it's useful to pick the point with the smallest value of $f(x)$, in which case $f(x_i^*)\Delta x$ will be a lower bound for the area of the strip. (If $f(x)$ is an increasing function on I_i, this is the same as picking the left endpoint, while if $f(x)$ is decreasing, it's the same as picking the right endpoint.) Other times it's useful to pick the point with the largest value of $f(x)$, in which case $f(x_i^*)\Delta x$ will be an upper bound for the area.

There is no single right answer for which point to choose. When computing derivatives, we could use forward differences, backward differences, or centered differences. Centered differences were the most accurate, but they all gave the same answer in the limit of smaller and smaller intervals. Likewise, we can compute areas using left endpoints, right endpoints, midpoints, or a host of other schemes. Some are more accurate than others, but as $N \to \infty$ (and the sizes of the intervals go to 0), they all approach the correct area.

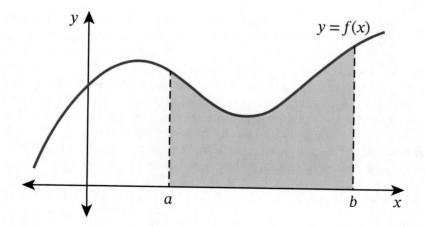

Figure 7.10. What is the area of the shaded region?

Now that we have our notation settled, we can summarize the procedure for finding the area under the curve $y = f(x)$ between $x = a$ and $x = b$, the shaded region in Figure 7.10.

> (1) Decide on the number N of pieces into which we are going to break the interval $I = [a, b]$. This is the same as the number of strips into which we are going to break the region under the curve $y = f(x)$ from $x = a$ to $x = b$.
>
> (2) Let $x_0 = a$, $x_N = b$, and pick points $x_1 < x_2 < \cdots < x_{N-1}$ in between. Most of the time we want these points to be spaced apart at an equal distance $\Delta x = (b-a)/N$, in which case we have $x_i = a + i\Delta x = a + \frac{i(b-a)}{N}$. However, other choices are possible, in which case the various widths $\Delta_i x = x_i - x_{i-1}$ may not all be the same.
>
> (3) In each interval $I_i = [x_{i-1}, x_i]$, pick a representative point x_i^*.
>
> (4) Approximate the ith strip (that is, the region over I_i) as a rectangle of height $f(x_i^*)$ and width $\Delta_i x$, so it has area $f(x_i^*)\Delta_i x$.
>
> (5) Add up the pieces. The total area under the curve is approximately
>
> (7.5) $$\sum_{i=1}^{N} f(x_i^*)\Delta_i x.$$
>
> This expression is called a **Riemann sum**.[1]
>
> (6) For more accuracy, increase N. The exact area under the curve is
>
> (7.6) $$\lim_{N \to \infty} \sum_{i=1}^{N} f(x_i^*)\Delta_i x.$$
>
> This limit does *not* depend on the choices you made. As long as $f(x)$ is continuous and the lengths of all the intervals approach 0, the Riemann sum (7.5) will get closer and closer to the true area as $N \to \infty$.

We are using area problems to develop intuition for Riemann sums, but Riemann sums apply to *much* more than just area problems. The same ideas apply to distance, to volume, to arc-length, to accumulated income, to present value, and to many other applications. You'll see some of these in the exercises, but the list of possible applications is way too long to fit in one book. The common feature of all of these applications is that there is a big quantity that can be broken into many little pieces, and that the contribution of the ith piece can be approximated as $f(x_i^*)\Delta x$ for some function $f(x)$ and some variable x (or as $f(y_i^*)\Delta y$ or $f(t_i^*)\Delta t$ or $f(r_i^*)\Delta r$ or whatnot). To understand what the function $f(x)$ is, we need to understand something about distance or volume or arc-length or income or present value. *There is no magic formula for that step!* But

---

[1] Bernhard Riemann (1826–1866) was one of the greatest mathematicians in history. He died of tuberculosis at age 39 and only published about a dozen papers (some posthumously), but every one of his papers was groundbreaking. Besides being the first person to define integrals rigorously, he invented much of differential geometry (sometimes called Riemannian Geometry), complex analysis (Riemann surfaces), and number theory (the Riemann zeta function).

7.2. Riemann Sums

once we've done that, the machinery of Riemann sums, and later of definite integrals, takes care of the rest.

Area Problems. Let's apply our procedure to the two area problems of Section 7.1.

First, we want the area under $y = f(x) = \sqrt{16 - x^2}$ between $x = 0$ and $x = 4$. We set $N = 4$ and break $[0, 4]$ into equal sized intervals, each of length $\Delta x = \frac{4-0}{4} = 1$. We then have

(7.7)
$$\begin{aligned} x_0 &= a & &= 0, \\ x_1 &= a + \Delta x & &= 1, \\ x_2 &= a + 2\Delta x & &= 2, \\ x_3 &= a + 3\Delta x & &= 3, \\ x_4 &= a + 4\Delta x & &= 4 = b. \end{aligned}$$

To do the calculation with left endpoints, we pick $x_1^* = x_0 = 0$, $x_2^* = x_1 = 1$, $x_3^* = x_2 = 2$, and $x_4^* = x_3 = 3$. In other words, $x_i^* = i - 1$. Our estimated area is then

(7.8)
$$\begin{aligned} \sum_{i=1}^{4} f(x_i^*)\Delta x &= \sum_{i=1}^{4} f(i-1)\Delta x \\ &= \sum_{i=1}^{4} \sqrt{16 - (i-1)^2} \\ &= 4 + \sqrt{15} + \sqrt{12} + \sqrt{7} \approx 13.98. \end{aligned}$$

In general, equal-sized intervals and left endpoints always give us

(7.9)
$$\begin{aligned} \sum_{i=1}^{N} f(x_i^*)\Delta x &= \sum_{i=1}^{N} f(x_{i-1})\Delta x \\ &= \Delta x (f(x_0) + \cdots + f(x_{N-1})). \end{aligned}$$

If instead we used right endpoints, then each x_i^* would be x_i instead of x_{i-1}. Our estimated area would then be

(7.10)
$$\begin{aligned} \sum_{i=1}^{4} f(x_i^*)\Delta x &= \sum_{i=1}^{4} f(i)\Delta x \\ &= \sum_{i=1}^{4} \sqrt{16 - (i)^2} \\ &= \sqrt{15} + \sqrt{12} + \sqrt{7} + \sqrt{0} \approx 9.98. \end{aligned}$$

In general, equal-sized intervals and right endpoints always give us

(7.11)
$$\begin{aligned} \sum_{i=1}^{N} f(x_i^*)\Delta x &= \sum_{i=1}^{N} f(x_i)\Delta x \\ &= \Delta x (f(x_1) + \cdots + f(x_N)). \end{aligned}$$

If the function is decreasing, as in this case, then using left endpoints gives too big an answer and using right endpoints gives too small an answer. If the function is increasing, then it's the other way around, with left endpoints giving an underestimate and right endpoints giving an overestimate. The **trapezoidal rule** says to take the average of the answers. It gives

$$\text{Area} \approx \frac{1}{2}\left(\sum_{i=1}^{N} f(x_{i-1})\Delta x + \sum_{i=1}^{N} f(x_i)\Delta x\right)$$

$$= \Delta x\left(\frac{1}{2}f(x_0) + f(x_1) + f(x_2) + \cdots + f(x_{N-1}) + \frac{1}{2}f(x_N)\right)$$

(7.12) $$= \frac{1}{2}4 + \sqrt{15} + \sqrt{12} + \sqrt{7} + \frac{1}{2}\sqrt{0} \approx 11.98.$$

Another obvious compromise between left and right endpoints is to use midpoints. Since $I_i = [x_{i-1}, x_i]$, the **midpoint rule** has

(7.13) $$x_i^* = \frac{1}{2}(x_{i-1} + x_i) = \frac{1}{2}(a + (i-1)\Delta x + a + i\Delta x) = a + \left(i - \frac{1}{2}\right)\Delta x.$$

In our example, that gives an estimated area of

$$\sum_{i=1}^{4} f(x_i^*)\Delta x = \sum_{i=1}^{4} f\left(i - \frac{1}{2}\right)\Delta x$$

$$= \sum_{i=1}^{4} \sqrt{16 - \left(i - \frac{1}{2}\right)^2}$$

(7.14) $$= \sqrt{15.75} + \sqrt{13.75} + \sqrt{9.75} + \sqrt{3.75} \approx 12.74.$$

The midpoint rule is slightly more accurate than the trapezoidal rule and is a *lot* more accurate than either the left endpoint rule or the right endpoint rule.

For the second area problem of Section 7.1, namely computing the area under the parabola $y = x^2$, we will program a computer to do our calculations for us using left endpoints. Since the algorithm is the same for all choices of starting point a, endpoint b, a number of intervals N, and even for all functions $f(x)$, we will leave these as input parameters for our program. In the exercises you will then be asked to modify the program to work with right endpoints, midpoints, or the trapezoidal rule and with functions other than $y = x^2$. The following program is written for MATLAB, but it can be easily adapted to work on other platforms. Note that the counter in this program is called k instead of i. That is because MATLAB also works with complex numbers, where i already means $\sqrt{-1}$.

```
function v = RIEMANN(f,a,b,N)     % f, a, b and N are
                                   % input parameters
    v = 0;                         % Initialize our sum at 0
    deltax = (b - a)/N;            % Compute Δx
    for k = 1:N
        xstar = a + (k-1)*deltax;  % Compute x_k^* = x_{k-1}.
        v = v + f(xstar);          % Add f(x_k^*) to v
    end                            % Finish the loop.
    v = deltax*v;                  % Multiply ∑_{k=1}^{N} f(x_k^*) by Δx.
```

Instructions for using this program can be found just above Exercise 7.22.

7.3. The Definite Integral

The Definition of an Integral. In Section 7.1 we saw that a lot of bulk quantities can be expressed as limits of sums. The procedure was the same whether we were talking about losses at a company, distance traveled by a projectile, the moment of inertia of a bar, the volume of a pyramid, or the area under a curve. In Section 7.2 we looked more closely at the area under a curve and developed notation and terminology to describe what we were doing in Section 7.1. With that notation and terminology in hand, we are finally ready to define the **definite integral**.

> **Definition.** Suppose that $f(x)$ is a function defined on the interval $[a, b]$. The quantity
>
> $$(7.15) \qquad \lim_{N \to \infty} \sum_{i=1}^{N} f(x_i^*) \Delta_i x$$
>
> is called the **definite integral of $f(x)$ with respect to x from a to b** and is denoted $\int_a^b f(x) dx$.

The **integral sign** $\int$, sometimes written $\smallint$, is designed to look like a stretched-out letter S. The S stands for "sum", as did the Greek letter Σ. The expression dx in the integral is analogous to the Δx in the Riemann sum. It tells us that our variable is x, and that we are adding up contributions from all values of x from a to b. *Do not leave it out!* The expression $\int_2^3 x^2$, without a dx in it, is meaningless.

There are a lot of details, and a hidden assumption, built into equation (7.15). To begin with, we need to build the Riemann sum $\sum_{i=1}^{N} f(x_i^*) \Delta_i x$ as in Section 7.2. That is, we have to break the interval $[a, b]$ into N subintervals of length $\Delta_1 x$, $\Delta_2 x$, etc., pick a representative point x_i^* in each subinterval, and add up all the values of $f(x_i^*) \Delta_i x$. As we have seen, this process involves choices. For the integral to make sense, we need the Riemann sum to approach the same limit as $N \to \infty$, regardless of those choices. If that happens, we say that the function $f(x)$ is **integrable** on the interval $[a, b]$.

Fortunately, *all continuous functions are integrable*, and functions with a finite number of jumps are also integrable. The only functions you're likely to encounter that aren't integrable are artificial examples cooked up for that purpose.

Here is one of those artificial examples. Let

$$(7.16) \qquad f(x) = \begin{cases} 0 & \text{if } x \text{ is rational,} \\ 1 & \text{if } x \text{ is irrational.} \end{cases}$$

If we try to compute $\int_0^1 f(x) dx$ by using N equal-sized intervals and left endpoints, then each x_i^* will be rational, each $f(x_i^*)$ will equal 0, and our Riemann sum $\sum_{i=1}^{N} f(x_i^*) \Delta x$ will be 0. However, if we pick each x_i^* to be an irrational number between x_{i-1} and x_i, then $f(x_i^*)$ will be 1 and our Riemann sum will be 1. This is true no matter how big N is. Since the limiting answer depends on how we set up our Riemann sums, there is no single limiting value. In this example, $f(x)$ is not integrable and $\int_0^1 f(x)dx$ is not defined.

Most of the time that we write $\int_a^b f(x)dx$, we imagine that $a < b$. However, expressions like $\int_3^2 x^2 dx$ make sense, too. We just have to break the interval from 3 to 2 into pieces, add up the $f(x_i^*)\Delta_i x$'s, and take a limit. This is exactly the same as breaking the interval from 2 to 3 into pieces, adding up the $f(x_i^*)\Delta_i x$'s, and taking a limit, except for one thing. When going from a bigger number to a smaller number, the quantity $\Delta_i x = x_i - x_{i-1}$ is *minus* the distance from x_i to x_{i-1}. This means that $\int_3^2 f(x)dx = -\int_2^3 f(x)dx$, and more generally

(7.17) $$\int_b^a f(x)dx = -\int_a^b f(x)dx.$$

Properties of Integrals.

- If $f(x) = c$ is a constant function, then $\int_a^b f(x)dx = c(b-a)$. No matter how we break up our interval $[a, b]$ and no matter how we pick the points x_i^*, we will have $f(x_i^*) = c$ for every i, so

(7.18) $$\sum_{i=1}^N f(x_i^*)\Delta_i x = \sum_{i=1}^N c\Delta_i x = c(b-a).$$

Since this is true for every N, the limit as $N \to \infty$ is also $c(b-a)$.

- If $f(x)$ and $g(x)$ are integrable functions on the interval $[a, b]$ and c_1 and c_2 are constants, then $c_1 f(x) + c_2 g(x)$ is integrable on $[a, b]$, and

(7.19) $$\int_a^b (c_1 f(x) + c_2 g(x))dx = c_1 \int_a^b f(x)dx + c_2 \int_a^b g(x)dx.$$

In particular,

$$\int_a^b (f(x) + g(x))dx = \int_a^b f(x)dx + \int_a^b g(x)dx, \quad \text{and}$$

(7.20) $$\int_a^b (f(x) - g(x))dx = \int_a^b f(x)dx - \int_a^b g(x)dx.$$

Equation (7.19) comes straight from the corresponding property of sums. If we break the interval $[a, b]$ into N pieces and pick our representative points x_i^*, then

$$\int_a^b (c_1 f(x) + c_2 g(x))dx \approx \sum_{i=1}^N (c_1 f(x_i^*) + c_2 g(x_i^*))\Delta_i x$$

$$= c_1 \sum_{i=1}^N f(x_i^*)\Delta_i x + c_2 \sum_{i=1}^N g(x_i^*)\Delta_i x$$

(7.21) $$\approx c_1 \int_a^b f(x)dx + c_2 \int_a^b g(x)dx.$$

7.3. The Definite Integral

The bigger N is, the better the approximations in the first and third lines. In the limit all the expressions are equal.

- If $a < b$ and if $f(x) > 0$ on the interval $[a, b]$, then $\int_a^b f(x)$ equals the area under the curve $y = f(x)$ and above the x axis between $x = a$ and $x = b$. However, if $f(x)$ is sometimes negative, then $\int_a^b f(x)dx$ is a **signed area**. That is, it is the area of the region above the x-axis and below the curve $y = f(x)$, minus the area of the region below the x-axis and above the curve. See Figure 7.11.

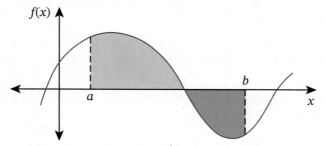

Figure 7.11. The integral $\int_a^b f(x)\,dx$ is a signed area.

In one sense this is obvious, since we already know that areas can be computed as integrals. However, it tells us that integrals can *always* be visualized as signed areas, even if they come from completely different problems, like distance or volume or corporate profit.

- Likewise, if $f(x)$ and $g(x)$ are integrable functions, then $\int_a^b f(x) - g(x)dx$ is the signed area between $y = f(x)$ and $y = g(x)$, as shown in Figure 7.12. If we cut the region between the two curves into vertical strips, each strip will have height $f(x) - g(x)$ and width Δx. Adding up the approximate areas $(f(x) - g(x))\Delta x$ of these strips and taking a limit gives us the integral $\int_a^b (f(x) - g(x))dx$.

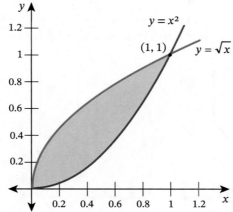

Figure 7.12. The area between the vertical parabola $y = x^2$ and the horizontal parabola $y = \sqrt{x}$ is $\int_0^1 (\sqrt{x} - x^2)\,dx$.

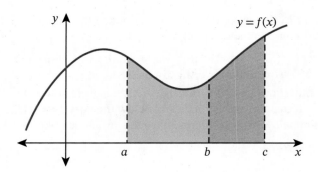

Figure 7.13. The area from a to c is the area from a to b plus the area from b to c.

- If a, b, and c are real numbers and the function $f(x)$ is integrable on the relevant intervals, then

(7.22) $$\int_a^c f(x)dx = \int_a^b f(x)dx + \int_b^c f(x)dx.$$

This is easy to see in terms of area. If $a < b < c$, as in Figure 7.13, then the area from a to c is the same as the area from a to b plus the area from b to c.

If $a < c < b$, as in Figure 7.14, then we also need a little algebra:

$$\int_a^b f(x)dx = \int_a^c f(x)dx + \int_c^b f(x)dx,$$

$$\int_a^b f(x)dx - \int_c^b f(x)dx = \int_a^c f(x)dx,$$

(7.23) $$\int_a^b f(x)dx + \int_b^c f(x)dx = \int_a^c f(x)dx.$$

That is, the area from a to c is the area from a to b minus the area from c to b, which is the integral from a to b *plus* the integral *from b to c*.

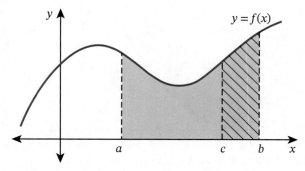

Figure 7.14. The area from a to c is the area from a to b minus the area from c to b.

7.4. The Accumulation Function

- All continuous functions are integrable. The reason is that if the function $f(x)$ is continuous, then the value of $f(x)$ can only change a tiny amount when x changes a tiny amount. This means that when we try to integrate $f(x)$ from a to b using a Riemann sum with N terms, with N large, then $f(x)$ can only change by a small amount in each interval I_i. (Mathematicians use the Greek letter ε to denote such small quantities.)

Since all of the choices of x_i^* give values of $f(x_i^*)$ that are within ε of one another, $f(x_i^*)\Delta_i x$ is within $\varepsilon \Delta_i x$ of the true area of the ith strip. We then have

$$\text{Error computing area of } I_i \leq \varepsilon \Delta_i x,$$
$$\text{Total error} \leq \sum_{i=1}^{N} \text{Error in } I_i$$
$$\leq \sum_{i=1}^{N} \varepsilon \Delta_i x$$
(7.24)
$$= \varepsilon(b-a).$$

As $N \to \infty$, ε goes to 0 and our total error disappears, regardless of how we pick the points x_i^*.

7.4. The Accumulation Function

So far we have talked a lot about what a definite integral *is*, and relatively little about how to compute one. In this section we introduce the **accumulation function** and we use it to compute some definite integrals.

As a warmup, consider $\int_2^5 x\,dx$, shown in Figure 7.15. This is the area of the trapezoid underneath the line $y = x$ between $x = 2$ and $x = 5$. The area of this trapezoid is the area $\frac{5^2}{2}$ of the right triangle of width 5 and height 5 minus the area $\frac{2^2}{2}$ of the right triangle of width 2 and height 2, or $\frac{5^2 - 2^2}{2} = \frac{21}{2}$.

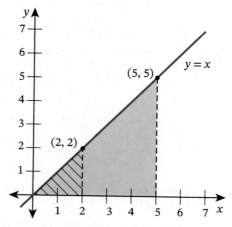

Figure 7.15. The area from 2 to 5 is the difference of the areas of two triangles.

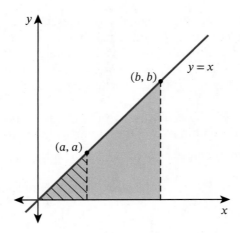

Figure 7.16. The area from a to b is the difference of two accumulations.

More generally, $\int_a^b x\,dx$ is the area of a trapezoid of width $b - a$, as shown in Figure 7.16. That is, the area of the triangle with vertices $(0, 0)$, $(b, 0)$, and (b, b) minus the area of the triangle with vertices $(0, 0)$, $(a, 0)$, and (a, a). In other words,

$$\text{(7.25)} \qquad \int_a^b x\,dx = \int_0^b x\,dx - \int_0^a x\,dx = \frac{b^2}{2} - \frac{a^2}{2} = \frac{b^2 - a^2}{2}.$$

The function $\frac{b^2}{2} = \int_0^b x\,dx$ is an example of an **accumulation function**. If $f(x)$ is an integrable function, we define the accumulation function $A(X)$ to be

$$\text{(7.26)} \qquad \int_0^X f(x)dx.$$

Geometrically, this represents a running total of all the area under the curve $y = f(x)$ from 0 to X. Note the difference between X and x! The variable of the A function is capital X, while the variable in the integral $\int_0^X f(x)dx$ is lower case x. An integral is a kind of sum, so x takes on *all* values between 0 and X, while X is a particular number.

An accumulation function is a running total, like an odometer or the reading on a water meter. See Figure 7.17. It measures the total accumulation of some quantity, whether area or distance or water usage, from the starting point (when your car rolled off the assembly line, or when your house was hooked up to the water main) until the point indicated by the variable X.

If we want to call the variable of the A function x, we have to give the variable in the integral a different name, like s or t:

$$\text{(7.27)} \qquad A(x) = \int_0^x f(s)ds = \int_0^x f(t)dt,$$

7.4. The Accumulation Function

where *s* and *t* are called *dummy variables*. They only serve to label the values of the f function that we are adding up to make our integral. Just as the expressions

(7.28) $$\sum_{s=1}^{100} s^2, \quad \sum_{t=1}^{100} t^2, \quad \text{and} \quad \sum_{w=1}^{100} w^2$$

all represent the same sum $1^2 + 2^2 + \cdots + 100^2$, the expressions

(7.29) $$\int_0^x f(s)\,ds, \quad \int_0^x f(t)\,dt, \quad \text{and} \quad \int_0^x f(w)\,dw$$

all represent limits of the same Riemann sums, and give the area under the f curve from 0 to x.

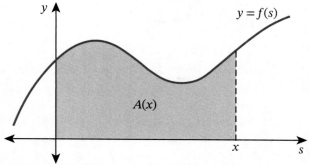

Figure 7.17. The accumulation function $A(x)$ is all the area from 0 up to x. Since x has a fixed value, the variable of integration must have a different name, in this case s.

If we can figure out the accumulation function $A(x)$, then it's easy to figure out any definite integral involving $f(x)$:

(7.30) $$\int_a^b f(x)\,dx = \int_0^b f(x)\,dx - \int_0^a f(x)\,dx = A(b) - A(a).$$

The total distance traveled between time a and time b is the odometer reading at time b minus the odometer reading at time a. The total water usage in March is the meter reading on April 1 minus the meter reading on March 1. Of course, this approach is only useful if we know $A(x)$!

In the example where $f(x) = x$, $A(x) = \frac{x^2}{2}$, since this is the area of a triangle with base x and height x. We could also write $A(b) = \frac{b^2}{2}$ or $A(z) = \frac{z^2}{2}$ or $A(X) = \frac{X^2}{2}$; the name of the variable doesn't matter. Eventually we will learn a lot of tricks for figuring out accumulation functions. Here are two more examples, one easy and one hard, that we can work out using geometry to identify areas.

(1) If $f(x) = 1$, then $A(x) = \int_0^x f(s)ds = x$. This is the area of a rectangle of height 1 and width x.

(2) If $f(x) = \sqrt{1-x^2}$, then $A(x) = \int_0^x f(s)ds$ is the area of the region shown in Figure 7.18. This region consists of a triangle and a pie wedge. The sine of the angle of

7. The Whole Is the Sum of the Parts

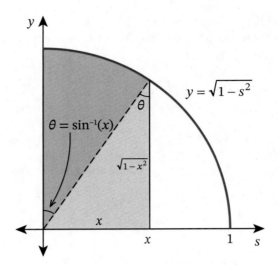

Figure 7.18. The accumulation function $A(x)$ is the area of a triangle plus the area of a pie wedge.

the pie wedge is x, so the angle itself is $\sin^{-1}(x)$. Since this is a fraction $\frac{\sin^{-1}(x)}{2\pi}$ of a circle of area π, the pie wedge has area $\sin^{-1}(x)/2$. Meanwhile, the triangle has base x and height $\sqrt{1-x^2}$, and so has area $x\sqrt{1-x^2}/2$. Putting this together, we have

$$(7.31) \qquad A(x) = \int_0^x \sqrt{1-s^2}\,ds = \frac{\sin^{-1}(x) + x\sqrt{1-x^2}}{2}.$$

We can also add accumulation functions. If $f(x) = 13 + 6x + 8\sqrt{1-x^2}$, then

$$\begin{aligned} S(x) &= \int_0^x (13 + 6s + 8\sqrt{1-s^2})\,ds \\ &= 13\int_0^x 1\,dx + 6\int_0^x s\,ds + 8\int_0^x \sqrt{1-s^2}\,ds \\ (7.32) &= 13x + 3x^2 + 4\sin^{-1}(x) + 4x\sqrt{1-x^2}. \end{aligned}$$

Do you notice anything about the derivatives of the accumulation functions that we have found so far? We'll have a lot more to say about them in the next chapter.

7.5. Chapter Summary

The Main Ideas.

- Integration means adding up the pieces to make a whole. It applies to any quantity that can be broken into pieces, not just to the area under a curve.
- Our strategy is to break the quantity into pieces, estimate the size of each piece, and add up the pieces. *The whole is the sum of the parts.*

7.5. Chapter Summary

- Most of the time, adding up the pieces results in an expression of the form $\sum_{i=1}^{N} f(x_i^*)\Delta_i x$. This is called a Riemann sum and can be visualized as the area under the curve $y = f(x)$.

- Setting up Riemann sums involves choices. We need to decide on the number N of pieces, on the sizes of the intervals I_i, and the choices of the representative points x_i^*.

- Some choices give much more accurate approximations than others. Both the trapezoidal rule and the midpoint rule are much more accurate than using left endpoints or right endpoints.

- A definite integral is the limit of a Riemann sum as the number of pieces goes to ∞. That is,

$$(7.33) \qquad \int_a^b f(x)dx = \lim_{N \to \infty} \sum_{i=1}^{N} f(x_i^*)\Delta_i x.$$

- If $f(x)$ is integrable, then all choices give the same limit as $N \to \infty$. (That's the definition of integrability!) All continuous functions, and most other functions we will encounter, are integrable.

- While integrals come from a variety of problems, all integrals can be *visualized* as signed areas. This allows us to apply geometric arguments to integrals.

- The accumulation function $A(X)$ is the running total of the integral of $f(x)$, namely $\int_0^X f(x)dx$. We often write $A(x) = \int_0^x f(s)\,ds$, where s is a dummy variable. So far we only know how to compute a few accumulation functions, but they will eventually turn out to be very useful.

- If $A(x)$ is the accumulation function of $f(x)$, then $\int_a^b f(x)dx = A(b) - A(a)$.

Expectations. You should be able to:

- Apply the strategy of breaking things into pieces, estimating the pieces, and adding up the pieces to approximate a variety of bulk quantities as Riemann sums.
- Understand and use $\sum$ notation for sums.
- Approximate bulk quantities numerically via Riemann sums, using either left endpoints, right endpoints, midpoints, or the trapezoidal rule.
- Take the limit as $N \to \infty$ to describe bulk quantities exactly as integrals.
- Convert back and forth between a Riemann sum and the integral that it approximates.
- Relate integrals to areas under, over, and between curves.
- Understand and use the main properties of integrals.
- Use accumulation functions to compute integrals of linear functions such as $f(x) = c_0 + c_1 x$.

7.6. Exercises

Note: The exercises for this chapter are about understanding bulk quantities by slicing and dicing them into pieces that we can estimate, adding up the pieces, and taking a limit to get an integral. Some exercises are about setting up those integrals, some are about evaluating Riemann sums numerically, and some are about accumulation functions. A few are about evaluating integrals that can be visualized as the area under a curve, but mostly the point is to understand how any bulk quantity, not just area, can be understood as an integral.

Riemann Sums

7.1. Express each sum using $\sum$ notation.
 (a) $3 + 6 + 9 + \cdots + 300$
 (b) $1 + \frac{1}{2} + \frac{1}{3} + \cdots + \frac{1}{N}$
 (c) $\tan(x_1^*)\Delta x + \tan(x_2^*)\Delta x + \cdots + \tan(x_8^*)\Delta x$
 (d) $\sqrt{x_1}\Delta x + \sqrt{x_2}\Delta x + \cdots + \sqrt{x_N}\Delta x$

7.2. Write out each sum explicitly and evaluate it.
 (a) $\sum_{i=1}^{10}(i-1)$.
 (b) $\sum_{j=1}^{5} e^{-j}$.
 (c) $\sum_{k=2}^{5}\left(\frac{1}{k-1} - \frac{1}{k}\right)$. Simplify your final answer as much as possible.
 (d) $\sum_{n=2}^{1000}\left(\frac{1}{n-1} - \frac{1}{n}\right)$. In this part you only have to evaluate the sum. You do not have to write out the 999 terms!

7.3. A CAT scan of a football-shaped tumor shows slices that are spaced 2.5 cm apart. The areas of the slices are 0, 20, 24, 38, 56, 40, 30, 23, and 0 square centimeters.
 (a) Write down a Riemann sum that approximates the volume of the tumor.
 (b) Evaluate that Riemann sum.

Present Value. We saw in Exercise 6.18 that the **present value** of a payment of A dollars t years from now is Ae^{-rt}, where r is the interest rate. Here we consider the present value of an *income stream*.

7.4. Suppose that an annuity pays \$10,000 per year for 10 years, with payments made continuously. We will compute the present value of this income stream, assuming an annual interest rate of 4% ($r = 0.04$/year).
 (a) Determine upper and lower bounds for the present value of the first year's payments. Bear in mind that a dollar paid immediately is worth \$1, but a dollar paid at the end of the year is worth e^{-r} dollars.
 (b) Repeat part (a) for years 2 through 10.

7.6. Exercises

(c) Add up the pieces for years 1 through 10 to get upper and lower bounds for the present value of the entire annuity.

(d) Average the upper and lower bounds from part (c) to get a good estimate for the present value of the annuity. This is (approximately) what an insurance company would charge you for this annuity.

7.5. Repeat Exercise 7.4, only now dividing the income stream into N pieces instead of 10. (For instance, if you were calculating things one month at a time, N would be 120.) Using $\sum$ notation, write down expressions for an upper bound on the present value of the annuity, a lower bound, and the average.

7.6. Now take the limit as $N \to \infty$ of the sum you wrote in Exercise 7.5. Write down a definite integral that gives the present value of the annuity. You are not expected to evaluate this integral.

7.7. Generalizing the scenario of Exercises 7.4–7.6, suppose that an annuity pays A dollars per year for T years and that the interest rate is r. Write down an integral involving A, T, and r that gives the present value of this annuity.

Area and Volume Problems. In Exercises 7.8–7.12, we use our slice-and-dice procedure to compute areas and volumes in unusual ways.

7.8. We are going to compute the area of the region between the parabola $y = 4 - x^2$ and the horizontal line $y = 2$ by slicing it into *horizontal* strips, estimating the area of each strip, adding them up, and taking a limit to get an integral *over y*. (Yes, there are ways to get this area by slicing the region into vertical strips, but we aren't doing that!) We imagine cutting the region into N strips of equal height. The figure shows how that looks when $N = 5$, but you should think of N as an unspecified parameter. We number the strips from the bottom of the region to the top.

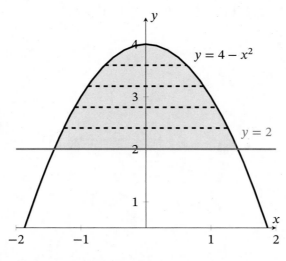

(a) What is the vertical thickness of each strip?

(b) Let y_i be the y-value at the top of the ith strip. Find a formula for y_i in terms of i and N.

(c) What is the *width* of the top of the ith strip? You can leave your answer in terms of y_i.

(d) In terms of i and N, what is the approximate area of the ith strip?

(e) Express the (approximate) total area of the region as a Riemann sum.

(f) Express the limit of the Riemann sum, as $N \to \infty$, as a definite integral over y. You do not have to evaluate this integral.

7.9. Now we are going to compute the volume of the bullet-shaped solid obtained by rotating the region of Exercise 7.8 around the y axis. We will slice the solid horizontally into roughly cylindrical slabs that we'll call "pancakes", estimate the volume of each pancake, and add them up.

(a) If there are N pancakes, each of equal thickness, what is the thickness of each pancake? Call this distance Δy.

(b) Let y_i be the height above the ground of the top of the ith pancake, where we're numbering pancakes starting at the bottom. Find an explicit formula for y_i.

(c) Find the cross-sectional area of the solid at height y from the ground.

(d) Based on your result in part (c), what is the approximate volume of the ith pancake? You can write your answer either in terms of i and N or in terms of y_i and Δy.

(e) Express the approximate total volume of the solid as a Riemann sum.

(f) Taking a limit of that Riemann sum, express the total volume of the solid as a definite integral over y. You do not have to evaluate that integral.

7.10. Imagine that we cut down a four-year-old tree, leaving a circular stump with a radius of 10 cm, as shown in the figure below. The tree has four equally spaced growth rings. We can estimate the area of each ring by multiplying its circumference by its width. However, we have to decide which circumference to use: the outside of the ring or the inside of the ring.

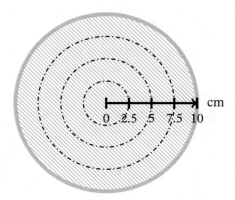

(a) Use the outer circumference of each ring to estimate its area. Add up these estimates to get an estimated area for the entire stump surface. Is your result an underestimate or an overestimate of the true total area?

7.6. Exercises

(b) Now imagine having N rings instead of just four, with the radius of the stump still being 10 cm. Write an expression for the area of the ith ring, counting outward, using outer circumferences.

(c) Add up the results from part (b) to get a Riemann sum that approximates the area of the stump surface. You do not have to evaluate this sum.

(d) Taking a limit of the Riemann sum as $N \to \infty$, write down a definite integral that computes the exact area of the stump surface.

(e) Evaluate the integral from part (d) using what you know about areas of triangles. Does the result agree with the usual formula for the area of a circle of radius 10 cm?

7.11. Repeat steps (b)–(e) of Exercise 7.10 for a stump of arbitrary radius R instead of radius 10 cm. The final answer should look very familiar.

7.12. Sometimes, an oyster ingests a speck of sand that gets stuck on its inside. To protect itself, the oyster coats the speck with a material that hardens, much like a coat of paint. Then it applies another coat, and another coat, and so on, until the speck of sand grows into a pearl. The volume of the pearl is then the sum of the volumes of all the coats of "paint".

(a) Suppose that a pearl of radius 5 mm consists of five coats, each a millimeter thick. Estimate the volume of the outermost coat as surface area times thickness, using the fact that the surface area of a sphere of radius r is $4\pi r^2$. (You have to decide whether to use the inner or outer radius of each coat. One gives an underestimate of the volume, while the other gives an overestimate.)

(b) Add up the estimated volumes of the five coats to get an approximate volume of the pearl.

(c) Now imagine that the pearl consists of N coats. Write down a Riemann sum that approximates the volume of the sphere. You do not need to evaluate this sum.

(d) Finally, take a limit of the Riemann sum as $N \to \infty$ to get a definite integral that gives the exact volume of the pearl. You do not need to evaluate the integral.

Distance. In the projectile problem of Section 7.1, we saw that distance traveled is the integral of velocity. We explore this idea further in Exercises 7.13–7.14.

7.13. You are going on a road trip to Big Bend National Park. At noon, you start the trip and set your trip odometer to 0. You drive at 60 miles/hour (mph) for two hours, until you reach the open road. You then drive at 75 mph for four hours. At 6:00 you turn onto a park road and drive one more hour at 40 mph, reaching your destination at 7:00.

Let $v(t)$ be your speed at time t, measured in miles per hour, where t is the number of hours since noon. Let $D(t)$ be the reading of your trip odometer at time t, indicating the total distance you have traveled since noon. This is the same as the accumulation function of V: $D(t) = \int_0^t v(s)\,ds$.

(a) Graph your velocity $v(t)$ (measured in mph) as a function of the time t (measured in hours since noon).

(b) What will your trip odometer read at 1:00? At 2:00? At 3:00? At 4:00? At 5:00? At 6:00? At 7:00?

(c) Find a formula for $D(t)$ that applies for $0 \leq t \leq 2$. Make sure that it agrees with your answers to part (b).

(d) Find formulas for $D(t)$ when $2 \leq t \leq 6$ and when $6 \leq t \leq 7$. Again, make sure that these formulas agree with your answers to part (b).

(e) Graph $D(t)$ for $0 \leq t \leq 7$.

7.14. After having fun at Big Bend, you decide to go bungee jumping from the Royal Gorge Bridge in Colorado, almost 1000 feet above the canyon below.[2] You jump off the bridge, attached to a 484 foot bungee cord. Your downward velocity $v(t)$ for the first part of your fall is $32t$, where the velocity $v(t)$ is measured in feet/second and t is the time in seconds since you jumped. This continues until you have fallen 484 feet, at which point the cord starts to stretch, slowing you down and eventually pulling you back up.

(a) Graph $v(t)$ as a function of t for the first part of your fall.

(b) Express the distance traveled between $t = 1$ and $t = 4$ as the area of a region under your graph. Using what you know about triangles, compute that area.

(c) Write an expression for how far you fall in the first T seconds that applies until your bungee cord starts to stretch.

(d) After how many seconds will your bungee cord start to stretch?

Work. Labor costs are based on how many people work at a task, and for how long. If three people work on a task for two hours, or if two people work on the task for three hours, that's six **person-hours** (also called **staff-hours** or just **hours**).

7.15. Landscaping can be done by several people working together, so it makes sense to measure the work done in staff-hours. The number of workers that a landscaping business can assign to a certain eight-hour job is given by the function

$$s(t) = \begin{cases} 4 & 0 \leq t < 2, \\ 2 & 2 \leq t < 4, \\ 3 & 4 \leq t \leq 8, \end{cases}$$

where t is the time in hours since the beginning of the job.

(a) Draw the graph of the step function defined here, and compute the total number of staff-hours for this job.

(b) The **accumulated work** function $W(t)$ gives the total number of staff-hours up to time t. Draw the graph of $W(t)$.

(c) How are the graphs of $s(t)$ and $W(t)$ related?

7.16. A different landscaping job requires twenty staff-hours of work. From 8:00 am until 10:00 am, three workers landscape the yard. From 10:00 am until noon, there are only two workers. After that, there are four workers until the job is finished.

(a) How long will the whole job take?

---

[2] From 2003 to 2008, bungee jumping at Royal Gorge was part of the annual Go Fast! games, but to the best of my knowledge it has not been allowed since 2008.

7.6. Exercises

(b) Graph the staffing function for this exercise. Mark on the graph the time that the job is finished.

(c) Graph the accumulated work function $W(t)$.

(d) How are the graphs of $s(t)$ and $W(t)$ related?

7.17. The **average staffing** for a job is the total number of staff-hours divided by the total number of hours.

(a) What is the average staffing of the job considered in Exercise 7.15?

(b) What is the average staffing of the job considered in Exercise 7.16?

Electrical Energy. Exercises 7.18–7.20 deal with electricity. The work done by electricity is usually referred to as **electrical energy**, and the rate at which this work is done is called **electric power**. Electric *power* is measured in **watts**. Electrical *energy* is measured in **watt-hours** (or kilowatt-hours or megawatt-hours, which are thousands and millions of watt-hours, respectively). Running a 20-watt light bulb for five hours consumes 100 watt-hours of electrical energy, as does running a 100-watt light bulb for one hour.

7.18. On a summer day, a 2000-watt air conditioner runs from 10:00 am until 10:00 pm. How many watt-hours of energy does it consume? How many kilowatt-hours does it consume?

7.19. A certain air conditioner has three settings: "economy" runs at 1200 watts, "normal" runs at 1600 watts, and "maximum" runs at 2000 watts. A homeowner programs the air conditioner to run on economy from midnight to 7:00 am, on normal from 7:00 am until noon, on maximum from noon until 7:00 pm, and on normal again from 7:00 until midnight.

(a) Let $p(t)$ be the rate of power consumption, where t is the number of hours since midnight. Sketch the graph of $p(t)$.

(b) Express the total energy consumption during the day (in watt-hours, or in kilowatt-hours if you prefer) as a definite integral. Evaluate that integral.

(c) Let $P(t)$ be the accumulated energy consumption in the first t hours of the day. Sketch the graph of $P(t)$.

7.20. Most air conditioners are controlled by thermostats. The hotter it is outside, the more they run and the more energy they consume. On a certain day, it got hot enough to trigger the air conditioner at 9:00 am, and it kept getting hotter until 4:00 pm. The rate of power consumption was $p(t) = 200t$ watts, where t is the number of hours since 9:00 am.

(a) How much energy was used in the first t hours? That is, compute an accumulation function.

(b) How much energy was used in the entire seven-hour period?

(c) How much energy was used between noon and 3:00 pm?

7.21. The following is actual data from Microsoft regarding capital expenditures (in billions of dollars per quarter) over a three-year period. The notation FY16Q1 means "Fiscal Year 2016 Quarter 1". Unfortunately, we only have data for six of the twelve quarters.

Capital Expenditure	Time Interval
1.5	FY16Q1
3.1	FY16Q4
2.5	FY17Q2
2.1	FY17Q3
3.3	FY18Q2
4.1	FY18Q4

(a) Write down and compute a Riemann sum that gives the approximate total capital expenditures for fiscal years 2016–2018. Because of the missing data, your Δt's may have different lengths.
(b) Will everyone have the same answers for part (a)? Why or why not?
(c) Good news! Some of the missing data has been found. Our chart now reads:

Capital Expenditure	Time Interval
1.5	FY16Q1
2.0	FY16Q2
3.1	FY16Q4
2.5	FY17Q2
2.1	FY17Q3
3.3	FY17Q4
2.7	FY18Q1
3.3	FY18Q2
4.1	FY18Q4

Use a Riemann sum to approximate total capital expenditures for the three-year period. Do you think this is a better approximation than what you computed in part (a)? Why or why not?

Using and Modifying the Riemann Program. Exercises 7.22–7.26 use the Riemann program, shown at the end of Section 7.2. This program tells MATLAB how to approximate $\int_a^b f(x)dx$ as a Riemann sum,

$$\sum_{i=1}^{N} f(x_i^*)\Delta x,$$

using N intervals, all of size $\Delta x = (b-a)/N$, with t_i^* being the left endpoint of the ith interval.

Note: Riemann is not a built-in part of MATLAB, but needs to be added by hand. If your instructor has not already provided you with an electronic copy of this file, you can simply type it in yourself to a file called Riemann.m. The .m ending tells MATLAB

7.6. Exercises

that it is a program called Riemann. For MATLAB to find this program, you must be working in the directory that contains Riemann.m.

To use the program, you first have to specify what the function $f(x)$ is. For instance, if you want to compute $\int_1^4 x^3 dx$ using $N = 100$, you would first type
`f = @(x) x^3;`
at the command prompt. You then type
`Riemann(f,1,4,100)`
and MATLAB returns the answer 62.8084.

7.22. (a) Using the Riemann program, calculate left-endpoint Riemann sums for the function e^{-x^2} on the interval $[0, 3]$ using $N = 50, 500, 5000$, and $50{,}000$. (Type `f = @(x) exp(-x^2);` at the command prompt to tell MATLAB what $f(x)$ is.)

(b) The left-endpoint Riemann sums that you calculated in part (a) seem to be approaching a limit as the number N of pieces increases without bound. Estimate that limit to four decimal places.

7.23. (a) Modify the Riemann program so it calculates the Riemann sum using right endpoints instead of left endpoints. Describe exactly how you changed the program to do this. Call your new function RiemannRit and save it as RiemannRit.m.

(b) Use your RiemannRit program to calculate right-endpoint Riemann sums for the function e^{-x^2} on the interval $[0, 3]$ using $N = 50, 500, 5000$, and $50{,}000$.

7.24. (a) Modify the Riemann program so it calculates the Riemann sum by taking t_i^* to be the midpoint of the ith interval, instead of the left or right endpoint. Describe exactly how you changed the program to do this. Call your new function RiemannMid and save it as RiemannMid.m.

(b) Use RiemannMid to calculate midpoint Riemann sums for the function e^{-x^2} on the interval $[0, 3]$ using $N = 50, 500, 5000$, and $50{,}000$.

(c) Compare your results from using RiemannMid to the results of using Riemann, the results of using RiemannRit, and the results of taking the average of Riemann and RiemannRit. Which method do you think gives the most accurate estimate of $\int_0^3 e^{-x^2} dx$? Which gives the least accurate estimate?

(d) For each method, estimate how big N needs to be to get within 0.0001 of that actual integral. How can you tell? (We will have a lot more to say about the accuracy of different numerical methods in Chapter 9.)

7.25. Suppose we want to compute the (signed) area "under" the curve $y = \sin(x^2)$ between $x = 0$ and $x = 4$.

(a) Use Riemann to estimate this area with Riemann sums using left endpoints and $N = 100, 1000$, and $10{,}000$.

(b) Use the RiemannRit program from Exercise 7.23 to estimate this area with Riemann sums using right endpoints and $N = 100, 1000$, and $10{,}000$.

(c) Use the RiemannMid program from Exercise 7.24 to estimate this area with Riemann sums using midpoints and $N = 100, 1000$, and $10{,}000$.

7.26. Suppose we want to compute the (signed) area "under" the curve $y = x^5 + x^3$ between $x = -2$ and $x = 2$.
   (a) Use the RiemannMid program from Exercise 7.24 to estimate this area with Riemann sums using midpoints and $N = 10$, 100, and 1000. Explain why each Riemann sum is 0.
   (b) Repeat part (a) using Riemann sums with left endpoints. Are the sums still 0? Explain how these estimates are (or aren't) different from those in part (a).

Probability. Most random quantities follow a bell-shaped distribution, also called a **normal distribution** or a **Gaussian distribution**. The most widely accepted measure of the width of the distribution is called the **standard deviation**. For instance, if you flip 100 coins many times, the number of heads in each set of 100 flips follows a normal distribution with an average (or **mean**) of 50 heads and a standard deviation of 5. If you flip 10,000 coins, the number of heads is described by a normal distribution with a mean of 5000 and a standard deviation of 50. Exercises 7.27 and 7.28 explore this distribution, and Exercise 7.29 explores another common distribution, the **exponential distribution**.

7.27. Estimate the area under the curve $y = e^{-x^2}$ between $x = 0$ and $x = 1$ to four decimal places. Sketch the curve and shade the area.

7.28. For a random quantity following a normal distribution, the probability of being between a standard deviations and b standard deviations above the mean is the area under the graph of

$$f(x) = \frac{1}{\sqrt{2\pi}} e^{-x^2/2}$$

between $x = a$ and $x = b$. That is, the probability is $\int_a^b f(x)\,dx$. Note that a and b don't have to be positive, although we do want $a \leq b$. The probability of being between three standard deviations below the mean and two standard deviations above the mean is $\int_{-3}^{2} f(x)\,dx$. This is almost (but not exactly) the same function as in Exercise 7.27.
   (a) With the help of MATLAB, compute to four decimal places the probability of being within one standard deviation of the mean (that is, the probability of being between one standard deviation below the mean and one standard deviation above the mean).
   (b) Compute the probability, to four decimal place accuracy, of being within two standard deviations of the mean.
   (c) Compute the probability, to four decimal place accuracy, of being between 0.5 standard deviations below the mean and 1.5 standard deviations above the mean.

7.29. The **exponential distribution** describes how long a piece of machinery lasts. The probability of lasting between a and b times the average lifetime is $\int_a^b e^{-x}\,dx$.

(a) A certain brand of light bulb has an average lifetime of 100 hours. What is the probability of its lasting less than 75 hours? Feel free to use MATLAB (and specifically the Riemann program) to compute this.

(b) What is the probability that the light bulb lasts less than 150 hours?

(c) What is the probability that the light bulb lasts between 75 and 150 hours? How is that related to your answers to parts (a) and (b)?

7.30. An oil field is currently producing oil at a rate of 3 million barrels of oil per year. The production rate typically falls at a rate proportional to the current rate of production, with rate constant 0.15/year. That is, if $P(t)$ (P for petroleum) represents the production rate in barrels/year after t years, then $P(t)$ solves the initial value problem

$$P'(t) = -0.15P(t), \qquad P(0) = 3{,}000{,}000.$$

(a) What is the solution to this initial value problem?

(b) If a and b are specific times, what real-world quantity does $\int_a^b P(t)\,dt$ represent?

(c) Using $\Delta t = 1$ year and right endpoints, estimate how much oil is produced between $t = 0$ and $t = 5$.

(d) Write down a definite integral that gives the exact total oil production over the five-year period of part (c).

7.31. Evaluate each integral.
(a) $\int_2^{10} 3\,dx$
(b) $\int_{10}^{2} 3\,dx$
(c) $\int_{-4}^{9} -2\,dz$
(d) $\int_9^{-4} -2\,dz$

7.32. This exercise uses the function $f(x) = \begin{cases} -2 & 0 \le x < 4, \\ 5 & 4 \le x \le 9. \end{cases}$

(a) Sketch the graph of $f(x)$.

(b) Evaluate $\int_0^6 f(x)\,dx$, $\int_6^9 f(x)\,dx$, and $\int_0^9 f(x)\,dx$.

7.33. A pyramid is 20 feet tall. The area of a horizontal cross-section x feet from the top of the pyramid is $4x^2$ square feet.

(a) What is the area of the base?

(b) What is the volume of the pyramid, to the nearest cubic foot? (*Hint*: Use Riemann or RiemannRit or RiemannMid.)

7.34. Find a formula for the accumulation function $A(x) = \int_0^x 2s + 5\,ds$. Then use that accumulation function to compute $\int_{-1}^{4} 2x + 5\,dx$.

Demand Curves and Consumer Surplus. Recall from Chapter 6 that the price p of an item and the demand (number x of items that are sold) are related. When we write x as a function of p, it's called the **demand function**. When we write $p = D(x)$, meaning that the price is a function of the demand, it's called the **demand curve**. Not only is $D(x)$ the price that could be charged if the manufacturer only wanted to sell x items, it's also the amount that the xth most enthusiastic consumer would be willing to pay for the item.

Each consumer who buys an item at price p thinks that it's worth more than p, or else he wouldn't have bought it. The difference between the item's value to the consumer (its **utility**) and the price the consumer actually paid is that individual consumer's **surplus**. The **consumer surplus** of a product is the sum of the surpluses of *all* the consumers buying that product.

7.35. Consider the demand curve shown below.

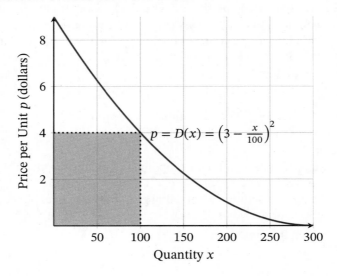

(a) Explain why demand curves (not just this example) are generally decreasing functions.

(b) The figure shows the demand curve $p = \left(3 - \frac{x}{100}\right)^2$ for a hypothetical product, where p is measured in dollars. If the product is priced at \$4/item and 100 items are sold, what does the area of shaded region represent? What is its value?

(c) Explain why the consumer surplus equals the area beneath the demand curve $p = D(x)$ and above line $p = p^*$, where p^* is the actual price for the quantity sold.

(d) For the given figure and demand curve, use four rectangles to estimate the area beneath the function $p = D(x)$ and above the line $p = 4$. Use the trapezoidal rule, which is the average of what you get using left endpoints and using right endpoints.

(e) Express the consumer surplus as a definite integral. (You do not have to evaluate this integral.)

(f) As the price of the product decreases, why does the consumer surplus increase?

Chapter 8

The Fundamental Theorem of Calculus (One Step at a Time)

As noted at the beginning of Chapter 5, there are three questions you should ask about every mathematical concept that you encounter:

- What is it?
- How do you compute it?
- What is it good for?

When it comes to integrals, we tackled the first and third question in Chapter 7, but we barely touched the second. At this point, we haven't even learned how to evaluate $\int_0^1 x^2\,dx$. In this chapter, we're going to see how anti-derivatives can help us evaluate integrals. At the end of this chapter and into Chapter 9, we'll develop techniques for finding those anti-derivatives.

The **Fundamental Theorem of Calculus**, or FTC, relates three quantities that come up a lot in integral calculus: the definite integral $\int_a^b f(x)dx$, the accumulation function $\int_0^x f(s)ds$, and the anti-derivative of $f(x)$. We already saw the first two in Chapter 7, and we learned how they are related. We will introduce the third in Section 8.1. The full situation is illustrated in Figure 8.1.

The FTC actually has two parts, called the first and second Fundamental Theorems of Calculus, or just FTC1 and FTC2 for short.[1] In Section 8.2 we will go over FTC2, which relates definite integrals and anti-derivatives. This gives us our most powerful tool for computing definite integrals, and with it we will see why, for instance,

---

[1] We follow the practice of most contemporary authors in calling the two parts FTC1 and FTC2, although we discuss them in the opposite order.

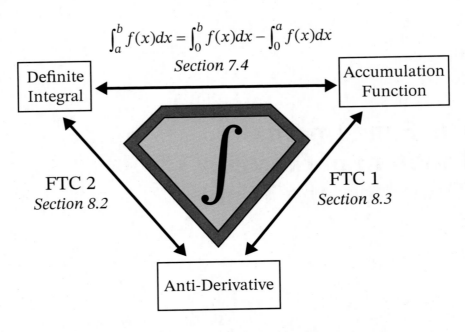

Figure 8.1. The main topics of this chapter

$\int_0^1 x^2 dx = \frac{1}{3}$. In Section 8.3 we will go over FTC1, which relates accumulation functions and anti-derivatives. We will also see how the two halves of the FTC fit together. In Section 8.4 we will treat anti-derivatives in their own right, and we'll see how they solve many problems involving basic physics. These physics problems will give us an additional perspective on why the FTC is true. Finally, in Section 8.5 we will develop some tools for computing anti-derivatives.

8.1. Three Different Quantities

Let's start by defining our three related concepts: definite integrals, accumulation functions, and anti-derivatives.

Definite Integrals.

The **definite integral** $\int_a^b f(x)dx$ is the limit of a Riemann sum:

$$(8.1) \qquad \int_a^b f(x)dx = \lim_{N \to \infty} \sum_{i=1}^{N} f(x_i^*) \Delta_i x.$$

A definite integral can always be visualized as the (signed) area under the curve $y = f(x)$ between $x = a$ and $x = b$, but it applies to lots of different things as well. For instance, if $v(t)$ is the velocity of a particle at time t, then $\int_a^b v(t)dt$ is the net distance traveled between time a and time b.

Defining a definite integral as the limit of a Riemann sum is more precise than just saying "the area under the curve", but making sense of that definition in Chapter 7 took

8.1. Three Different Quantities

a lot of machinery. We needed $\sum$ notation, we needed to understand what $\Delta_i x$ means, we needed to pick the representative points x_i^*, and we needed to understand how to take a limit as $N \to \infty$. Most of all, we had to learn how to estimate bulk quantities with Riemann sums, and how to get them exactly as limits of Riemann sums. The upshot of Chapter 7 is that lots and lots of real-world quantities can be represented as definite integrals.

Note that a definite integral is a *number*, not a function. Between 1:00 pm and 3:00 pm we might travel 105 miles (by car), or we might travel 5 miles (on foot), but we don't travel $f(t)$ miles. Also, there isn't any single value of x (or t) to associate with the integral. The integral $\int_a^b f(x)dx$ adds up the contributions of *all* values of x between a and b.

Accumulation Functions. The **accumulation function** $\int_0^x f(s)ds$ is a running total, and it is a function of the endpoint x. For instance, if the speed of your car is $v(t)$, then $\int_0^t v(s)ds$ is the reading of your odometer at time t. It gives the total distance that the car drove between time 0 (the day the car was made) and time t. The dummy variable s in the integral runs from 0 to t, since the odometer reading today is the result of all of the driving that you did between time 0 and time t. Note that the accumulation function is a *function*, not a number. It takes x or t as the input and outputs the number:

$$(8.2) \qquad A(x) = \int_0^x f(s)ds \quad \text{or} \quad A(t) = \int_0^t f(s)ds.$$

Accumulation functions are closely related to definite integrals. Partly, this is because the output of the accumulation function *is* a definite integral. Partly, it is because we can use accumulation functions to compute definite integrals. As we saw in Chapter 7,

$$(8.3) \qquad \int_a^b f(x)dx = \int_0^b f(x)dx - \int_0^a f(x)dx = A(b) - A(a).$$

So far, we have defined our accumulation functions starting at 0, but we could just as well have used a different starting point or starting time. For instance, a trip odometer is an accumulation function, started at whenever you last pressed the reset button. To figure out how far we drove in a certain period of time (that is, $\int_a^b v(t)dt$), we can compare the odometer readings at time a and b or we can compare the trip odometer readings at a and b.

In particular, if $A_1(x)$ and $A_2(x)$ are accumulation functions for the same function $f(x)$ with different starting points, then $A_1(x)$ and $A_2(x)$ just differ by a constant. For instance, if A_1 starts at $x = 0$ and A_2 starts at $x = 3$, then

$$(8.4) \qquad \begin{aligned} A_1(x) &= \int_0^x f(s)ds = \int_0^3 f(s)ds + \int_3^x f(s)ds \\ &= A_2(x) + \int_0^3 f(s)ds. \end{aligned}$$

The *function* $A_1(x)$ equals the *function* $A_2(x)$ plus the *constant* $\int_0^3 f(s)ds$. No matter what a and b are, we can compute $\int_a^b f(x)dx$ using either function:

$$\int_a^b f(x)dx = A_1(b) - A_1(a)$$

$$= \left(A_2(b) + \int_0^3 f(s)ds\right) - \left(A_2(a) + \int_0^3 f(s)ds\right)$$

(8.5) $$= A_2(b) - A_2(a).$$

The only thing wrong with accumulation functions is that we don't know how to actually compute them! That will change once we learn FTC1.

Anti-Derivatives.

An **anti-derivative** of $f(x)$ is a function $F(x)$ whose derivative (with respect to x) is $f(x)$. For instance, the derivative of x^3 is $3x^2$, so we say that x^3 is an anti-derivative of $3x^2$.

We already know a lot about derivatives, so we know lots of pairs of functions where the second is the derivative of the first. For instance, $5x^4$ is the derivative of x^5, $2e^{2x}$ is the derivative of e^{2x}, and $\sec^2(x)$ is the derivative of $\tan(x)$. That means we know lots of pairs of functions where the first is the anti-derivative of the second! x^5 is an anti-derivative of $5x^4$, e^{2x} is an anti-derivative of $2e^{2x}$, and $\tan(x)$ is an anti-derivative of $\sec^2(x)$.

Another example comes from motion. The derivative of the position $x(t)$ is the velocity $v(t)$, so the position is an anti-derivative of the velocity. The derivative of the velocity $v(t)$ is the acceleration $a(t)$, so the velocity is an anti-derivative of the acceleration.

In the last paragraph I wrote *an* anti-derivative, rather than *the* anti-derivative, for a reason. The anti-derivative of a function isn't unique. The derivative of $x^3 + 7$ is $3x^2$, so $x^3 + 7$ and x^3 are both anti-derivatives of $3x^2$. In general, adding a constant to any anti-derivative gives another anti-derivative. However, on connected regions that is the only ambiguity.

Here's why: If $F_1(x)$ and $F_2(x)$ are both anti-derivatives of $f(x)$, then

(8.6) $$\frac{d}{dx}(F_1(x) - F_2(x)) = F_1'(x) - F_2'(x) = f(x) - f(x) = 0.$$

Since the derivative of $F_1(x) - F_2(x)$ is always 0, $F_1(x) - F_2(x)$ never changes. That is, $F_1(x) - F_2(x)$ is a *constant*, and we can write

(8.7) $$F_1(x) = F_2(x) + C,$$

where C is a constant. In other words, every anti-derivative of $3x^2$ is x^3 plus a constant, and we say that *the* anti-derivative of $3x^2$ is

(8.8) $$x^3 + C,$$

where now C is an *arbitrary* constant.

To summarize,

(1) Checking whether $F(x)$ is an anti-derivative of $f(x)$ is easy. Just take the derivative of $F(x)$ and see if it equals $f(x)$.

(2) Once we have one anti-derivative $F(x)$ of $f(x)$, perhaps by guessing and checking, we know that every other anti-derivative of $f(x)$ is of the form $F(x) + C$, and we might say that $F(x) + C$ is *the* anti-derivative of $f(x)$.

Unfortunately, we still don't know what anti-derivatives are good for, so there isn't anything for us to do with this knowledge. That's about to change.

As a hint, consider the way that position and velocity are related. The position $x(t)$ is an *anti-derivative* of velocity. The change in position from time a to time b, namely $x(b) - x(a)$, is the distance traveled, which is the *definite integral* of the velocity from time a to time b. This suggests that definite integrals and anti-derivatives have something to do with one another. FTC2 will tell us what.

8.2. FTC2: The Integral of the Derivative

The second Fundamental Theorem of Calculus relates anti-derivatives and definite integrals.

> **Theorem 8.1** (FTC2). *Suppose that $f(x)$ and $F(x)$ are functions defined on all of $[a, b]$ and that $F(x)$ is an anti-derivative of $f(x)$. Then $f(x)$ is integrable on $[a, b]$ and*
>
> (8.9) $$\int_a^b f(x)dx = F(b) - F(a).$$

In other words, computing the definite integral $\int_a^b f(x)dx$ boils down to finding an anti-derivative $F(x)$, evaluating it at the endpoints, and taking the difference. We often use the notation $F(x)\Big|_a^b$ to mean $F(b) - F(a)$. For instance, $x^3/3$ is an anti-derivative of x^2, so

(8.10) $$\int_0^1 x^2 dx = \frac{x^3}{3}\Big|_0^1 = \frac{1^3}{3} - \frac{0^3}{3} = \frac{1}{3}.$$

Similarly, $-e^{-x}$ is an anti-derivative of e^{-x} (as you can check by taking the derivative of $-e^{-x}$ with the help of the chain rule), so

(8.11) $$\int_0^3 e^{-x} dx = -e^{-x}\Big|_0^3 = -e^{-3} - (-e^0) = 1 - e^{-3}.$$

That's it! No messy sums, no worrying about left vs. right endpoints, and no limits. Those sums were *extremely* important for understanding why $\int_0^1 x^2\, dx$ is the volume of a pyramid and the moment of inertia of a rod, but when it comes to *computing* $\int_0^1 x^2\, dx$ (or $\int_0^3 e^{-x}\, dx$), using FTC2 is much, much easier.

This begs two questions:

(1) Why is FTC2 true? That's the content of the rest of this section.

(2) How do we go about finding anti-derivatives? We will start on that in Section 8.5, and follow it up more seriously in Sections 9.1 and 9.2.

The reason that FTC2 works can be summed up in the 5th Pillar of Calculus:

One step at a time.

If we want to know how much something changes in a calendar year, we can figure out how much it changes on January 1, how much it changes on January 2, all the way through December 31, and add up those little changes. Likewise, if we have points

(8.12) $$a = x_0 < x_1 < x_2 < \cdots < x_{N-1} < x_N = b$$

and we ask how much a function $F(x)$ changes between $x = a$ and $x = b$, then the whole change is the change from x_0 to x_1, plus the change from x_1 to x_2, all the way up to the change from x_{N-1} to x_N. That is,

$$\begin{aligned}\sum_{i=1}^{N} F(x_i) - F(x_{i-1}) &= -F(x_0) + F(x_1) \\ & \quad - F(x_1) + F(x_2) \\ & \quad \quad - F(x_2) + F(x_3) \\ & \quad \quad \quad \ddots \\ & \quad \quad \quad \quad - F(x_{N-1}) + F(x_N) \\ &= -F(x_0) + F(x_N) \\ &= F(b) - F(a),\end{aligned}$$

(8.13)

since all the terms in the middle cancel in pairs.

Next we look at the change $F(x_i) - F(x_{i-1})$ that occurs in each step of the sum (8.13). By the microscope equation, this is approximately $F'(x_{i-1})(x_i - x_{i-1}) = f(x_{i-1})\Delta_i x$. That is,

(8.14) $$F(b) - F(a) \approx \sum_{i=1}^{N} f(x_{i-1})\Delta_i x.$$

However, the right-hand side is exactly the Riemann sum we would get if we tried to estimate $\int_a^b f(x)dx$ using left endpoints! As $N \to \infty$, the approximation gets better and better, and the right-hand side gets closer and closer to that integral. In the limit, we have

(8.15) $$F(b) - F(a) = \int_a^b f(x)dx,$$

which is exactly FTC2.

For another perspective on FTC2, imagine that a rock moves upward with velocity $v(t) = 160 - 32t$ feet/second, where t is the time measured in seconds. That is, at time

8.2. FTC2: The Integral of the Derivative

$t = 0$ it is shot upward at 160 feet/second, and then it slows down thanks to gravity. How much does the rock rise between time $t = 1$ and $t = 4$?

On the one hand, the distance traveled is the integral of velocity, so our answer must be $\int_1^4 160 - 32t\, dt$. On the other hand, the distance traveled is the change in position, so our answer must be $z(4) - z(1)$, where $z(t)$ is the height at time t. Equating our two answers tells us that

$$(8.16) \qquad \int_1^4 v(t)dt = z(t)\Big|_1^4.$$

The derivative of $z(t)$ is $v(t)$, so $z(t)$ is an anti-derivative of $v(t)$ and this equation is *exactly* what FTC2 says.

We can go further with anti-derivatives. Since the derivative of $160t - 16t^2$ is $160 - 32t$, our position function must be

$$(8.17) \qquad z(t) = 160t - 16t^2 + C,$$

for some constant C. The problem doesn't give us enough information to figure out what C is, but that doesn't stop us, because the *change* in z doesn't depend on C.

$$\int_1^4 160 - 32t\, dt = \int_1^4 v(t)dt \;=\; z(t)\Big|_1^4$$
$$= 160t - 16t^2 + C\Big|_1^4$$
$$= 160t - 16t^2\Big|_1^4 + C\Big|_1^4$$
$$(8.18) \qquad\qquad\qquad\qquad\qquad\qquad\; = 160t - 16t^2\Big|_1^4.$$

To find the integral of $v(t)dt$ from 1 to 4, just take *any* anti-derivative of $v(t)$, plug in the values $t = 4$ and $t = 1$, and take the difference to get a final answer of 240 feet.

There are three ways of thinking about FTC2, and all three are useful.

(1) From the perspective of $f(x)$, FTC2 is a statement about *integrals*, and in particular it's a recipe for computing them. To compute $\int_a^b f(x)\,dx$, just find an anti-derivative of $f(x)$, plug in the beginning and ending values of x, and take the difference.

(2) From the perspective of $F(x)$, FTC2 is a statement about *changes*. The change in the quantity $F(x)$ is the integral of the derivative of $F(x)$. The change in position is the integral of velocity. The change in velocity is the integral of acceleration. The change in the national debt over a period of time is the integral of the national budget deficit over that period of time.[2] In fact, some texts treat this perspective as a theorem in its own right and call it the **net change theorem**.

(3) If we start with a function $F(x)$, take its derivative, and then integrate, we get something involving $F(x)$ again. (Specifically, we get $F(x)\Big|_a^b$.) So in some sense

---

[2] Or at least it would be if we used the same accounting procedures for both, as we saw in Section 4.1.

differentiation and integration are inverse operations. We will explore this further in Section 8.3.

8.3. FTC1: The Derivative of the Accumulation

Suppose that $f(x)$ is a function, that $A(x)$ is its accumulation function, and that $F(x)$ is an anti-derivative of $f(x)$. The second Fundamental Theorem of Calculus tells us that

$$(8.19) \qquad \int_a^b f(x)\,dx = F(x)\Big|_a^b.$$

However, we already know that

$$(8.20) \qquad \int_a^b f(x)\,dx = A(x)\Big|_a^b.$$

This suggests that $A(x)$ and $F(x)$ must have something to do with each other. The first Fundamental Theorem of Calculus tells us what.

> **Theorem 8.2** (FTC1). *If $f(x)$ is an integrable function and $A(x)$ is its accumulation function, then $A'(x) = f(x)$. In an equation,*
>
> $$(8.21) \qquad \frac{d}{dx}\int_0^x f(s)\,ds = f(x).$$

In other words, accumulation functions are anti-derivatives! If $F(x)$ is any other anti-derivative, then we must have $A(x) = F(x) + C$ for some constant C.

We have already seen this in a couple of examples. The accumulation function of $f(x) = 1$ is $A(x) = x$, whose derivative is 1. The accumulation function of $f(x) = x$ is $A(x) = x^2/2$, whose derivative is x. The accumulation function of $\sqrt{1-x^2}$ is complicated (specifically, we worked it out to be $\frac{1}{2}(\sin^{-1}(x) + x\sqrt{1-x^2})$), and if we take the derivative of *that*, we get $\sqrt{1-x^2}$.

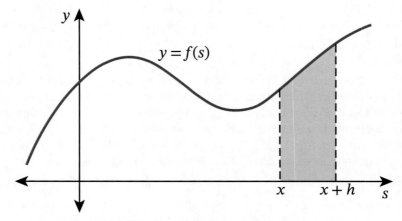

Figure 8.2. $A(x+h) - A(x)$ is the area of the striped region.

8.3. FTC1: The Derivative of the Accumulation

To see why this works for an arbitrary function f, think about the graph of $y = f(s)$, shown in Figure 8.2. (Note that we are using s as a dummy variable, as with all accumulation functions.) We have

$$A(x) = \text{area under the curve from 0 to } x,$$
$$A(x+h) = \text{area under the curve from 0 to } x+h,$$
(8.22) $$A(x+h) - A(x) = \text{area under the curve from } x \text{ to } x+h.$$

When h is small, the region under the curve $y = f(s)$ between $s = x$ and $s = x + h$ is the shaded strip of width h and approximate height $f(x)$, so its area is approximately $hf(x)$. That is, $A(x+h) - A(x) \approx hf(x)$. Dividing by h, we get that

(8.23) $$\frac{A(x+h) - A(x)}{h} \approx f(x).$$

By definition, $A'(x)$ is the limit of the left-hand side of (8.23) as $h \to 0$. But the smaller h is, the better our approximation becomes, and the closer the left-hand side gets to $f(x)$. In other words,

(8.24) $$\frac{d}{dx}\int_0^x f(s)\,ds = \frac{d}{dx}A(x) = f(x).$$

This completes the picture of how integration and differentiation can be viewed as inverse operations. If you start with $f(x)$, integrate to get its accumulation function $A(x)$, and then take the derivative of $A(x)$, you get exactly $f(x)$. If you start with $F(x)$, compute its derivative $f(x)$, and then integrate $f(x)$ to get its accumulation function $A(x)$, you wind up with $F(x)$ plus a constant. In words,

(1) The derivative of the accumulation function (integral) is the original function.

(2) The accumulation function (integral) of the derivative is the original function plus a constant.

The Graph of $A(x)$. Armed with FTC1, we can sketch accumulation functions. For instance, consider the function

(8.25) $$f(x) = \begin{cases} -1 & x < -1, \\ x & -1 \leq x \leq 1, \\ 1 & x > 1, \end{cases}$$

shown in Figure 8.3. At each point on the graph of $A(x)$, the slope of $A(x)$ is $f(x)$. This tells us that the graph of the accumulation function $A(x)$ must:

- Have slope -1 when $x < -1$. This part of the graph of $A(x)$ must be a straight line of slope -1.
- Have a slope that gradually increases from -1 to 1 as x goes from -1 to 1. This part of the graph must be concave up, and there must be a local minimum at $x = 0$, since that is where $f(x) = A'(x)$ goes from negative to positive.
- Have slope 1 when $x > 1$. This part of the graph of A is a straight line of slope 1.
- Be differentiable everywhere (since $A'(x) = f(x)$, which is defined everywhere) and, in particular, not have any jumps.
- Go through $(0,0)$, since $A(0) = \int_0^0 f(s)ds = 0$.

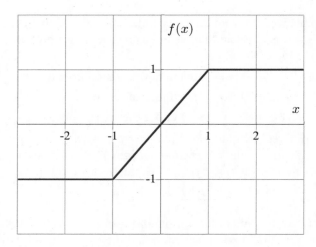

Figure 8.3. A function $f(x)$ defined in pieces

Based on these clues, we see that the graph of $A(x)$ has to more-or-less look like Figure 8.4.

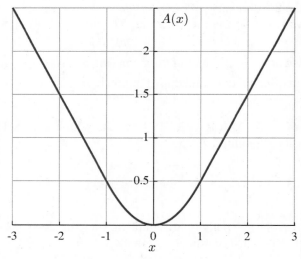

Figure 8.4. The accumulation function $A(x)$

Notation for Integrals and Anti-Derivatives. Integrals are so closely tied to anti-derivatives that we use the same notation for both. The expression

$$\int_a^b f(x)\,dx \tag{8.26}$$

8.3. FTC1: The Derivative of the Accumulation

denotes a definite integral, which as we have seen is the limit of a sum. The expression

$$\int f(x)\,dx, \tag{8.27}$$

without limits of integration, means the anti-derivative of $f(x)$. This notation makes FTC2 especially easy to write down:

$$\int_a^b f(x)\,dx = \left(\int f(x)\,dx\right)\Big|_a^b. \tag{8.28}$$

Not only is the notation for anti-derivatives and integrals the same, but so is the language. The process of finding anti-derivatives is usually called **integration**. In fact, most people call $\int f(x)\,dx$ an **indefinite integral**.[3] In this language, equation (8.28) says that a definite integral (the left-hand side) is the difference of the values of the indefinite integral at the endpoints. For both definite and indefinite integrals, we call the quantity inside the integral sign the **integrand**. Note that the integrand of $\int f(x)\,dx$ is $f(x)\,dx$, not just $f(x)$!

This unified terminology *completely* obscures the fact that integration is all about adding up little parts to make a whole! However, resisting it is hopeless. For the rest of this book, I will raise the white flag and write things like "the indefinite integral of $x^2\,dx$ is $x^3/3$ plus a constant".

Relating FTC1 and FTC2. We proved the two halves of the Fundamental Theorem of Calculus separately, but it's easy to get each half from the other. Suppose that $f(x)$ is a function, $F(x)$ is an anti-derivative, and $A(x)$ is the accumulation function.

- If FTC1 is true, then $A'(x) = f(x) = F'(x)$, so $A(x) = F(x) + C$ for some constant C. But then

$$\begin{aligned}\int_a^b f(x)\,dx &= A(b) - A(a) \\ &= F(b) + C - (F(a) + C) \\ &= F(b) - F(a),\end{aligned} \tag{8.29}$$

which is FTC2.

- If FTC2 is true, then

$$A(x) = \int_0^x f(s)\,ds = F(x) - F(0). \tag{8.30}$$

Since $F(0)$ is a constant, this means that $A'(x) = F'(x) = f(x)$, which is FTC1.

---

[3] A few curmudgeons, including the author, believe that the term "indefinite integral" should only be applied to accumulation functions, since an accumulation function is an integral whose upper limit is unspecified, hence indefinite. But it doesn't really matter, since FTC1 says that accumulation functions are anti-derivatives.

Evaluating Derivatives of More Complicated Integrals. A student's knowledge of FTC1 is often tested with computations like

$$(8.31) \qquad \frac{d}{dx}\int_0^x \cos(s^2)\,ds, \quad \frac{d}{dx}\int_x^7 e^s \tan^{-1}(s)\,ds, \quad \text{and} \quad \frac{d}{dx}\int_{\ln(x)}^{x^2} \frac{e^s\,ds}{s+1}.$$

FTC1 applies directly to the first problem, and there are tricks for applying it to the other two. However, all three of these problems can be solved easily with FTC2.

In the first problem, we are studying the derivative of $\int_0^x f(s)\,ds$, where $f(s) = \cos(s^2)$. But $\int_0^x f(s)\,ds = F(x) - F(0)$, so the derivative of this with respect to x is $F'(x) = f(x) = \cos(x^2)$. Notice that we never had to figure out what $F(x)$ actually was! All we needed was its derivative, which we already knew.

The second problem is similar. If we let $f(x) = e^x \tan^{-1}(x)$ and let $F(x)$ be an anti-derivative, then

$$\int_x^7 f(s)\,ds = F(7) - F(x),$$

$$\frac{d}{dx}\int_x^7 f(s)\,ds = \frac{d}{dx}F(7) - \frac{d}{dx}F(x)$$

$$= 0 - F'(x)$$

$$= -f(x)$$

$$(8.32) \qquad\qquad\qquad\qquad = -e^x \tan^{-1}(x).$$

The first line is FTC2. In the third line, the derivative of $F(7)$ is *not* $F'(7)$! $F(7)$ is a number that doesn't change as we vary x, so its derivative with respect to x is 0. Meanwhile, $F'(x) = f(x)$, since $F(x)$ is an anti-derivative of $f(x)$. Once again, we never needed to actually figure out what $F(x)$ was.

The third problem is the same story, plus the chain rule. Let $f(x) = e^x/(x+1)$, and let $F(x)$ be its anti-derivative, which we don't need to ever compute. Then

$$\int_{\ln(x)}^{x^2} f(s)\,ds = F(x^2) - F(\ln(x)),$$

$$\frac{d}{dx}\int_{\ln(x)}^{x^2} f(s)\,ds = \frac{d}{dx}F(x^2) - \frac{d}{dx}F(\ln(x))$$

$$= F'(x^2)\frac{d(x^2)}{dx} - F'(\ln(x))\frac{d(\ln(x))}{dx}$$

$$= 2xf(x^2) - \frac{1}{x}f(\ln(x))$$

$$= \frac{2xe^{x^2}}{1+x^2} - \frac{1}{x}\frac{e^{\ln(x)}}{1+\ln(x)}$$

$$(8.33) \qquad\qquad\qquad\qquad = \frac{2xe^{x^2}}{1+x^2} - \frac{1}{1+\ln(x)}.$$

The first four lines apply to any function $f(x)$. In the fifth line we entered the values of $f(x^2)$ and $f(\ln(x))$, and on the last line we used the fact that $e^{\ln(x)} = x$. We never needed to know what $F(x)$ or $F(x^2)$ or $F(\ln(x))$ were. We just needed their derivatives, and those came from the chain rule.

In general, whenever we're faced with a problem of the form $\frac{d}{dx}\int_{g(x)}^{h(x)} f(s)\,ds$, we can write

$$\int_{g(x)}^{h(x)} f(s)\,ds = F(h(x)) - F(g(x)),$$

$$\frac{d}{dx}\int_{g(x)}^{h(x)} f(s)\,ds = \frac{d}{dx}F(h(x)) - \frac{d}{dx}F(g(x))$$

$$= F'(h(x))h'(x) - F'(g(x))g'(x)$$

(8.34)
$$= f(h(x))h'(x) - f(g(x))g'(x).$$

Please do not attempt to memorize equation (8.34). Remember the *method* instead.

8.4. Anti-Derivatives and Ballistics

Anti-derivatives (a.k.a. indefinite integrals) are essential for using the Fundamental Theorem of Calculus to compute definite integrals. However, they are also very important in their own right. In particular, Newton's second law says that the acceleration of a particle is proportional to the force applied to it. If we know the force, we know the acceleration. To get from the acceleration to the velocity, we just need to integrate. To get from the velocity to the position, we just need to integrate a second time.

For instance, in Section 6.1 we considered the problem of a falling rock. At time $t = 1$ the rock was 41 feet above the ground and falling at 12 feet/second. What is the rock's elevation $z(t)$ as a function of time? When will it hit the ground?

Under gravity (and ignoring air resistance), all objects near the surface of the Earth accelerate downward at 32 feet/second^2. That is, if the velocity is $v(t) = z'(t)$ and the acceleration is $a(t) = v'(t) = z''(t)$, and if we measure z in feet, v in feet/second, and a in feet/second^2, then

(8.35)
$$a(t) = -32,$$

where the minus sign indicates that the acceleration is downward. Since $a(t) = v'(t)$, we must have $v(t) = \int a(t)dt = \int -32 dt$. We already know that an anti-derivative of 1 is t, so we must have

(8.36)
$$v(t) = -32t + c_1,$$

where c_1 is a constant.

Every time we compute an indefinite integral, there is an unknown **constant of integration** like c_1. To figure out what that constant is, we have to compare our formulas to actual data. In this case, we know that $v(1) = -12$, since the rock was falling

at 12 feet/second at time $t = 1$. This means that
$$v(1) = -12,$$
$$-32(1) + c_1 = -12,$$
$$c_1 = 20,$$
(8.37) $$v(t) = 20 - 32t.$$

Once we have $v(t)$, we can integrate again to get $z(t)$. An anti-derivative of t is $t^2/2$, so we must have

(8.38) $$z(t) = 20t - 16t^2 + c_2,$$

where once again we have a constant of integration c_2. We determine the constant by using the known value of $z(1) = 41$:
$$z(1) = 41,$$
$$20(1) - 16(1)^2 + c_2 = 41,$$
$$c_2 = 37,$$
(8.39) $$z(t) = 37 + 20t - 16t^2.$$

The position, velocity, and acceleration are shown in Figure 8.5.

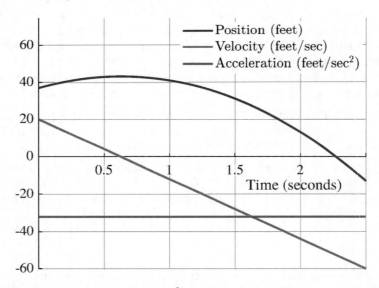

Figure 8.5. The position $z(t) = -16t^2 + 20t + 37$, velocity $v(t) = -32t + 20$, and acceleration $a(t) = -32$ of a falling rock. Note that the slope of $z(t)$ is always equal to $v(t)$, and the slope of $v(t)$ is always equal to $a(t)$.

Finally, setting $z(t) = 0$, we get the time when the rock hits the ground. By the quadratic formula, the solutions to $37 + 20t - 16t^2 = 0$ are approximately $t = 2.2691$ and $t = -1.0191$, so the rock will hit the ground at time $t = 2.2691$. (If the rock was thrown from the ground, then the other solution represents the time that it was thrown.

However, it's also possible that the rock was thrown or dropped from a different height at a different time.)

Even if we don't know the position and velocity at time $t = 1$, we can use integration to find the general form of a rock's trajectory:

(8.40)
$$\begin{aligned} a(t) &= -32, \\ v(t) &= -32t + c_1, \\ z(t) &= -16t^2 + c_1 t + c_2. \end{aligned}$$

The height of a projectile is always a quadratic function of time. The coefficients of 1 and t depend on the initial position and velocity, but the coefficient of t^2 is always -16. Similarly, in any situation where a particle has constant acceleration a, its position will be of the form $\frac{1}{2}at^2 + c_1 t + c_2$, while its velocity will be $at + c_1$. That formula applies equally well to falling rocks, to drag racing (where cars undergo more-or-less constant acceleration for the first few seconds), and to slamming on the brakes (where the braking force is more-or-less constant).

In the last problem we considered height as a function of time. We can use the same sort of reasoning to figure out the arc of a thrown object, such as a baseball thrown from center field to home plate, or a cannon ball fired at a castle.

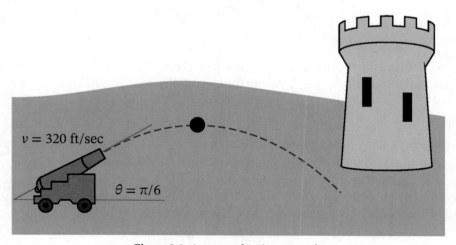

Figure 8.6. A cannon shooting at a castle

Suppose that a cannon is pointed at an angle of $\pi/6$ (30 degrees) above the horizontal, and it fires a cannon ball at an initial speed of 320 feet/second, as in Figure 8.6.[4] How far does the cannon ball fly before it hits the ground?

---

[4]Technically, an artillery piece that shoots at a high angle is called a *mortar* rather than a cannon. A velocity of 320 feet/second is realistic for 17th and 18th century mortars. Modern artillery pieces are much more powerful.

This is actually two problems rolled into one, since we have to separately keep track of the height $z(t)$ of the cannon ball and its horizontal distance $x(t)$ downrange. Our strategy is as follows:

(1) Figure out the initial height $z(0)$ and the initial vertical velocity $v_z(t)$.
(2) Use that to compute the height $z(t)$ as a function of time.
(3) Compute the time t_f at which the cannon ball hits the ground.
(4) Compute the horizontal position $x(t)$ as a function of time.
(5) Plug t_f into the formula for $x(t)$.

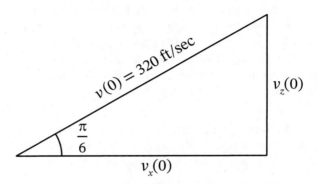

Figure 8.7. The vertical and horizontal components of the initial velocity

Since the cannon is fired at an angle of $\pi/6$, the initial horizontal velocity $v_x(0)$, the initial vertical velocity $v_z(0)$, and the speed form a triangle (as shown in Figure 8.7), we then have

(8.41) $\qquad v_z(0) = 320\sin(\pi/6) = 160, \qquad v_x(0) = 320\cos(\pi/6) \approx 277,$

where both v_x and v_z are measured in feet/second.

Since $a_z(t) = -32$, we must have $v_z(t) = c_1 - 32t$ for some constant c_1. Comparing to our initial velocity $v_z(0) = 160$, we see that $c_1 = 160$, so $v_z(t) = 160 - 32t$. $z(t)$ is the anti-derivative of $v_z(t)$, so $z(t) = c_2 + 160t - 16t^2$. Since the cannon ball essentially started at the ground, we have $z(0) = 0$, so we must have $c_2 = 0$. In other words,

(8.42) $\qquad\qquad\qquad z(t) = 160t - 16t^2.$

To figure out when the cannon ball hits the ground, we just solve $z(t) = 0$. The two solutions are $t = 0$ and $t = 10$. The first is when the cannon is fired. The second is when the cannon ball hits the ground. In other words, the cannon ball hits the ground 10 seconds after it is fired.

Next we consider the horizontal motion. Since there is no horizontal force on the cannon ball, the horizontal acceleration is 0, so the horizontal velocity is constant. We already found that $v_x(0) = 277$, so $v_x(t) = 277$. We then integrate $v_x(t)\,dt$ to get $x(t) = c_3 + 277t$. Since $x(0) = 0$, we have $c_3 = 0$, so

(8.43) $\qquad\qquad\qquad x(t) = 277t.$

8.5. Computing Anti-Derivatives

Finally, we plug in $t_f = 10$ to get $x(10) = 2770$. The cannon ball will land at a spot about 2770 feet downrange, or a little over half a mile away.

8.5. Computing Anti-Derivatives

We already know the derivatives of a lot of functions. To create a table of indefinite integrals, we just take a table of derivatives, swap the columns, and write $+C$ a few times. See Tables 8.1 and 8.2.

Table 8.1. A table of derivatives

Function	Derivative
x^n	nx^{n-1}
e^x	e^x
$\ln(x)$	$1/x$
$\sin(x)$	$\cos(x)$
$\cos(x)$	$-\sin(x)$
$\tan(x)$	$\sec^2(x)$
etc.	etc.

Table 8.2. A table of integrals

Function	Integral
nx^{n-1}	$x^n + C$
e^x	$e^x + C$
$1/x$	$\ln(x) + C$
$\cos(x)$	$\sin(x) + C$
$-\sin(x)$	$\cos(x) + C$
$\sec^2(x)$	$\tan(x) + C$
etc.	etc.

Table 8.2 is correct, but not as useful as it could be. We usually aren't interested in the integral of $5x^4$. We'd rather just know the integral of x^4, and we'd rather describe that as $n = 4$ than as $n = 5$. Likewise, we'd rather know the integral of $\sin(x)$ than of $-\sin(x)$. By tweaking this table and adding a few more rows, we get a list of common integrals, shown in Table 8.3.

Table 8.3. A better table of integrals

Function	Integral	Comment
x^n	$x^{n+1}/(n+1) + C$	for $n \neq -1$
$1/x$	$\ln(x) + C$	which is the $n = -1$ case
e^x	$e^x + C$	world's easiest integral
e^{kx}	$e^{kx}/k + C$	
$\sin(x)$	$-\cos(x) + C$	note the minus sign
$\cos(x)$	$\sin(x) + C$	
$\sec^2(x)$	$\tan(x) + C$	
$\sec(x)\tan(x)$	$\sec(x)$	
$f'(g(x))g'(x)$	$f(g(x)) + C$	substitution
$f(x)g'(x) + f'(x)g(x)$	$f(x)g(x) + C$	integration by parts

Those aren't all the functions we will need, but they're a good start. The last two lines are just the chain rule and the product rule. The two most powerful techniques for finding additional anti-derivatives, called **integration by substitution** and **integration by parts**, are based on these two rules. We will learn these techniques, and use them to expand our table further, in Chapter 9.

8.6. Chapter Summary

The Main Ideas.

- Definite integrals, accumulation functions, and anti-derivatives are different things:
  - A **definite integral** is the limit of a sum:

  (8.44)
  $$\int_a^b f(x)\,dx = \lim_{N \to \infty} \sum_{i=1}^N f(x_i^*)\Delta_i x.$$

  It is a number, not a function.
  - The **accumulation function** $A(x) = \int_0^x f(s)\,ds$ is a running total, like an odometer or a bar tab. It is a function, not a number.
  - An **anti-derivative** $F(x)$ is a function with $F'(x) = f(x)$. Anti-derivatives are unique, up to adding a constant.

8.6. Chapter Summary

- Accumulation functions and definite integrals are related by

(8.45) $$\int_a^b f(x)\,dx = A(b) - A(a).$$

- Anti-derivatives and definite integrals are related by FTC2:

(8.46) $$\int_a^b f(x)\,dx = F(b) - F(a).$$

Depending on your perspective, this is either a great way to compute the definite integral $\int_a^b f(x)\,dx$ or a great way to compute the change in the function $F(x)$.

- FTC1 says that the accumulation function is an anti-derivative:

(8.47) $$\frac{d}{dx}\int_0^x f(s)\,ds = f(x).$$

- Together, FTC1 and FTC2 say that differentiation (that is, computing rates of change) and integration (getting whole quantities by adding up the pieces) are, in some sense, inverse operations.
- Because of this, the process of finding anti-derivatives is usually called **integration**. Anti-derivatives are often called **indefinite integrals** and are denoted $\int f(x)\,dx$.
- Anti-derivatives are important in their own right. If you know the acceleration of an object, you can get its velocity by integration. If you know the velocity, you can get the position by integration.
- Every time you compute an indefinite integral, there is an arbitrary additive constant that must be included in the answer. The value of this constant can be determined from the initial conditions.
- Every fact about derivatives is also a fact about anti-derivatives. From this, we already know the anti-derivatives of x^n, e^x, $\cos(x)$, $\sin(x)$, and $\sec^2(x)$.

Expectations. You should be able to:

- Explain the difference between definite integrals, accumulation functions, and anti-derivatives.
- Explain the key ideas behind FTC1 and FTC2.
- Use your knowledge of derivatives to sketch an accumulation function.
- Use correct notation for definite and indefinite integrals. This involves the uses of the integral sign $\int$, the limits of integration, using correct variable names, and including dx (or dt or ds, etc.) inside the integral.
- Find the anti-derivatives of functions such as $x^3 - 3x + \frac{2}{x} + \cos(x) - 5e^x$.
- Use FTC2 to compute definite integrals of such functions.
- Use FTC2 and the chain rule to compute quantities such as $\frac{d}{dx}\int_{x^2}^{x^3}\sin(s^2)\,ds$.

8.7. Exercises

8.1. Clearly and concisely explain the difference between an anti-derivative, an accumulation function, and a definite integral.

Present Value and Mortgages. In the Chapter 7 exercises we expressed the present value of an income stream as an integral. If the interest rate is r and payments are made at a constant rate of A dollars/year for T years, then the present value after T years can be expressed as an integral

$$(8.48) \qquad PV = \int_0^T Ae^{-rt}\, dt.$$

In Exercises 8.2 and 8.3, we evaluate that integral with the help of the Fundamental Theorem of Calculus.

8.2. First we need to compute an anti-derivative.
   (a) What is the derivative of $f(t) = e^{-.02t}$?
   (b) Find a function whose derivative is $f(t)$.
   (c) Using FTC2, find the accumulation function $\int_0^t f(s)\,ds$. Although FTC1 says that this is an anti-derivative of $f(t)$, it probably isn't the same as what you found in part (b).

8.3. Using what you learned in Exercise 8.2, find the accumulation function of Ae^{-rt}, where r is an arbitrary constant.

The answer to Exercise 8.3 is the formula that banks use to set the terms of a mortgage. If we know three of the four quantities PV, A, r, and T, we can compute the fourth. In practical terms, PV is the size of the mortgage, A is the yearly payment (usually broken up into 12 monthly payments), T is the length of the mortgage (usually 20 or 30 years), and r is the interest rate.

8.4. Rina is buying a house. If she takes out a 30-year $200,000 mortgage at 3% interest, how much will her monthly payments be?

8.5. Rina decides that she can only afford to spend $1500/month on mortgage payments. If she can find a bank that offers 30-year mortgages at 2.5% interest, how big a mortgage can she afford?

8.6. In the end, Rina finds a house that she can buy with a $150,000 mortgage. If the interest rate is 2.5% and she makes payments of $1500/month, how long will it take her to pay off the mortgage?

Accumulation Functions and Their Derivatives.

8.7. Let $f(x) = 1 + 2x$.
   (a) Sketch the graph of $f(x)$ on the interval $-2 \leq x \leq 4$.
   (b) Using the methods we learned in Chapter 7, find a formula for the accumulation function

$$A(x) = \int_0^x f(s)\, ds.$$

   Sketch the graph of $A(x)$ on the interval $-2 \leq x \leq 4$.

8.7. Exercises

(c) Verify that $A'(x) = f(x)$. This implies that, at any point x, the *slope* of the graph of $A(x)$ is the same as the *height* of the graph of $f(x)$.

8.8. Consider the accumulation function

$$A(x) = \int_0^x s^4 \, ds.$$

(a) Using the fact that $A'(x) = x^4$, write a formula that expresses A in terms of x.

(b) Now consider a modified accumulation function $\tilde{A}$ that begins at $x = 2$ instead of at $x = 0$. That is,

$$\tilde{A}(x) = \int_2^x s^4 \, ds.$$

It is still true that $\tilde{A}'(x) = x^4$, but now we have $\tilde{A}(2) = 0$ instead of $A(0) = 0$. Write an explicit formula for $\tilde{A}(x)$ in terms of x. How do the formulas for $\tilde{A}(x)$ and $A(x)$ differ?

8.9. Consider the accumulation function $A(x) = \int_0^x g(t) \, dt$, where $g(t)$ is the function whose graph is given below.

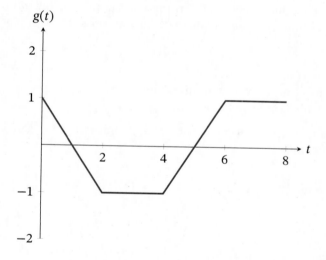

(a) Evaluate $A(0)$, $A(1)$, $A(4)$, and $A(8)$.
(b) On what interval(s) is $A(x)$ increasing?
(c) On what interval(s) is $A(x)$ decreasing?
(d) On what interval(s) is the graph of $A(x)$ concave up?
(e) On what interval(s) is the graph of $A(x)$ concave down?
(f) Where does $A(x)$ have a local maximum? Where does $A(x)$ have a local minimum?
(g) Sketch a graph of $A(x)$.

8.10. In each part, use the Fundamental Theorem of Calculus to find $A'(x)$.

(a) $A(x) = \int_1^x \sin^3(s)\,ds$

(b) $A(x) = \int_x^1 \sin^3(s)\,ds$

(c) $A(x) = \int_e^x \sqrt{t}\ln(t)\,dt$

(d) $A(x) = \int_e^{x^3} \sqrt{t}\ln(t)\,dt$

(e) $A(x) = \int_{x^2}^{x^4} e^s \sin^3(s)\,ds$

8.11. Find all critical points of the function $A(x) = \int_1^x s^3 - 16s^5\,ds$.

8.12. Find an anti-derivative $F(t)$ of $f(t) = t + \cos(t)$ for which $F(0) = 3$.

8.13. Find a formula for the solution of each initial value problem (IVP).
(a) $y' = 2x^3 - 5x^4$, $\quad y(1) = 3$.
(b) $y' = \sin(5x)$, $\quad y(\pi/2) = -2$.

8.14. Express the solution to each IVP using an accumulation function. That is, each solution should include an integral with a variable upper limit of integration.
(a) $y' = \sin(x^4)$, $\quad y(0) = 3$.
(b) $y' = e^{x^4}$, $\quad y(1) = 0$.
(c) $y' = f(x)$, $\quad f(x_0) = y_0$.
Leave your answer in terms of the (unknown) function $f(x)$ and the unspecified constants x_0 and y_0.

8.15. Determine the exact value of each definite integral. See Exercise 8.16 for the anti-derivative of $\cos(4x)$.

(a) $\int_0^2 1 + 2x + 6x^2\,dx$

(b) $\int_2^3 \dfrac{4}{x^2}\,dx$

(c) $\int_0^{3\pi} \sin(x) + 3\,dx$

(d) $\int_0^{\pi/8} \cos(4x)\,dx$

(e) $\int_0^2 4^t\,dt$

(f) $\int_{\ln(3)}^5 1 + e^t\,dt$

(g) $\int_{-2}^2 t^4\,dt$

(h) $\int_{-2}^2 t^5\,dt$

8.16. Evaluate each indefinite integral in terms of the given parameters. Parts (b), (c), and (d) can all be done using integration by substitution, which we'll cover in Section 9.1, but that's overkill. You should be able to just guess an antiderivative, and then adjust your guess if it isn't quite right. *Guess and check* is an important method of integration!

(a) $\int rx^3\, dx$

(b) $\int 2^{kt}\, dt$

(c) $\int \cos(\alpha x)\, dx$

(d) $\int 3 + \sec^2(Cz)\, dz$

(e) $\int \alpha x^n\, dx, \quad n \neq -1$

(f) $\int \lambda x^4 - x^5\, dx$

8.17. Suppose that bacteria are growing in a Petri dish at rate $b(t)$, where t is measured in hours and b is measured in colonies/hour. The sample begins with $B(0) = B_0$ colonies.
(a) Provide a clear and concise interpretation for the quantity $\int_1^2 b(t)\, dt$.
(b) Provide a clear and concise interpretation for the quantity $B_0 + \int_0^T r(t)\, dt$.

8.18. A medical engineering company has designed a high resolution ultrasound device. Records indicate that 10,000 staff-hours were used for the design and assembly of the first eight units. As production continued, the production team found that the function

$$L'(x) = 1200x^{-1/3}$$

worked well to describe the staff-hours per item needed to produce additional devices. That is, after the production of eight units of this device, it takes approximately $1200(8)^{-1/3} = 600$ staff-hours to produce the next device.
(a) How many staff-hours $L(x)$ are required for the design and assembly of a total of x units? (*Note*: $L(0)$ is not 0, since it takes many staff-hours of design work to even begin assembly.)
(b) How many staff-hours are required for the design and assembly of the first 27 units?

Comparing Depreciation Methods. When accounting for the cost of a piece of equipment, like a tractor or a company car, it makes sense to think of the value of the equipment being gradually used up during its useful lifetime L. That way, a certain fraction of the cost of buying the equipment can be assigned to each year of use (and counted as a business expense for tax purposes). There are a number of ways to account for this depreciation (or *to depreciate* the equipment). Exercises 8.19–8.21 explore and compare three methods allowed by the IRS.

8.19. The simplest type of depreciation is called **straight-line depreciation**. As the name implies, straight-line depreciation assumes that the piece of equipment loses the same amount of value each year until it is worthless.
   (a) If the equipment has a useful life of L years, what fraction of its original value I_0 does it lose each year?
   (b) What is the rate of change in the value of the equipment?
   (c) Write down an IVP for the depreciated value $V_{SL}(t)$ of the equipment at time t. Be sure to include the time interval for which the IVP is valid.
   (d) Solve the IVP in part (c) with $L = 5$ years and $I_0 = 30{,}000$ to find a formula for $V_{SL}(t)$.
   (e) Using the values given in part (d), what is $V_{SL}(4)$?

8.20. Another type of depreciation is called **remaining years depreciation**. In this method, the rate of change in the depreciated value $V_{RY}(t)$ of a piece of equipment after t years is proportional to the amount of useful life that the equipment has left. This is different from straight-line depreciation because the loss of value in the piece of equipment decreases from year to year. For example, a car loses more value in its first year of use (when it has a large remaining lifetime) than in its last year of use.
   (a) Using L for the expected life of the equipment, I_0 for its initial value, and k for the proportionality constant, write down an IVP for $V_{RY}(t)$.
   (b) If $L = 5$ and $I_0 = \$30{,}000$, what is the solution to the IVP of part (a)?
   (c) Your answer in part (b) still has the parameter k. Using the fact that $V_{RY}(5) = 0$ (why?), find k.
   (d) What is $R_{RY}(4)$?

8.21. A third type of depreciation is called **double-declining balance depreciation**. In this type of depreciation, rate of loss in the depreciated value $V_{DD}(t)$ of the piece of equipment is proportional to the current value of the equipment, with a proportionality constant r. The constant r is chosen so that the equipment depreciates by a fraction $2/L$ of its purchase price I_0 in the first year. (This is twice what it would lose in its first year with straight-line depreciation.)
   (a) Write down an IVP for the depreciated value $V_{DD}(t)$ of the piece of equipment at time t, using $L = 5$ and $I_0 = \$30{,}000$.
   (b) Solve the IVP in part (a). Determine the value for r by noting that $V_{DD}(1) = (3/5)I_0$.
   (c) Show that your solution in part (b) can be rewritten as $V_{DD}(t) = 30{,}000\left(1 - \frac{2}{5}\right)^t$. Then hypothesize a generalized formula for $V_{DD}(t)$ in terms of the parameters I_0 and L.
   (d) What is $V_{DD}(4)$?

8.22. Use MATLAB to create a single graph that includes plots of $V_{SL}(t)$, $V_{RY}(t)$, and $V_{DD}(t)$. Your figure should include a legend that specifies which curve corresponds to each method.

Describing Motion. Exercises 8.23 and 8.24 explore the relationship between velocity, position, and distance traveled.

8.23. The following graphs represent the velocity of a particle moving on the x-axis. The particle starts at $x = 2$ when $t = 0$. Time is measured in seconds, velocity is measured in feet/second, and distance is measured in feet.

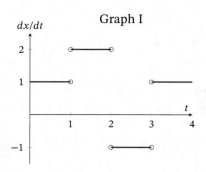
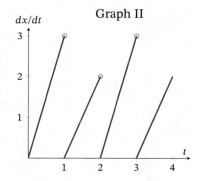

(a) For each graph, find where the particle is at the end of the trip four seconds later.

(b) For each graph, find the total (as opposed to net) distance traveled.

Food for Thought: Real-world objects can't change velocity instantaneously, as that would require infinite acceleration and infinite force. However, a baseball bat can apply a huge force to a ball, changing the ball's velocity *very* quickly! Physicists refer to such sudden intense forces as **impulses** and treat the resulting changes in velocity as if they were instantaneous. In other words, the graphs in Exercise 8.23 can't be taken literally, but they are exactly how a physicist would approximate a situation with impulsive forces.

8.24. The following graph represents the velocity of a particle moving on the x-axis. The particle starts at $x = 2$ when $t = 0$. Time is measured in seconds, and distance is measured in meters.

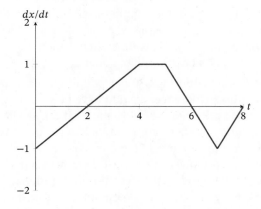

(a) Find where the particle is at the end of the trip eight seconds later.

(b) Find the total (as opposed to net) distance traveled.

8.25. **Baby Boom.** After World War II ended in 1945, the birth rate in the United States increased dramatically. Suppose the birth rate $b(t)$ (in millions of babies per year) for the first 20 years after the end of the war was

$$b(t) = 4 + 0.025t,$$

where t is the number of years since the end of the war.
(a) In all, how many babies were born in the 20 years after the end of World War II?
(b) What was the average number of babies born each year during these 20 years?
(c) Find an accumulation function $B(t)$ that gives a running total number of babies born since the end of the war.
(d) Starting at the end World War II, how long did it take for 20 million babies to be born?

Producer Surplus. In Exercise 7.35, we considered the **consumer surplus** of a product, which is the total difference between the value that consumers place on the items they buy (i.e., what they would have been willing to pay for them) and the price they actually pay for them. In Exercises 8.26–8.28, we look at commerce from the producer's perspective and compute the **producer surplus**. We then combine the supply and demand perspectives in Exercises 8.29–8.31.

Imagine a camping cooperative stocking up on locally sourced beef jerky, which it sells to its members at cost. The co-op didn't find any suppliers willing to sell jerky for $10 per pound or less. However, they found that for each $4 increase in price per pound (starting at $10), they could find suppliers willing to provide an additional 100 pounds of jerky. This can be expressed as $x(p) = \frac{100}{4}(p - 10)$, where x represents the supply of beef jerky at a price p. This function is called the supply **function**. Its inverse, $p(x) = 10 + 0.04x$, is called the supply **curve**.

Suppose that the co-op is offering $30 per pound for jerky and manages to buy $x = 500$ pounds at that price. Producers would have been willing to supply much of that jerky for less than $30 per pound. For instance, we can see from the supply function that 400 pounds would have been supplied for $26 per pound or less and that 300 pounds would have been supplied for $22 per pound or less. To the jerky manufacturers that would have accepted a lower price, the jerky is worth less than $30 per pound. The difference between what producers receive for their products and what the products are worth to them (i.e., the difference between the price they were paid and the minimum price they would have accepted) is called the **producer surplus**. This is illustrated in the figure below, showing the supply curve $S(x) = 10 + 0.04x$ and the offering price $p = \$30$.

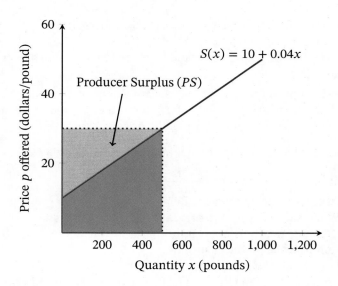

8.26. (a) What does the area of the blue region in the figure represent?
 (b) Explain why the producer surplus is the area of the gray region between the supply curve and the horizontal line $p = 30$.
 (c) Compute the producer surplus in this example.
 (d) If the co-op's offering price went up, would that result in a bigger or smaller producer surplus? Explain.

8.27. In general, why are supply curves always increasing functions of x?

8.28. Suppose that the supply curve for a particular product is $p = S(x) = 2 + \sqrt{x}$, where x runs from 0 to 100. Suppose that consumers offer a price just high enough to obtain $x = 64$ units of the product.
 (a) What is the price of the product?
 (b) Graph the supply curve and shade the region that corresponds to the producer surplus.
 (c) Compute the producer surplus.

Equilibrium Supply and Demand. When supply equals demand, the prices determined by the supply curve and the demand curve must be the same. The point where the demand curve and supply curve intersect is called the **equilibrium point**, which we denote by (x^*, p^*). That is, p^* is the equilibrium price and x^* is the quantity of goods sold at that price. We explore the balance between supply and demand in Exercises 8.29–8.31.

8.29. Suppose that, for our camping co-op in Exercises 8.26–8.28, the demand function is $x = 1000 - \frac{50}{3}p$.
 (a) Compute the demand *curve* $p = D(x)$. That is, find p as a function of x.
 (b) Using the supply curve $p = S(x) = 10 + 0.04x$, find the equilibrium point.
 (c) Use MATLAB to graph the supply curve $S(x)$ and the demand curve $D(x)$. Mark the equilibrium point.

(d) On the graph that you created in part (c), shade and label the region whose area is the consumer surplus.

(e) On the same graph, shade and label the region whose area is the supplier surplus.

8.30. In the situation of Exercise 8.29, what would happen if the co-op set a price that was higher than p^*? What would happen if they set the price lower than p^*? What incentives do they have to set the price exactly at p^*?

8.31. Suppose that a particular product has the demand curve $D(x) = \left(8 - \frac{x}{1000}\right)^2$ and the supply curve $S(x) = \frac{x^2}{1{,}000{,}000} + 16$.

(a) At equilibrium, how many items are sold?

(b) What is the equilibrium price?

(c) Compute the consumer surplus.

(d) Compute the producer surplus.

(e) Use MATLAB to graph $D(x)$ and $S(x)$.

(f) Print out the graph that you made in part (e). Label and shade the regions whose areas represent the consumer surplus and producer surplus.

8.32. **Ballistics.** A projectile is fired straight up from a 60-foot platform at an initial velocity of 28 feet/second. Ignoring air resistance,

(a) Find a formula for $z(t)$, the height (in feet) of the projectile t seconds after launch.

(b) What is the maximum height the projectile attains?

(c) What is the velocity of the projectile when it reaches the ground?

8.33. Imagine that you are an NFL kicker. With 5 seconds left in the Super Bowl, your team is behind by 2 points. You line up to attempt a 60-yard field goal. The center snaps the ball, and you give the ball a mighty kick!

(a) Once kicked, the ball has an initial velocity of 80 feet/second and an initial angle of 30 degrees, and it is aimed straight toward the middle of the goal posts. Ignoring air resistance, do you win? (*Note*: The crossbar that the ball must clear is 10 feet high, and a successful kick is worth 3 points.)

(b) Wait! The opposing team called a time out just before the snap, so your previous attempt doesn't count! You redo the kick, but this time you kick the ball (with perfect aim) at an angle of 40 degrees, with an initial velocity of 80 feet/second. Again ignoring air resistance, do you win?

Chapter 9

Methods of Integration

In this chapter we complete our survey of one-dimensional integral calculus. In Chapter 7 we learned about representing bulk quantities as definite integrals. In Chapter 8 we learned how anti-derivatives, also known as indefinite integrals, can help us compute definite integrals. In this chapter we'll go over some of the main techniques for computing both definite and indefinite integrals. The two most powerful techniques for finding indefinite integrals are called **integration by substitution** and **integration by parts**. These are covered in Sections 9.1 and 9.2. If we can find an anti-derivative, then the Fundamental Theorem of Calculus makes computing a definite integral easy.

But what if we can't find an anti-derivative? There are some functions, like e^{-x^2}, whose anti-derivatives cannot be expressed in terms of elementary functions.[1] In other situations, we only have numerical data for our original function. In those cases, we can't compute definite integrals using FTC2. We have to compute them by directly estimating the area under the curve or otherwise summing the different contributions.

There are three important numerical methods for estimating definite integrals, all of which are covered in Section 9.3. We have already seen two of them, the **trapezoidal rule** and the **midpoint rule**. While the naive left-endpoint and right-endpoint rules have errors that typically scale[2] as $1/N$, the trapezoidal and midpoint rules have errors that scale as $1/N^2$. By taking a weighted average of the trapezoidal and midpoint rules, we get an estimate whose error scales as $1/N^4$. This is called **Simpson's rule**.

---

[1] Some people even say that "we can't integrate $e^{-x^2} dx$." Of course that isn't true, since there *is* a function, namely the accumulation function, whose derivative is e^{-x^2}. We just can't find a *formula* for it!

[2] When we say that an error "scales as" $1/N$ or $1/N^2$ or $1/N^4$, we mean that it is roughly proportional to $1/N$ or $1/N^2$ or $1/N^4$. If we make N ten times larger, then our error will become roughly 10 or 100 or 10,000 times smaller.

9.1. Integration by Substitution

How can we compute

(9.1) $$\int_1^3 2x(x^2-1)^3 \, dx \, ?$$

We could multiply out $(x^2-1)^3$, multiply by $2x$, and then integrate the resulting polynomial term by term, but that's a mess. A better way involves the **chain rule**.

If u is any function of x, and if $F'(x) = f(x)$, then

(9.2) $$\frac{d}{dx}F(u) = F'(u)\frac{du}{dx} = f(u)\frac{du}{dx}.$$

In other words, $F(u)$ is an anti-derivative of $f(u)\frac{du}{dx}$, so we can write the following.

(9.3) $$\int f(u)\frac{du}{dx}\,dx = F(u) + C = \int f(u)\,du.$$

There's more to this equation than meets the eye! The first indefinite integral means "a function whose derivative with respect to x is $f(u)\frac{du}{dx}$", and the second indefinite integral means "a function whose derivative with respect to u is $f(u)$". Those are very different ideas! However, the equation says that it's legal to replace $\frac{du}{dx}dx$ with du, as if $\frac{du}{dx}$ was actually the ratio of quantities du and dx. So from now on, we will simply write

(9.4) $$du = \frac{du}{dx}dx.$$

In our example, we want to find an anti-derivative of $2x(x^2-1)^3$. If we let $u = x^2 - 1$, then $\frac{du}{dx} = 2x$, and we have

(9.5) $$\int 2x(x^2-1)^3 \, dx = \int u^3 \frac{du}{dx}\,dx = \int u^3 \, du = \frac{u^4}{4} + C = \frac{(x^2-1)^4}{4} + C.$$

Let's check:

(9.6) $$\frac{d}{dx}\frac{(x^2-1)^4}{4} = (x^2-1)^3 \frac{d(x^2-1)}{dx} = 2x(x^2-1)^3.$$

Our definite integral is then

(9.7) $$\int_1^3 2x(x^2-1)^3 \, dx = \frac{(x^2-1)^4}{4}\Big|_1^3 = \frac{8^4 - 0^4}{4} = 1024.$$

Since we substituted the variable u for some expression involving x, this process is called **integration by substitution**. It is also often called u-**substitution**, even if we don't actually call the second variable u! Here are several more examples of how this method works.

- To compute $\int e^{kx} \, dx$, we set $u = kx$, so $du = k \, dx$ and $dx = du/k$. Our integral is then $\int e^u \frac{du}{k} = \frac{e^u}{k} + C = \frac{e^{kx}}{k} + C$. When we take a derivative of e^{kx}, we multiply by k. When we integrate $e^{kx}dx$, we divide by k.

9.1. Integration by Substitution

- The derivative of $\ln(x)$ is $1/x$, so $\int \frac{dx}{x} = \ln(x) + C$. But that only makes sense when $x > 0$, since we can't take the log of a negative number. What is $\int_{-3}^{-1} \frac{dx}{x}$?
  When $x < 0$, we define $u = -x$, which is positive. Then $du = -dx$ and $\frac{dx}{x} = \frac{du}{u}$, so

(9.8) $$\int \frac{dx}{x} = \int \frac{du}{u} = \ln(u) + C = \ln(-x) + C.$$

In other words, $\int \frac{dx}{x}$ equals $\ln(x) + C$ when $x > 0$ and equals $\ln(-x) + C$ when $x < 0$. That is, $\int \frac{dx}{x} = \ln(|x|) + C$, no matter whether x is positive or negative. In particular, $\int_{-3}^{-1} \frac{dx}{x} = \ln(|x|)\Big|_{-3}^{-1} = \ln(1) - \ln(3) = -\ln(3)$.

- To compute $\int \tan(x) dx = \int \frac{\sin(x)}{\cos(x)} dx$, we let $u = \cos(x)$. Then $du = -\sin(x) dx$, and we have

$$\int \tan(x)\, dx \;=\; \int \frac{\sin(x)}{\cos(x)} dx = \int \frac{-du}{u} = -\ln(|u|) + C$$

(9.9) $$= -\ln(|\cos(x)|) + C = \ln(|\sec(x)|) + C.$$

In the last step we used the fact that $|\sec(x)| = |\cos(x)|^{-1}$, so $\ln(|\sec(x)|) = -\ln(|\cos(x)|)$. Another way to get the same result is to let $u = \sec(x)$, so $du = \sec(x)\tan(x) dx$, and

$$\int \tan(x)\, dx \;=\; \int \frac{\sec(x)\tan(x)}{\sec(x)} dx = \int \frac{du}{u}$$

(9.10) $$= \ln(|u|) + C = \ln(|\sec(x)|) + C.$$

Similarly, we can compute $\int \cot(x)\, dx$ by doing a substitution with $u = \sin(x)$ or by doing a substitution with $u = \csc(x)$. Either way we get

(9.11) $$\int \cot(x)\, dx = \ln(|\sin(x)|) + C = -\ln(|\csc(x)|) + C.$$

As the last example showed, we can sometimes get the answer in more than one way. *Integration by substitution is a tool, not a recipe.* Likewise, sometimes we can't get the answer by substitution at all. When faced with the integral

(9.12) $$\int (x^2 - 1)^3\, dx,$$

it's tempting to try $u = x^2 - 1$, but that doesn't help at all. The integrand contains u^3, but it isn't $u^3\, du$. It's $u^3\, dx$, which is something else entirely. That's why it's so important to keep track of the dx's and dt's and du's in our integrals. They really matter!

When a method doesn't work, we just have to give up and try something else. The only way to do this integral is to multiply out the expression $(x^2 - 1)^3\, dx = (x^6 - 3x^4 + 3x^2 - 1)\, dx$ and integrate each term separately to get $\frac{x^7}{7} - \frac{3x^5}{5} + x^3 - x + C$.

With the help of integration by substitution, we can expand our table of known integrals to obtain Table 9.1.

Table 9.1. A revised and expanded table of integrals

Integrand	Integral	Comment
$x^n \, dx$	$x^{n+1}/(n+1) + C$	for $n \neq -1$
dx/x	$\ln(|x|) + C$	for both $x > 0$ and $x < 0$
$e^x \, dx$	$e^x + C$	world's easiest integral
$e^{kx} \, dx$	$e^{kx}/k + C$	
$\sin(x) \, dx$	$-\cos(x) + C$	note the minus sign
$\cos(x) \, dx$	$\sin(x) + C$	
$\sec^2(x) \, dx$	$\tan(x) + C$	
$\sec(x)\tan(x) \, dx$	$\sec(x)$	
$\tan(x) \, dx$	$\ln(|\sec(x)|) + C$	
$dx/(1+x^2)$	$\tan^{-1}(x) + C$	
$u^n \, du$	$u^{n+1}/(n+1) + C$	for $n \neq -1$
du/u	$\ln(|u|) + C$	for both $u > 0$ and $u < 0$
$e^u \, du$	$e^u + C$	
$\sin(u) \, du$	$-\cos(u) + C$	note the minus sign
$\cos(x) \, du$	$\sin(u) + C$	
$du/(1+u^2)$	$\tan^{-1}(u) + C$	

Substitutions and Definite Integrals. We return to our original example of computing $\int_1^3 2x(x^2 - 1)^3 \, dx$. Our anti-derivative was $u^4/4$, which we had to evaluate at our endpoints. The trouble is that 1 and 3 are the starting and ending values of x, not the starting and ending values of u. To avoid mixing apples and oranges, we either have to express everything in terms of x or express everything in terms of u. Our two options are

- convert $u^4/4$ into a function of x and plug in $x = 1$ and $x = 3$, or
- convert $x = 1$ and $x = 3$ into values of u, namely $u = 0$ and $u = 8$.

That is, we can get our answer either as

$$(9.13) \quad \int_{x=1}^{3} 2x(x^2-1)^3 \, dx = \int_{x=1}^{3} u^3 \, du = \left.\frac{u^4}{4}\right|_{x=1}^{3} = \left.\frac{(x^2-1)^4}{4}\right|_{x=1}^{3} = \frac{8^4 - 0^4}{4} = 1024$$

9.1. Integration by Substitution

or as

$$(9.14) \quad \int_{x=1}^{3} 2x(x^2-1)^3\, dx = \int_{x=1}^{3} u^3\, du = \int_{u=0}^{8} u^3\, du = \frac{u^4}{4}\Big|_{u=0}^{8} = \frac{8^4 - 0^4}{4} = 1024$$

but not as

$$(9.15) \quad \int_{1}^{3} 2x(x^2-1)^3\, dx = \int_{1}^{3} u^3\, du = \frac{u^4}{4}\Big|_{1}^{3} = \frac{3^4 - 1^4}{4} = 20. \quad \text{(Wrong!)}$$

To avoid that mistake, it's a good idea to explicitly indicate what variable we're talking about in our limits of integration, as in the previous two lines.

We derived the rules for u-substitution by thinking about anti-derivatives and the chain rule. However, we can also understand u-substitution directly in terms of definite integrals and Riemann sums. If $u = x^2 - 1$, then every value of x corresponds to a value of u. If we break the interval $[1, 3]$ into N pieces, with $x_0 = 1 < x_1 < x_2 \cdots < x_N = 3$, then we have also broken the interval $[0, 8]$ into N pieces, with $u_0 = 0 < u_1 < u_2 < \cdots < u_N = 8$, where each $u_i = (x_i^2 - 1)$. We then have

$$\int_0^8 u^3\, du \approx \sum_{i=1}^{N} (u_i^*)^3 \Delta_i u$$

$$\approx \sum_{i=1}^{N} (u_i^*)^3 u'(x_i^*) \Delta_i x$$

$$= \sum_{i=1}^{N} 2x_i^* ((x_i^*)^2 - 1)^3 \Delta_i x$$

$$(9.16) \qquad\qquad \approx \int_1^3 2x(x^2-1)^3\, dx.$$

The first line is the definition of a definite integral with respect to u. On the second line we used the microscope equation for the function $u(x)$, centered at $x = x_i^*$, to say that $\Delta_i u = u_i - u_{i-1} \approx u'(x_i^*)\Delta_i x = 2x_i^* \Delta_i x$. On the third line we just rewrote everything in terms of x, and the fourth line is the definition of a definite integral with respect to x. All the approximations get better and better as $N \to \infty$, and we get equality in the limit.

In summary, the procedure for integration by substitution is summarized in the formal equation

$$(9.17) \qquad\qquad du = \frac{du}{dx} dx.$$

We derived this by thinking about the chain rule, and the difference between taking a derivative with respect to u and taking a derivative with respect to x. However, it can also be viewed as a statement about the intervals $\Delta_i u$ and $\Delta_i x$ that appear in our Riemann sums. In fact, that is pretty much how Leibniz thought of (9.17) when he invented the whole dx, du notation. To Leibniz, du and dx represented infinitesimally small intervals in u and x whose ratio is the conversion factor du/dx.

9.2. Integration by Parts

How can we integrate $xe^x\, dx$? We know that $\int x\, dx = \frac{x^2}{2} + C$ and that $\int e^x\, dx = e^x + C$, but we don't have a product rule for integrals. There is no obvious combination of x, x^2, and e^x whose derivative is xe^x.

The Method. Instead of going straight to the answer, let's go back to *close is good enough*. We'll guess a function whose derivative is close to xe^x, and then we'll adjust by how much we're off by. Since the derivative of e^x is e^x, we can try xe^x as a function whose derivative is close to xe^x.

The product rule tells us that

$$\frac{d}{dx} xe^x = xe^x + e^x. \tag{9.18}$$

Our guess has a derivative that is e^x too big. By subtracting off something whose derivative is e^x, we get our answer:

$$\frac{d}{dx}(xe^x - e^x) = xe^x + e^x - e^x = xe^x, \quad \text{so}$$

$$\int xe^x\, dx = xe^x - e^x + C. \tag{9.19}$$

In general, whenever we are trying to do an integral of the form $\int f(x)g'(x)dx$, we can guess $f(x)g(x)$. By the product rule, the derivative of $f(x)g(x)$ has two parts:

$$\frac{d}{dx}(f(x)g(x)) = f(x)g'(x) + f'(x)g(x). \tag{9.20}$$

The first part is what we want. The second part is extra, and we have to subtract off something whose derivative is $f'(x)g(x)$. That is, we have the following.

Integration by Parts Formula (long version):

$$\int f(x)g'(x)dx = f(x)g(x) - \int g(x)f'(x)\, dx. \tag{9.21}$$

Integration by parts doesn't solve our integration problem in one step. It just turns one integration problem (computing $\int f(x)g'(x)\, dx$) into another integration problem (computing $\int g(x)f'(x)\, dx$). However, that's often real progress. If the second integral is something that we know how to do, or at least is simpler than the first one, we win.

Integration by parts is often described in terms of expressions u and v instead of $f(x)$ and $g(x)$. If $u = f(x)$ and $v = g(x)$, then $du = f'(x)\, dx$ and $dv = g'(x)\, dx$, and we have the following.

Integration by Parts Formula (short version):

$$\int u\, dv = uv - \int v\, du. \tag{9.22}$$

9.2. Integration by Parts

Here are some examples of hard problems that can be solved by integration by parts.

- To compute $\int x^2 e^x \, dx$, we can take $u = x^2$ and $dv = e^x \, dx$, so $du = 2x \, dx$ and $v = e^x$. (We could also take $v = e^x + 3$ or $e^x + 17$, but that's needlessly complicated. We usually pick v to be the simplest anti-derivative of dv.) Then

$$(9.23) \qquad \int x^2 e^x \, dx = x^2 e^x - \int 2x e^x \, dx.$$

We didn't finish the problem, but we reduced it to a simpler problem. To finish the problem, we either integrate by parts a second time or remember equation (9.19) and get

$$(9.24) \qquad \int x^2 e^x \, dx = x^2 e^x - \int 2x e^x \, dx = x^2 e^x - 2x e^x + 2e^x + C.$$

- To compute $\int \ln(x) \, dx$, we take $u = \ln(x)$ and $dv = dx$, so $du = dx/x$ and $v = x$. Equation (9.22) becomes

$$(9.25) \qquad \int \ln(x) \, dx = x \ln(x) - \int x \frac{dx}{x} = x \ln(x) - \int dx = x \ln(x) - x + C.$$

- Computing $\int \tan^{-1}(x) \, dx$ is similar. We take $u = \tan^{-1}(x)$ and $dv = dx$, hence $du = \frac{dx}{1+x^2}$ and $v = x$, and we get

$$(9.26) \qquad \int \tan^{-1}(x) \, dx = x \tan^{-1}(x) - \int \frac{x \, dx}{1 + x^2}.$$

We can solve the second integral by substitution, setting $w = 1 + x^2$ and $dw = 2x \, dx$. (We already defined u, so we'd better use another letter for our substitution.) Then $\frac{x \, dx}{1+x^2} = \frac{dw}{2w}$. The integral of $\frac{dw}{2w}$ is $\ln(|w|)/2$, so we get

$$(9.27) \qquad \int \tan^{-1}(x) \, dx = x \tan^{-1}(x) - \int \frac{x \, dx}{1 + x^2} = x \tan^{-1}(x) - \frac{1}{2} \ln(1 + x^2) + C.$$

(We could have written $\ln(|1 + x^2|)$, but $1 + x^2$ is always positive, so $|1 + x^2|$ is just $1 + x^2$.)

The Hierarchy. At this point, integration by parts may seem mysterious. Yes, it worked in the examples we just did, but the way that we picked u and dv probably felt like black magic. For instance, when integrating $xe^x \, dx$, we set $u = x$ and $dv = e^x dx$ so that $du = dx$ and $v = e^x$. But we could have tried $u = e^x$ and $dv = x \, dx$, in which case we would have had $du = e^x \, dx$ and $v = x^2/2$, and (9.22) would have read

$$(9.28) \qquad \int x e^x \, dx = \frac{x^2}{2} e^x - \int \frac{x^2}{2} e^x \, dx.$$

That's a true statement, but it doesn't do us any good. We've just replaced a hard integral with an even harder one.

In this subsection we'll learn how to avoid that trap by picking u and dv effectively. Our goal is *to make $\int v \, du$ simpler than $\int u \, dv$*. Ideally, we want to pick u and dv so that du is simpler than u and v is simpler than dv. That isn't always possible, since integration usually makes a function more complicated, but we can often set things up so that what we gain by differentiating u outweighs what we lose by integrating dv.

To develop our strategy, we divide functions into four classes:

(1) Some functions either get much simpler when differentiated (such as $\ln(x)$ and $\tan^{-1}(x)$) or get much more complicated when integrated (such as $1/x$ and $1/(1+x^2)$). These functions should almost always be part of u.

(2) Some functions, such as positive powers of x, get a little simpler when differentiated and a little more complicated when integrated. These are the next priority for including in u.

(3) Some functions, such as $\sin(x)$, $\cos(x)$, and e^x, don't care. They look pretty much the same whether integrated or differentiated, and they usually belong in dv.

(4) A few functions, such as $1/x^2$ and $2xe^{x^2}$, actually get simpler when integrated. These almost always belong in dv.

When dealing with the product of two functions that are at different levels of this hierarchy, we want to make u the higher function and make dv the lower function (times dx). For instance, positive powers of x are higher than e^x, so when computing $\int x^n e^x \, dx$, we want to take $u = x^n$ and $dv = e^x \, dx$. However, powers of x are lower than $\ln(x)$, so when computing $\int x^3 \ln(x) \, dx$, we want to take $u = \ln(x)$ and $dv = x^3 \, dx$. In this case we have $v = x^4/4$ and $du = dx/x$, so

(9.29)
$$\int x^3 \ln(x) \, dx = \frac{x^4}{4} \ln(x) - \int \frac{x^4}{4} \frac{dx}{x} = \frac{x^4}{4} \ln(x) - \int \frac{x^3}{4} \, dx = \frac{x^4}{4} \ln(x) - \frac{x^4}{16} + C.$$

Converting x^3 to $x^4/4$ (making our job a little harder) was a small price to pay for converting $\ln(x)$ to $1/x$ (making our job a lot easier).

Remember that 1 is a function! We can think of $\ln(x)$ as $\ln(x)$ times 1. When integrating functions at the top of the hierarchy, like $\ln(x)$ and $\tan^{-1}(x)$, we can let u be the entire function being integrated, in which case $dv = dx$.

Integration by parts is often combined with other methods, particularly integration by substitution. For instance, if we wanted to compute

(9.30)
$$\int \frac{4x^2}{(2x+7)^2} \, dx,$$

we might take $u = 4x^2$ and $dv = (2x+7)^{-2} \, dx$. Computing v requires a substitution with $w = 2x + 7$, with the result that $v = -\frac{1}{2}(2x+7)^{-1}$. We then have

$$\int \frac{4x^2}{(2x+7)^2} \, dx = -\frac{2x^2}{2x+7} - \int \frac{-2x}{2x+7} \, dx$$

$$= -\frac{2x^2}{2x+7} + \int \left(1 - \frac{7}{2x+7}\right) dx$$

(9.31)
$$= -\frac{2x^2}{2x+7} + x - \frac{7}{2} \ln(|2x+7|) + C.$$

In the second line we used the fact that $1 - \frac{7}{2x+7} = \frac{2x}{2x+7}$, and then we did another substitution with $w = 2x + 7$.

9.2. Integration by Parts

Going in Circles. Integration by parts can help us when we have functions at different levels of the hierarchy. In those cases we replace the harder integral $\int u\, dv$ with an easier integral $\int v\, du$, which we can attack with all of our integration methods, including additional rounds of integration by parts. However, sometimes we have two functions at the same level of the hierarchy, so $\int v\, du$ is just as hard as $\int u\, dv$. What do we do then?

In those cases, we can often use integration by parts to turn our integral $I = \int u\, dv$ into a multiple of itself. It may seem like we're going in circles, trading a hard integral for the exact same hard integral, but there's a method to our madness. After one or more rounds of integration by parts, we hope to wind up with something like

(9.32) $$I = (\text{some function of } x) + kI,$$

where k is some constant. If $k \neq 1$, then we just put kI on the other side of the equation and solve

$$I = (\text{some function of } x) + kI,$$
$$I - kI = (\text{some function of } x) + C,$$
(9.33) $$I = \frac{\text{some function of } x}{1-k} + C.$$

Note the placement of the arbitrary constants. On the first line, both I and kI involve arbitrary constants. They aren't necessarily the same, so on the second line we have to include $+C$ on the right-hand side. The constant in the third line is really $\frac{1}{1-k}$ times the constant in the second line, but it's still arbitrary, so we just write $+C$.

For instance, to compute $\int \sin^2(x)\, dx$ we can take $u = \sin(x)$ and $dv = \sin(x)\, dx$, so $du = \cos(x)\, dx$ and $v = -\cos(x)$. We then have

$$\int \sin^2(x)\, dx = -\sin(x)\cos(x) - \int -\cos^2(x)\, dx$$
$$= -\sin(x)\cos(x) + \int (1 - \sin^2(x))\, dx$$
$$= x - \sin(x)\cos(x) - \int \sin^2(x)\, dx,$$
$$2\int \sin^2(x)\, dx = x - \sin(x)\cos(x) + C,$$
(9.34) $$\int \sin^2(x)\, dx = \frac{x - \sin(x)\cos(x)}{2} + C.$$

In this example we were able to go in a circle in a single step, but it usually takes two steps, and sometimes more. For instance,

$$\int e^x \sin(x)\, dx = e^x \sin(x) - \int e^x \cos(x)\, dx$$
$$= e^x \sin(x) - e^x \cos(x) - \int e^x \sin(x)\, dx,$$
$$2 \int e^x \sin(x)\, dx = e^x(\sin(x) - \cos(x)) + C,$$
(9.35) $$\int e^x \sin(x)\, dx = \frac{\sin(x) - \cos(x)}{2} e^x + C.$$

In the first line, we did an integration by parts with $u = \sin(x)$ and $dv = e^x\, dx$. In the second line we did an integration by parts with $u = \cos(x)$ and $dv = e^x\, dx$. That is, *we kept going in the direction we started.* In the first round we differentiated $\sin(x)$ to get $\cos(x)$, so in the second round we differentiated $\cos(x)$. In the first round we integrated $e^x\, dx$ to get e^x, so in the second round we integrated $e^x\, dx$. If in the second round we had integrated $\cos(x)\, dx$ and differentiated e^x, that would have undone what we did in the first round.

In the first round, we didn't have to take $u = \sin(x)$ and $dv = e^x\, dx$. We could just as well have taken $u = e^x$ and $dv = \sin(x)\, dx$. But then we would have differentiated e^x and integrated $\cos(x)$ in the second round, and the calculation would have looked like

$$\int e^x \sin(x)\, dx = -e^x \cos(x) + \int e^x \cos(x)\, dx$$
$$= e^x \sin(x) - e^x \cos(x) - \int e^x \sin(x)\, dx,$$
$$2 \int e^x \sin(x)\, dx = e^x(\sin(x) - \cos(x)) + C,$$
(9.36)
$$\int e^x \sin(x)\, dx = \frac{\sin(x) - \cos(x)}{2} e^x + C.$$

A different calculation, but the same answer.

In general, when we have two functions at the same level of the hierarchy, we can start by differentiating either one and integrating the other. It doesn't much matter which we pick. However, once we pick a direction, we have to stick with it until we get back to a multiple of the original integral.

Definite Integrals. So far we have talked about indefinite integrals. To apply integration by parts to definite integrals, we just evaluate at the endpoints.

Integration by Parts for Definite Integrals:
$$\int_a^b f(x) g'(x)\, dx = f(x)g(x)\Big|_a^b - \int_a^b g(x) f'(x)\, dx,$$
(9.37)
$$\int_a^b u\, dv = uv\Big|_a^b - \int_a^b v\, du.$$

The term $uv\Big|_a^b$ is called the **boundary term** (or the boundary *terms*, since we have both $u(b)v(b)$ and $u(a)v(a)$), while $-\int_a^b v\, du$ is called the **bulk term**. This terminology comes from physics. In many settings, we want to integrate a quantity $v\, du$ over some region in space, like the bulk of a crystal. The term $uv\Big|_a^b$ doesn't depend on what is happening inside the crystal, just on what is happening at the boundary.

Integration by parts is especially useful when the boundary term is 0, either because u and/or v vanish at the boundary, or because u and v are periodic and take on

9.2. Integration by Parts

the same values at a and b. For instance,

$$(9.38) \qquad \int_0^\pi x \cos(x)\,dx = x\sin(x)\Big|_0^\pi - \int_0^\pi \sin(x)\,dx.$$

Since $\sin(0) = \sin(\pi) = 0$, the boundary term vanishes and we are left with just the bulk term

$$(9.39) \qquad -\int_0^\pi \sin(x)\,dx = \cos(x)\Big|_0^\pi = -2.$$

In other cases, the boundary term is only approximately 0, and we can say that the original integral and the bulk term are approximately equal. For instance, consider $\int_0^{100} x^5 e^{-x}\,dx$. We integrate by parts with $u = x^5$ and $dv = e^{-x}\,dx$, hence $du = 5x^4\,dx$ and $v = -e^{-x}$, to get

$$(9.40) \qquad \int_0^{100} x^5 e^{-x}\,dx = -x^5 e^{-x}\Big|_0^{100} + 5\int_0^{100} x^4 e^{-x}\,dx.$$

Since $0^5 e^{-0} = 0$ and $100^5 e^{-100} \approx 3.7 \times 10^{-34}$ is incredibly small,

$$\int_0^{100} x^5\,dx \approx 5\int_0^{100} x^4\,dx.$$

Similarly,

$$\int_0^{100} x^4 e^{-x}\,dx \approx 4\int_0^{100} x^3 e^{-x}\,dx,$$

$$\int_0^{100} x^3 e^{-x}\,dx \approx 3\int_0^{100} x^2 e^{-x}\,dx,$$

$$\int_0^{100} x^2 e^{-x}\,dx \approx 2\int_0^{100} x^1 e^{-x}\,dx,$$

$$\int_0^{100} x^1 e^{-x}\,dx \approx 1\int_0^{100} e^{-x}\,dx$$

$$= 1 - e^{-100} \approx 1, \quad \text{so}$$

$$(9.41) \qquad \int_0^{100} x^5 e^{-x}\,dx \approx 5 \times 4 \times 3 \times 2 \times 1 = 5! = 120.$$

In fact, if b is any large number and n is a positive integer, then

$$\int_0^b x^n e^{-x}\,dx \approx n! \quad \text{and} \quad \lim_{b \to \infty} \int_0^b x^n e^{-x}\,dx = n!.$$

From the perspective of bulk and boundary terms, integration by parts is more than just a trick for computing anti-derivatives. It allows us to relate one bulk quantity to another. As such, it is one of the cornerstones of mathematical physics and mathematical analysis.

9.3. Numerical Integration

So far we have been talking about formulas. If somebody gave us a formula for $f(x)$, we wanted to find a formula for $\int f(x)\,dx$, and we developed two powerful techniques (integration by substitution and integration by parts) for finding one. However, these techniques don't always work. No matter how clever we are, and no matter how hard we try, we will never find a formula (involving only elementary functions) for the anti-derivative of e^{-x^2}. If we want to compute $\int_0^1 e^{-x^2}\,dx$, we can't plug the anti-derivative into the Fundamental Theorem of Calculus. We have to do something completely different.

Sometimes our functions are defined by data and we don't even have a formula. If we measured the speed of a car every minute for an hour (say, by glancing at the speedometer), how well could we estimate the distance traveled in that hour? Similarly, if we didn't know the formula for a function $f(x)$ but only knew $f(x_0)$, $f(x_1)$, ..., $f(x_N)$ at some equally spaced points $a = x_0 < x_1 < \cdots < x_N = b$, how well could we estimate $\int_a^b f(x)\,dx$?

The Trapezoidal Rule. We have already seen several ways to estimate $\int_a^b f(x)\,dx$. All of these involve dividing the interval $[a, b]$ into N subintervals $[x_{i-1}, x_i]$, each of width $\Delta x = (b - a)/N$, with right endpoints at $x_i = a + i\Delta x$.

(1) We can do a Riemann sum using left endpoints. We take $x_i^* = x_{i-1}$, so our Riemann sum is

(9.42) $$L_N = \sum_{i=1}^{n} f(x_{i-1})\Delta x = \Delta x(f(x_0) + f(x_1) + \cdots + f(x_{N-1})).$$

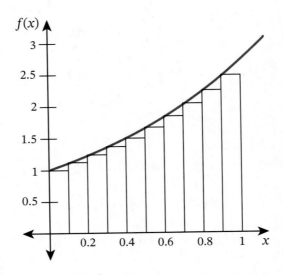

Figure 9.1. The approximation L_{10} with left endpoints

The trouble with this method is that it gives too low an estimate for increasing functions and too high an estimate for decreasing functions. As you can see in Figure 9.1, for increasing functions this estimate misses the area of N (roughly) triangular regions, each of width Δx and height $\Delta y \approx f'(x)\Delta x$, hence, with area about $\Delta x \Delta y/2$. The total error is then roughly

$$\sum (\Delta x \Delta y)/2 = \Delta x (\sum \Delta y)/2 = \Delta x (f(b) - f(a))/2.$$

Since $\Delta x = (b-a)/N$, this scales as $1/N$. To get twice as much accuracy, we need twice as many data points. To get ten times as much accuracy, we need ten times as many points.

(2) We can do a Riemann sum using right endpoints, as in Figure 9.2. Now we take $x_i^* = x_i$ and our Riemann sum is

(9.43) $$R_N = \sum_{i=1}^{n} f(x_i)\Delta x = \Delta x(f(x_1) + f(x_2) + \cdots + f(x_N)).$$

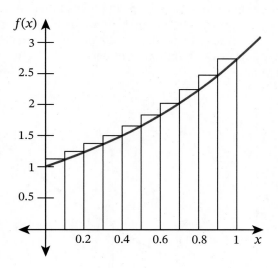

Figure 9.2. The approximation R_{10} with right endpoints

This has the opposite problem as using left endpoints, as it gives too high an estimate for increasing functions and too low an estimate for decreasing functions. For an increasing function it counts N roughly triangular regions of approximate total area $\Delta x(f(b) - f(a))/2$ that shouldn't have been counted.

Note that the difference between R_N and L_N is

$$\begin{aligned} R_N - L_N &= \Delta x (f(x_1) + \cdots + f(x_{N-1}) + f(x_N)) \\ & -\Delta x (f(x_0) + f(x_1) + \cdots + f(x_{N-1})) \end{aligned}$$

(9.44) $$= \Delta x (f(b) - f(a)).$$

(3) If using left endpoints gives an answer that is about $\Delta x(f(b) - f(a))/2$ too low, and using right endpoints gives an answer that is about the same amount too

high, then it makes sense to either add $\Delta x(f(b) - f(a))/2$ to the answer with left endpoints, or to subtract $\Delta x(f(b) - f(a))/2$ from the answer with right endpoints, or to average the results of using left and right endpoints. All of these procedures give exactly the same answer, namely

$$T_N = \sum_{i=1}^{N} \frac{f(x_{i-1}) + f(x_i)}{2} \Delta x$$

(9.45)
$$= \frac{\Delta x}{2}(f(x_0) + 2f(x_1) + \cdots + 2f(x_{N-1}) + f(x_N)).$$

This is called the **trapezoidal rule** because the area it assigns to the ith subinterval $I_i = [x_{i-1}, x_i]$ is $\Delta x(f(x_{i-1}) + f(x_i))/2$, which is the area of the trapezoid with vertices $(x_{i-1}, 0)$, $(x_{i-1}, f(x_{i-1}))$, $(x_i, 0)$, and $(x_i, f(x_i))$. It does fine with increasing and decreasing functions, but it isn't perfect. When the graph is concave up, as in Figure 9.3, the trapezoids overestimate the area under the curve, and when the graph is concave down they underestimate it. The over/under-estimates are small, and are barely visible in the figure, but they're there. We will figure out the size of this error in Exercise 9.43. For now, we merely note that the total error scales as $1/N^2$. Doubling the number of data points gives us four times as much accuracy, and multiplying the number of data points by 10 gives us 100 times more accuracy.

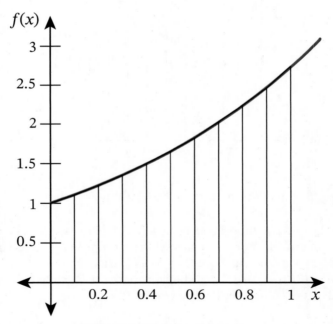

Figure 9.3. The trapezoidal rule T_{10}

To see how these methods work, let's try them out to compute $\int_0^1 f(x)\,dx$ for a function whose values are listed in Table 9.2.

9.3. Numerical Integration

Table 9.2. Values of a mystery function

x	$f(x)$
0.0	1
0.1	1.105171
0.2	1.221403
0.3	1.349859
0.4	1.491825
0.5	1.648721
0.6	1.822119
0.7	2.013753
0.8	2.225541
0.9	2.459603
1.0	2.718282

The sum of the first ten values is 16.337995, the sum of the last ten values is 18.056277, and $\Delta x = 0.1$, so we have

(9.46) $\qquad L_{10} = 1.6337995, \qquad R_{10} = 1.8056277, \qquad T_{10} = 1.7197136.$

So which of these is closest? The function in Table 9.2 is actually e^x, and the exact value of the integral is $\int_0^1 e^x \, dx = e^x \big|_0^1 = e - 1 \approx 1.7182818$. We summarize the different results and their errors in Table 9.3. It takes slightly more work to compute T_{10} than to compute L_{10} or R_{10}, but the result is a factor of $6N = 60$ better.

Table 9.3. Comparing different numerical methods of integration

Approximation	Value	Error
$\int_0^1 e^x \, dx$	1.7182818	none
L_{10}	1.6337955	-0.0844868
R_{10}	1.8056277	$+0.0873459$
T_{10}	1.7197136	$+0.0014318$

The Midpoint Rule and Simpson's Rule. When we introduced Riemann sums, we talked about using left endpoints, right endpoints, and midpoints, but there wasn't a single mention of midpoints in the first half of this section. You should be wondering: *what's wrong with midpoints?*

If we have a formula for $f(x)$, there is absolutely nothing wrong with using midpoints! It doesn't have any error associated with slope, and its error from curvature is only half as big as the error from using trapezoids.

For instance, suppose that we were estimating the area under $y = x^2$ from 1 to $1+h$ using a single interval ($N = 1$), where we imagine that h is a small number, such as 0.1. We have $x_0 = 1$, $x_1 = 1 + h$, and our midpoint is $x_1^* = 1 + \frac{h}{2}$. The exact answer (as you can work out yourself) is $\int_1^{1+h} x^2 \, dx = h + h^2 + \frac{1}{3}h^3$. The other approximations are given in Table 9.4.

Table 9.4. Comparing errors for a single interval

Approximation	Value	Error
$\int_1^{1+h} x^2 \, dx$	$h + h^2 + \frac{h^3}{3}$	none
L_1	h	$-h^2 - \frac{h^3}{3}$
R_1	$h + 2h^2 + h^3$	$h^2 + \frac{2h^3}{3}$
T_1	$h + h^2 + \frac{h^3}{2}$	$h^3/6$
M_1	$h + h^2 + \frac{h^3}{4}$	$-h^3/12$

While L_1 and R_1 are off by an amount that scales as h^2, T_1 and our midpoint estimate M_1 are only off by amounts proportional to h^3. In fact, T_1 is off by -2 times the amount that M_1 is off. If we can gather data at midpoints as easily as at endpoints, the midpoint rule is a little better than the trapezoidal rule. See Figure 9.4.

The trouble is that when we were integrating our function from a table, we didn't have the values of $f(0.05)$ or $f(0.15)$ or $f(0.25)$! The only way we could implement the midpoint rule was by making our intervals bigger. If we divided $[0, 1]$ into five intervals instead of ten, then the midpoints of the intervals $[0, 0.2]$, $[0.2, 0.4]$, etc., would be 0.1, 0.3, 0.5, 0.7, and 0.9. But while M_5 is a better estimate than T_5, it isn't as good an estimate as T_{10}.

This situation should remind you of what we saw with L_N and R_N. Both L_N and R_N had errors, but by taking their average, we eliminated the error that comes from slope. Likewise, M_N and T_N both have errors proportional to $f''(x)/N^2$, with the error in T_N being approximately -2 times the error in M_N. By looking at $(T_N + 2M_N)/3$, we can eliminate that quadratic error entirely.

So let's go back to our numerically defined function. The trapezoidal and midterm rules with $N = 5$ and their weighted average $S_{10} = (T_5 + 2M_5)/3$ are

$$T_5 = 0.1\left(f(0) + 2f(0.2) + 2f(0.4) + 2f(0.6) + 2f(0.8) + f(1)\right),$$

$$M_5 = 0.2\left(f(0.1) + f(0.3) + f(0.5) + f(0.7) + f(0.9)\right),$$

$$S_{10} = \frac{0.1}{3}\Big(f(0) + 4f(0.1) + 2f(0.2) + 4f(0.3) + 2f(0.4) + 4f(0.5)$$
$$+ 2f(0.6) + 4f(0.7) + 2f(0.8) + 4f(0.9) + f(1)\Big)$$

(9.47) $\qquad = 1.7182828.$

9.3. Numerical Integration

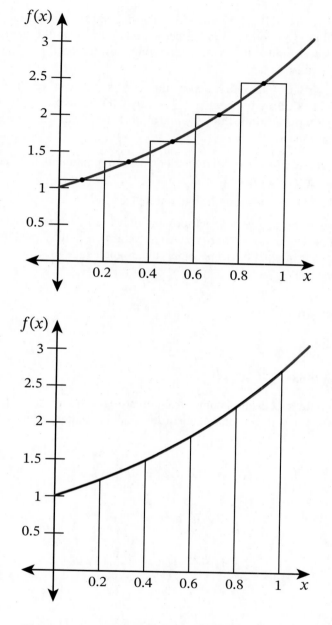

Figure 9.4. The midpoint rule M_5 and the trapezoidal rule T_5

S_{10} is called **Simpson's rule** with $N = 10$. We say $N = 10$ because we're using ten data points, even though we are only using five intervals. In general, if N is even, then

$$S_N = \frac{\Delta x}{3}\Big[f(x_0) + f(x_N) + 2(f(x_2) + f(x_4) + \cdots + f(x_{N-2}))$$
$$+ 4(f(x_1) + f(x_3) \cdots + f(x_{N-1}))\Big],$$

(9.48)

where $\Delta x = (b-a)/N$ and each $x_i = a + i\Delta x$. The beginning and endpoints each count once, all the other even points count twice, the odd points count four times, and the whole thing gets multiplied by $\Delta x/3$. Put another way, the coefficients for Simpson's rule follow the pattern $1, 4, 2, 4, \ldots, 2, 4, 1$.

If $f(x)$ is an arbitrary smooth function, then Simpson's rule gives an answer whose error scales like $1/N^4$. In our example, S_{10} is only 0.000001 too high! This makes S_{10} about 1500 times more accurate than T_{10}, and almost 100,000 times more accurate than L_{10} or R_{10}. In summary:

- The trapezoidal rule is easy to remember, easy to implement, and has good accuracy, with errors scaling as $1/N^2$.
- Simpson's rule is more complicated, but it gives $1/N^4$ accuracy.
- There do exist more accurate methods, but only hard-core numerical analysts use them. In most situations where you need to compute a numerical integral, Simpson's rule, or even just the trapezoidal rule, is good enough.

9.4. Chapter Summary

The Main Ideas.

- Integration by substitution comes from inverting the chain rule. If $F'(x) = f(x)$, then the rule for integration by substitution can be written in any of the following ways:

$$\int f(g(x))g'(x)\, dx = F(g(x)) + C,$$

$$\int f(u)\frac{du}{dx}\, dx = F(u) + C,$$

$$\int f(u)\frac{du}{dx}\, dx = \int f(u)\, du,$$

(9.49)
$$\frac{du}{dx} dx = du.$$

- Using integration by substitution, we obtain several more *standard* integrals, including

$$\int \frac{dx}{x} = \ln(|x|) + C \quad \text{even when } x < 0,$$

(9.50)
$$\int \tan(x)\, dx = -\ln(|\cos(x)|) + C = \ln(|\sec(x)|) + C.$$

- When doing definite integrals, the limits of integration have to match the variable being integrated. If $F'(x) = f(x)$, then

$$\int_{x=a}^{b} f(u)\frac{du}{dx}\,dx = F(u(x))\Big|_{x=a}^{b}$$

(9.51)
$$= F(u)\Big|_{u=u(a)}^{u(b)}$$
$$\neq F(u)\Big|_{u=a}^{b}.$$

- Integration by parts comes from inverting the product rule for derivatives. There are two common ways of writing it:

$$\int f(x)g'(x)\,dx = f(x)g(x) - \int g(x)f'(x)\,dx,$$

(9.52)
$$\int u\,dv = uv - \int v\,du.$$

- To decide what to use as u and dv, we pick u from higher in the following hierarchy of functions and dv from lower in the hierarchy, if possible. That makes $\int v\,du$ simpler than $\int u\,dv$.
   (1) Functions that get much simpler when differentiated or much more complicated when integrated, such as logs and inverse trig functions,
   (2) Functions that get a little simpler when differentiated, such as polynomials,
   (3) Function like $\sin(x)$, $\cos(x)$, and e^x that don't care, and
   (4) Functions that get simpler when integrated.
- When dealing with products of functions at the same level of the hierarchy, try going in circles.
- Integration by parts for definite integrals

(9.53)
$$\int_a^b u\,dv = uv\Big|_a^b - \int_a^b v\,du$$

is especially useful when the boundary term $uv\Big|_a^b$ is 0 or approximately 0.
- When we can't find an anti-derivative, or our function is only defined numerically, we have to evaluate our definite integrals numerically.
- The trapezoidal rule,

(9.54)
$$T_N = \frac{\Delta x}{2}\Big(f(x_0) + 2f(x_1) + \cdots + 2f(x_{N-1}) + f(x_N)\Big),$$

is accurate to within an error that scales like $1/N^2$.

- We can get $1/N^4$ accuracy with Simpson's rule with even values of N with

$$S_N = \frac{\Delta x}{3}\Big[f(x_0) + f(x_N) + 2(f(x_2) + f(x_4) + \cdots + f(x_{N-2}))$$
$$+ 4(f(x_1) + f(x_3) \cdots + f(x_{N-1}))\Big]$$
$$= \frac{\Delta x}{3}\Big[f(x_0) + 4f(x_1) + 2f(x_2) + 4f(x_3)$$
(9.55)
$$+ \cdots + 2f(x_{N-2}) + 4f(x_{N-1}) + f(x_N)\Big].$$

Expectations. You should be able to:

- Find anti-derivatives of basic functions.
- Use integration by substitution and integration by parts to convert harder integrals into easier ones.
- Apply limits of integration properly to definite integrals.
- Know how to persevere, repeating these methods as many times as necessary to evaluate the harder integrals.
- Also know when to quit. Some integrals, such as $\int_0^1 e^{-x^2}\,dx$, can't be evaluated exactly with these or any other methods.
- Estimate integrals numerically with the trapezoidal rule and Simpson's rule.
- Modify the Riemann program (Section 7.2) to instruct a computer to apply the trapezoidal rule and Simpson's rule.

9.5. Exercises

Integration by Substitution

9.1. Evaluate each integral using substitution.

(a) $\int 3y^2(y^3 - 4)\,dy$

(b) $\int \cos(t)\sqrt{2\sin(t)}\,dt$

(c) $\int \cos(4x)\,dx$

(d) $\int e^t \cos(1 + e^t)\,dt$

(e) $\int \frac{4x^2}{1 + x^2}\,dx$

(f) $\int \tan^2(2t)\sec^2(2t)\,dt$

(g) $\int_0^1 \frac{\sin(\pi\sqrt{x})}{\sqrt{x}}\,dx$

(h) $\int_1^3 \frac{dt}{3t + 1}$

(i) $\int_0^{\ln(e)} \frac{e^x}{e^x + 1}\,dx$

(j) $\int_0^2 \frac{dt}{1 + (t^2/4)}$

(k) $\int_0^{\pi/2} 3\sin(x)\cos^4(x)\,dx$

(l) $\int_0^1 \frac{t}{\sqrt{1 + 2t^2}}\,dt$

9.5. Exercises

9.2. Let $f(x) = x^2 \sin(x^3)$.
   (a) Compute $\int f(x)\,dx$.
   (b) Find an anti-derivative $F(x)$ of $f(x)$ with $F(0) = 8$.
   (c) Find an anti-derivative $\tilde{F}(x)$ of $f(x)$ with $\tilde{F}(\sqrt[3]{\pi}) = 5$.

9.3. An anti-derivative of $f(t)$ is the same thing as the solution to the differential equation $y' = f(t)$. Find three solutions to the differential equation $y' = 4t + 3\ln(t)$.

9.4. An auto parts store has determined that the marginal price of synthetic motor oil at a supply level of x quarts per week is

$$p'(x) = \frac{1}{3\sqrt{2x+25}}.$$

Find the supply curve (that is, the price as a function of supply) for the motor oil if the producer is willing to supply 100 quarts a week at a price of $6.00 per quart. How many quarts would the producer be willing to supply at a price of $7.00 per quart?

9.5. A small company installs artificial turf football fields. Suppose the marginal profit function for installing x fields in a given year is

$$P'(x) = 10{,}000\left(\frac{64x}{x^2+15} - 4\right).$$

   (a) How many fields should be installed in order to maximize profit?
   (b) What is the profit function $P(x)$ for the year if the fixed costs (that is, the cost of doing business while installing no fields) are $50,000?
   (c) What is the maximum profit?

9.6. Suppose that you have two investment opportunities to choose between. Each opportunity requires the same initial investment I_0, but the two investments differ in the expected flow of income. For the first investment opportunity, the income flow is expected to be $f(t) = 20{,}000$ dollars per year, while for the second investment opportunity the income flow is expected to be $g(t) = 5{,}000e^{0.25t}$ dollars per year. Which investment opportunity yields more money (total) over a ten-year period? Explain.

9.7. We previously considered the idea of **present value**, where a dollar collected t years from now is considered equivalent to e^{-rt} dollars today. If the interest rate is 10%, which of the two investments of Exercise 9.6 gives an income stream with a greater present value?

9.8. (a) Sketch the graph of the function $f(x) = x^2 e^{-x^3}$ on the interval $[0, 2]$.
   (b) Find the area under the graph of $f(x)$ between $x = 0$ and $x = 2$.
   (c) Find the area under the graph of $f(x)$ between $x = 0$ and $x = b$. That is, find the accumulation function $A(b) = \int_0^b f(x)\,dx$. Is $A(2)$ the same number you found in part (b)? What are the values of $A(10)$, $A(100)$, and $A(1000)$?
   (d) The area between the graph of $f(x)$ and the entire positive x-axis can be said to be $1/3$. Explain why.

Improper integrals. Even when a function has an unbounded domain, the area under the graph may still be finite. Exercise 9.8 showed us how to handle the integral of such a function. Exercises 9.9–9.12 explore such functions further.

The expression $\int_0^\infty f(x)\,dx$ does *not* mean "take the interval from 0 to ∞, break it into N pieces, build a Riemann sum, and take the limit as $N \to \infty$". That doesn't work because one of those intervals would have to be infinitely wide. Instead, we set a cutoff at a number b and compute $\int_0^b f(x)\,dx$. That involves breaking the *finite* interval $[0,b]$ into N pieces, setting up a Riemann sum, and taking a limit as $N \to \infty$. Only then do we take a limit as $b \to \infty$. That is,

(9.56) $$\int_0^\infty f(x)\,dx \quad \text{means} \quad \lim_{b \to \infty} \int_0^b f(x)\,dx.$$

Similarly, we can integrate from $-\infty$ to 0.

(9.57) $$\int_{-\infty}^0 f(x)\,dx \quad \text{means} \quad \lim_{b \to \infty} \int_{-b}^0 f(x)\,dx.$$

Finally, we can integrate from $-\infty$ to ∞ by combining these operations:

(9.58) $$\int_{-\infty}^\infty f(x)\,dx = \int_{-\infty}^0 f(x)\,dx + \int_0^\infty f(x)\,dx.$$

Integrals of the form $\int_a^\infty f(x)\,dx$, $\int_{-\infty}^a f(x)\,dx$, and $\int_{-\infty}^\infty f(x)\,dx$ are called **improper integrals** because their definitions are a little different from what we usually mean by integrals. But when $f(x)$ is well behaved, they still represent the total area under the curve!

In most cases, we can integrate from $-\infty$ to ∞ by setting cutoffs at $-b$ and b:

(9.59) $$\int_{-\infty}^\infty f(x)\,dx = \lim_{b \to \infty} \int_{-b}^b f(x)\,dx.$$

Be careful, however. The right-hand side sometimes gives an answer even when the improper integral doesn't makes sense. $\int_{-\infty}^\infty x\,dx$ does not exist, since there's an infinite amount of area under the line $y = x$ to the right of the origin, and there's an infinite amount of area over the line to the left of the origin. But $\lim_{b\to\infty} \int_{-b}^b x\,dx = \lim_{b\to\infty} 0 = 0$. The *definition* of $\int_{-\infty}^\infty f(x)\,dx$ is (9.58); equation (9.59) is just a useful-but-not-100%-reliable shortcut.

Improper integrals come up a lot in probability. Many random quantities are described by a **probability density function** $f(x)$, often just called a **pdf**. The probability that a particular random quantity is between a and b is then $\int_a^b f(x)\,dx$. Since the quantity has to have *some* value, we must have $\int_{-\infty}^\infty f(x)\,dx = 1$.

9.9. The **exponential distribution** is often used to describe how long a piece of equipment lasts. (We saw a version of this in Exercise 7.29.) Let x be the lifetime of a machine, in hours, and suppose that its pdf is

$$f(x) = \begin{cases} Ce^{-x/1000} & x \geq 0, \\ 0 & x < 0, \end{cases}$$

where C is a constant chosen so that $\int_0^\infty f(x)\,dx = 1$.
(a) Find C.
(b) What is the probability that the machine will last less than 1000 hours? Less than 2000 hours? More than 3000 hours?
(c) Which is more likely, the machine failing in the first 100 hours or failing between the 1000th and 1100th hour?

9.10. The lifetime of a different piece of equipment might be described by a pdf that decays exponentially but at a different rate. Let λ be a positive number, and suppose that
$$f(x) = \begin{cases} Ce^{-x/\lambda} & x \geq 0, \\ 0 & x < 0, \end{cases}$$
where C is a constant chosen so that $\int_0^\infty f(x)\,dx = 1$. As in Exercise 9.9, x is the equipment's lifetime in hours.
(a) Find C in terms of λ.
(b) What is the probability that the equipment will last less than λ hours? Less than 2λ hours? More than 3λ hours?

9.11. The average lifetime of a piece of equipment is given by the improper integral $\int_0^\infty x f(x)\,dx$. Compute this average for an exponential distribution with parameter λ, as in Exercise 9.10. (*Hint*: Integrate by parts.)

9.12. If a random quantity is described by a pdf $f(x)$, then the **cumulative probability function**, or **cdf**, is the function $F(X) = \int_{-\infty}^X f(x)\,dx$. This gives the probability that the quantity is X or less. Compute the cdf of an exponential random variable with parameter λ, as in Exercise 9.10. (Note that $f(x) = 0$ if $x < 0$, so we can start our integral at 0 instead of $-\infty$.) If a and b are positive numbers with $a < b$, what is the probability that the random quantity is between $a\lambda$ and $b\lambda$?

Food for Thought: The parameter λ for an exponential distribution is usually called the **mean lifetime**, for reasons that shouldn't surprise you after doing Exercise 9.11. One advantage of the cdf over the pdf is that it is a function of x/λ. If you know the cdf for one quantity that is exponentially distributed, then you essentially know the cdf for every quantity that is exponentially distributed. The **standard exponential distribution** is the distribution with $\lambda = 1$.

Integration by Parts

9.13. For each integral, pick u and dv so that $\int v\,du$ is simpler than $\int u\,dv$. Indicate what v and du are. You do not have to evaluate $\int v\,du$.

(a) $\int (x^3 + 7)\sin(x)\,dx$

(b) $\int (x^3 + 7)\ln(x)\,dx$

(c) $\int t\tan^{-1}(t)\,dt$

(d) $\int t\sec^2(t)\,dt$

(e) $\int x\sin(x^2)\ln(x)\,dx$

(f) $\int x^3 e^{x^2}\,dx$

9.14. Evaluate each integral using integration by parts. Some examples may require two rounds, or may require a different trick after integrating by parts.

(a) $\int 2xe^{-x}\,dx$

(b) $\int x\cos(x)\,dx$

(c) $\int x^2 e^x\,dx$

(d) $\int x^3 e^{x^2}\,dx$

(e) $\int_{1/2}^{1} x^4 \ln(2x)\,dx$

(f) $\int_{1}^{9} \frac{\ln(x)}{\sqrt{x}}\,dx$

9.15. Use the going-in-circles trick to evaluate each integral.

(a) $\int e^{-x}\cos(x)\,dx$

(b) $\int_{0}^{\pi/2} \sin(x)\sin(2x)\,dx$

9.16. Consider the definite integral $\int_{0}^{\pi/2} \sin^n(x)\,dx$, where $n \geq 2$.

(a) Integrate by parts using $u = \sin^{n-1}$ and $dv = \sin(x)\,dx$ to show that

$$\int_{0}^{\pi/2} \sin^n(x)\,dx = \frac{n-1}{n} \int_{0}^{\pi/2} \sin^{n-2}(x)\,dx.$$

This involves both the going-in-circles trick and the identity $\cos^2(x) = 1 - \sin^2(x)$.

(b) Using the recursion from part (a), compute

$$\int_{0}^{\pi/2} \sin^5(x)\,dx \quad \text{and} \quad \int_{0}^{\pi/2} \sin^6(x)\,dx.$$

9.17. The **average value** of the function $f(x)$ over the interval $[a, b]$ is defined to be $\frac{1}{b-a}\int_{a}^{b} f(x)\,dx$.

(a) Explain in words why that is a reasonable definition.

(b) What is the average value of the function $\ln(x)$ on the interval $[1, e]$?

(c) What is the average value of $\ln(x)$ on $[1, b]$? Express your answer in terms of b.

(d) Explain why, for large values of b, the average value of $\ln(x)$ on $[1, b]$ is approximately $\ln(b) - 1$.

9.18. What is the average value of $\sin^2(x)$ (as defined in Exercise 9.17) on the interval $[0, 2\pi]$? What is the average value of $\cos^2(x)$ on the same interval?

9.19. At a large oil field, oil is produced at a rate of $b(t)$ thousand barrels per month, where t is measured in months. Suppose the function

$$b(t) = 10te^{-0.1t}$$

describes the production rate.

(a) Use MATLAB to graph $b(t)$. Be sure to label the axes appropriately.

(b) On your graph, shade the region that corresponds to the oil production during the first six months.

(c) Find the total oil production in the first six months of operation.

9.5. Exercises

9.20. When sales of an older model of a product drop below a certain level, stores will stop carrying the model and the manufacturer will stop making it. Suppose that sales of an old TV are currently at 2000/month, and that the model will be discontinued when sales drop to 200/month. Sales of the TV (in units per month) are decreasing at rate

$$S'(t) = -4te^{0.25t},$$

where t is measured in months.

    (a) Compute the function $S(t)$ that gives sales as a function of time.

    (b) To the nearest month, when will the model be discontinued? (*Hint*: There is no formula for the solution to $S(t) = 200$. However, you can solve that equation accurately with a variety of numerical methods, including Newton's method, the bisection method, or just graphing $S(t)$ on MATLAB and zooming in on the graph.)

9.21. (This exercise uses the concept of **producer surplus**, which was introduced in the Chapter 8 Exercises.) Find the producer surplus, to the nearest dollar, at a price of $p = \$20$ for the supply curve

$$p = S(x) = x2^{0.1x}.$$

9.22. The yeast in a tank at a brewery is producing alcohol at a rate of

$$R(t) = 80te^{-0.2t}$$

liters per day, where t is measured in days. Find the total amount of alcohol produced in the first week.

Future Value. Instead of talking about present value, some economists talk about future value. Since a dollar received at time t and invested at interest rate r turns into $e^{r(T-t)}$ dollars at time T, a dollar received at time t as said to have a **future value** of $e^{r(T-t)}$.

9.23. (a) If you receive income at a rate $f(t)$ from time $t = 0$ until time $t = T$, what is the future value (FV) of your income stream? Express your answer as an integral.

    (b) Recall that the present value of the income stream is $PV = \int_0^T f(t)e^{-rt}dt$. Find a simple formula that relates PV and FV.

    (c) Which is greater, FV or the result of investing PV for T years at interest rate r, compounded continuously?

9.24. Find the future value, at 4% interest compounded continuously for 5 years, of a continuous income stream with a flow rate of

$$f(t) = 2,000 - 300t,$$

where t is measured in years.

Means, Variances, and Standard Deviations. Probability density functions (pdf's) were introduced in Exercises 9.9–9.11. In Exercises 9.25–9.30, we see how we can find the mean and standard deviation of some common random variables from their pdf's, with the help of improper integrals and integration by parts.

If $f(x)$ is the pdf of a random variable, then the **mean** or the **average** of the probability distribution, often denoted μ or $\langle x \rangle$, is given by the integral

$$\mu = \langle x \rangle = \int_{-\infty}^{\infty} x f(x)\,dx.$$

This is a weighted average of each possible value of x, where the weight is the probability $f(x)\,\Delta x$ of being between x and $x + \Delta x$. Likewise, the average value of x^2 is

$$\langle x^2 \rangle = \int_{-\infty}^{\infty} x^2 f(x)\,dx.$$

The **variance**, often written σ^2, is $\langle x^2 \rangle - \langle x \rangle^2$. Finally, the **standard deviation**, usually written σ, is the square root of the variance. The standard deviation measures, in a technical sense, how wide the probability distribution is.

9.25. The **standard uniform distribution** has the pdf

$$f(x) = \begin{cases} 1 & 0 < x < 1, \\ 0 & \text{otherwise.} \end{cases}$$

This is what people usually mean by "pick a random number between 0 and 1". Find the mean, variance, and standard deviation of this distribution.

9.26. A uniform distribution on the interval $[a, b]$, where $a < b$, has the pdf

$$f(x) = \begin{cases} \frac{1}{b-a} & a < x < b, \\ 0 & \text{otherwise.} \end{cases}$$

This is what people usually mean by "pick a random number between a and b". Find the mean, variance, and standard deviation of this distribution. Notice that the standard deviation only depends on the width $b - a$ of the interval.

9.27. The **standard exponential distribution** has the pdf

$$f(x) = \begin{cases} e^{-x} & x \geq 0, \\ 0 & x < 0. \end{cases}$$

Compute the mean and standard deviation. This requires integrating by parts and using the fact that $\lim_{b \to \infty} b e^{-b} = \lim_{b \to \infty} b^2 e^{-b} = 0$.

9.28. An exponential random variable with parameter λ has the pdf

$$f(x) = \begin{cases} \frac{e^{-x/\lambda}}{\lambda} & x \geq 0, \\ 0 & x < 0. \end{cases}$$

Compute the mean and standard deviation. For this calculation, you can use the fact that $\lim_{b \to \infty} b e^{-b} = \lim_{b \to \infty} b^2 e^{-b} = 0$.

9.29. The **standard normal distribution** has the pdf

$$f(x) = C e^{-x^2/2},$$

where C is a constant chosen so that $\int_{-\infty}^{\infty} f(x)\,dx = 1$. (The actual value happens to be $C = 1/\sqrt{2\pi}$, but that isn't relevant here.)

(a) Compute the mean.

(b) Compute $\langle x^2 \rangle$. You do not need to know C or the anti-derivative of $e^{-x^2/2}$ (which can't be expressed as a formula) to do this! Instead, integrate by parts to convert the integral for $\langle x^2 \rangle$ into a multiple of $\int_{-\infty}^{\infty} f(x)\, dx$, which equals 1.

(c) Compute the standard deviation.

9.30. The pdf of a normal distribution with mean μ and standard deviation σ is given by the formula

(9.60) $$f(x) = \frac{C}{\sigma} e^{-(x-\mu)^2/(2\sigma^2)},$$

where C is the same constant as for the standard normal distribution.

Let X be a random variable described by a normal distribution with mean μ and standard deviation σ, and let Z be a random variable described by a standard normal distribution. Let a and b be real numbers, with $a < b$. Show that the probability of X being between $\mu + a\sigma$ and $\mu + b\sigma$ is the same as the probability of Z being between a and b. (*Hint*: Do a u-substitution with $z = (x - \mu)/\sigma$.)

Food for Thought: Since the anti-derivative of $e^{-x^2/2}$ cannot be expressed as a formula involving elementary functions, most probability textbooks contain tables of the cdf of the standard normal distribution; these tables are easily accessed online. Computing things involving a general normal distribution almost always involves converting to *z-scores* and using the standard normal distribution.

Numerical Integration

9.31. Consider the integral $\int_0^{\pi/2} \frac{dx}{1+\sin^2(x)}$.
  (a) Approximate this integral numerically using the trapezoidal rule and $N = 4$.
  (b) Approximate this integral numerically using Simpson's rule and $N = 4$.

9.32. A bicycle courier needs to deliver an important package quickly! The courier leaves the pick-up location with the package promptly at 9:00 am and rides down a straight road but is slowed by traffic. The courier needs to deliver the package to a location 1.5 miles away by 9:08 am. The following table contains GPS estimates of the velocity $v(t)$ (in mph) of the courier at one-minute intervals with $t = 0$ corresponding to 9:00 am.

t	$v(t)$	t	$v(t)$
0	0	5	8
1	12	6	15
2	10	7	20
3	14	8	8
4	10		

Here t is the number of minutes since 9:00 am.
  (a) Use the midpoint rule with $\Delta t = 2$ to estimate the distance the courier traveled from 9:00 am to 9:08 am. That is, find M_4. Based on this estimate, was the delivery made on time?

(b) Use the trapezoidal rule with $\Delta t = 2$ to estimate the distance the courier traveled from 9:00 am to 9:08 am. That is, find T_4. Based on this estimate, was the delivery made on time?

(c) Use Simpson's rule with $\Delta t = 2$ to estimate the distance the courier traveled from 9:00 am to 9:08 am. That is, find S_4. Based on this estimate, was the delivery made on time?

9.33. The area of the shaded region in the graph below represents the definite integral $\int_0^2 f(x)\, dx$.

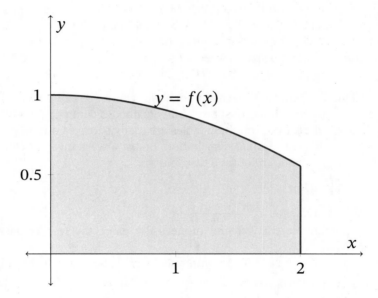

To estimate the definite integral, the left endpoint, right endpoint, midpoint, and trapezoidal rule approximations were found (using the same N for each estimate). The estimates, from smallest to largest, were 1.6577, 1.7022, 1.7044, and 1.7466. Which rule produced which estimate?

Numerical Integration by Computer. Exercises 9.34–9.41 involve first creating MATLAB programs to apply the trapezoidal rule and Simpson's rule, and then using those programs to approximate some integrals.

9.34. Create a MATLAB program named Trap that will numerically estimate $\int_a^b f(x)\, dx$ using the trapezoidal rule. You can modify the Riemann program, or you can write your own program from scratch.

9.35. The following is a MATLAB program named Simp that will numerically estimate $\int_a^b f(x)\, dx$ using Simpson's method.

9.5. Exercises

```
function S = SIMP(f,a,b,N )
% This function uses Simpson's rule to approximate the
% integral of the function f from a to b where N is an
% even number of equally spaced subintervals.
deltax = (b - a)/N;
S = f(a) + f(b);
for k = 1:2:N-1
    x = a +k*deltax;
    S = S + 4*f(x);
end
for k = 2:2:N-2
    x = a +k*deltax;
    S = S + 2*f(x);
end
S=S*deltax/3;
```

Carefully explain what each line of the program is doing to implement Simpson's rule.

9.36. Consider the integral $\int_0^2 \frac{8}{x^2+4} dx$.
   (a) Evaluate this integral exactly, using the substitution $u = x/2$.
   (b) Approximate this integral numerically, using the trapezoidal rule with $N = 1000$. How accurate is the approximation? (Use the `format long` command in MATLAB to display more than four digits of your answer.)
   (c) Approximate this integral numerically, using Simpson's rule with $N = 100$. How accurate is the approximation?

9.37. Consider the integral $\int_0^3 x^2 e^{-x} dx$.
   (a) Evaluate this integral exactly, using integration by parts.
   (b) Approximate this integral numerically, using the trapezoidal rule with $N = 100$.
   (c) Approximate this integral numerically, using Simpson's rule with $N = 10$.
   (d) Which numerical approximation is more accurate?

9.38. (a) Use Simp to find numerical estimates for $\int_0^1 f(x) dx$, where $f(x) = x$, $f(x) = x^2$, $f(x) = x^3$, and $f(x) = x^4$ when $N = 10$, $N = 100$, and $N = 1000$. Compare these estimates to the exact results found using the Fundamental Theorem of Calculus.
   (b) Why would you expect the answer found using Simp for $f(x) = x^2$ to be the same as the exact answer?
   (c) Repeat part (a), only this time using Trap. Is it true that the error in these estimates is of order $1/N^2$?
   (d) Use Riemann to find L_{10}, L_{100}, and L_{1000} for $f(x) = x^2$. Is it true that the error in these estimates is of order $1/N$?

9.39. Use Simpson's rule with $N = 6$ to estimate the area under the curve $y = 2^x/x$ from $x = 1$ to $x = 4$.

9.40. At an auto parts store, the demand curve for synthetic motor oil is given by

$$p = D(x) = \frac{150}{\sqrt{100 + 10\sqrt{x}}},$$

where x is the number of cans of oil that can be sold at a price of p dollars. Using $\Delta x = 25$, do the following.
   (a) Estimate the average price over the demand interval $[50, 200]$ using the trapezoidal rule.
   (b) Estimate the average price over the demand interval $[50, 200]$ using the midpoint rule.
   (c) Estimate the average price over the demand interval $[50, 200]$ using Simpson's rule.

9.41. Suppose the probability density function for adult female height (in inches) in a particular country is approximately

$$h(x) = \frac{1}{2.5\sqrt{2\pi}} e^{-(x-63)^2/12.5}.$$

(This is an example of a normal distribution, as in Exercise 9.30.)
   (a) Use Simpson's rule with $N = 12$ to estimate the probability that a woman in that country is between 5 and 6 feet tall.
   (b) Use Simpson's rule to estimate the probability that a woman in that country is over 6.5 feet. You will have to be somewhat creative as Simpson's rule can't be applied to infinite intervals. A way around this is to make the reasonable assumption that there are no women in the country over the height of 7.5 feet.

Improving the Trapezoidal Rule. We noted earlier that the accuracy of the trapezoidal rule goes as $1/N^2$. In Exercises 9.42 and 9.43, we will see why, and we will develop a better approximation, called the **corrected trapezoidal rule**, that provides $1/N^4$ accuracy without the complication of the 1-4-2-4$\cdots$2-4-1 pattern of Simpson's rule.

9.42. Let $f(x) = c_0 + c_1 + c_2 x^2$ be an arbitrary quadratic function, let a and b be arbitrary numbers, and let $\Delta x = b - a$.
   (a) Compute $\int_a^b f(x)dx$ exactly using the Fundamental Theorem of Calculus.
   (b) Compute the area of the trapezoid with vertices $(a, 0)$, $(a, f(a))$, $(b, f(b))$, and $(b, 0)$. (You can assume that $f(a)$ and $f(b)$ are both positive.)
   (c) Show that the area of the trapezoid, minus the integral, is exactly

$$c_2(\Delta x)^3/6 = f''(a)(\Delta x)^3/12.$$

9.43. Now suppose that $f(x)$ is an arbitrary function, and that we are trying to estimate $\int_a^b f(x)dx$ with the trapezoidal rule using N intervals. Write down a Riemann sum, involving $f''(x)$, that approximates the difference between the trapezoidal rule and the actual integral. (*Hint*: Apply the results of Exercise 9.42 to each subinterval and add up the errors.)

The Corrected Trapezoidal Rule. By now you should be used to approximating integrals with Riemann sums, and approximating Riemann sums with integrals. The answer to Exercise 9.43 is approximately

$$(9.61) \qquad \frac{(\Delta x)^2}{12} \int_a^b f''(x)dx = \frac{(\Delta x)^2}{12} f'(x)\Big|_a^b = \frac{(\Delta x)^2}{12}(f'(b) - f'(a)).$$

Since Δx is proportional to $1/N$, this shows that the accuracy of the trapezoidal rule goes as $1/N^2$.

Now that we have a good approximation to the error, we can subtract this (approximate) error to get an even better approximation. The **corrected trapezoidal rule** is

$$(9.62) \qquad C_N = T_N - \frac{(\Delta x)^2}{12}(f'(b) - f'(a)).$$

9.44. Let $f(x) = \cos(x)$. We will approximate $\int_0^{\pi/2} f(x)\,dx$ numerically in several ways: using the left endpoint rule L_N, the right endpoint rule R_N, the trapezoidal rule T_N, Simpson's rule S_N, and the corrected trapezoidal rule C_N. We already know that this integral is exactly $\sin(x)\Big|_0^{\pi/2} = 1$, so it's easy to see how accurate each estimate is.
  (a) Compute L_2, R_2, T_2, S_2, and C_2.
  (b) Now compute L_4, R_4, T_4, S_4, and C_4. How much additional accuracy did we get by doubling the number of points?
  (c) For each method, indicate roughly how big N would have to be to give an answer that is accurate to six decimal places.

9.45. Let $f(x) = e^{-x}$. We will approximate $\int_0^1 f(x)dx$.
  (a) L_{10}, T_{10}, S_{10} and C_{10}. You may want to use the MATLAB programs Riemann, Trap, and Simp.
  (b) Now compute $\int_0^1 f(x)\,dx$ exactly using the Fundamental Theorem of Calculus. Which approximation was the most accurate, and by how much?

9.46. Consider the integral $\int_{-5}^5 e^{-x^2}\,dx$. This integral cannot be evaluated exactly, but it is within 10^{-10} of $\sqrt{\pi}$. (See Exercise 10.31 to understand why.)
  (a) Evaluate this integral numerically, using the trapezoidal rule with $N = 10$, 100, and 1000.
  (b) Evaluate this integral numerically, using Simpson's rule with $N = 10$, 100, and 1000.
  (c) The answers using the trapezoidal rule are almost as accurate as the answers using Simpson's rule. Why?

9.47. Suppose that $f(x)$ is a periodic function, with $f(x + 1) = f(x)$, such as $f(x) = \dfrac{1}{1 + \sin^2(2\pi x)}$.

(a) Explain why estimating $\int_0^1 f(x)\,dx$ with left endpoints gives the same answer as using right endpoints, the trapezoidal rule, or the corrected trapezoidal rule.

(b) Explain why, for a fixed value of N, using left endpoints to estimate $\int_0^1 f(x)\,dx$ is just as accurate as using midpoints or Simpson's rule.

9.48. Let $f(x) = c_0 + c_1 x + c_2 x^2 + c_3 x^3$, where $c_0, c_1, c_2,$ and c_3 are arbitrary constants. Show that, when estimating $\int_a^b f(x)\,dx$, both C_2 and S_2 give *exactly* the right answer. (So do C_N and T_N for larger values of N.)

Food for Thought: The errors for both Simpson's rule and the corrected trapezoidal rule come from the *fourth* derivative of f and scale roughly as $1/N^4$. It's possible to get even more accurate rules by taking a weighted average of C_N and S_N (specifically, $\frac{1}{5}(C_N + 4S_N)$), or by adding a term proportional to $(f'''(b) - f'''(a))/N^4$ to Simpson's rule, but that's overkill. In practice, the most common methods for numerical integration are the trapezoidal rule, Simpson's rule, and the corrected trapezoidal rule. When integrating a smooth periodic function over a complete period, as in Exercise 9.47, all of the methods discussed in this chapter are especially accurate, with errors that shrink faster than any power of N.

Chapter 10

One Variable at a Time!

So far we have developed the key tools of calculus, namely derivatives and integrals, for functions of a single variable. The derivative of $f(x)$ is

$$(10.1) \qquad f'(x) = \lim_{h \to 0} \frac{f(x+h) - f(x)}{h},$$

and the definite integral is

$$(10.2) \qquad \int_a^b f(x)\,dx = \lim_{N \to \infty} \sum_{i=1}^N f(x_i^*)\Delta_i x.$$

However, many quantities in the real world depend on two or more inputs. The boiling point of salt water depends on your elevation and the salinity of the water. The wind chill on a winter day depends on the temperature and the wind speed. The heat index on a summer day depends on the temperature and the relative humidity. All sorts of weather data, from temperature to atmospheric pressure to average annual rainfall, depend on location, which is described by two coordinates, such as latitude and longitude. The rate of a chemical reaction depends on the temperature and on the concentrations of all of the chemicals involved in the reaction.

The key idea for handling functions of several variables is the 6th and last Pillar of Calculus:

> *One variable at a time!*

When studying how a quantity depends on several variables, we never fiddle with all of the variables at once. Instead, we hold all but one of the variables fixed and see how the quantity depends on the one variable that we are studying. Once we know both how the boiling point of water depends on elevation and how it depends on salinity, we can put the two effects together to approximate the boiling point by using a linear function of elevation and salinity.

The derivative of a function with respect to one variable, while holding all the other variables fixed, is called a **partial derivative**. Partial derivatives are developed in Section 10.1. In Section 10.2 we use partial derivatives to build linear approximations to our functions.

In Section 10.3 we turn to integrals. Not anti-derivatives, but genuine whole-is-the-sum-of-the-parts integrals! If we know the rate of wheat production at every point in a state, how can we compute the total amount of wheat that is grown in the state? As with one-dimensional integrals, we break the state into little pieces, estimate the contribution of each piece, add up the pieces, and take a limit. Since we are working in two dimensions, this is called a **double integral**. In practice, we add the pieces one variable at a time to get something called an **iterated integral**.

10.1. Partial Derivatives

Don't Sweat the Details. The heat index I_H describes how hot it feels as a function of the actual temperature T (usually measured in the USA in degrees Fahrenheit) and the relative humidity R (measured as a percentage between 0 and 100). Since the human body cools itself by the evaporation of sweat, and since sweat evaporates less when the humidity is higher, increased humidity makes it feel hotter, even if the actual temperature doesn't change.[1]

There isn't any formula for the heat index. Instead, it is computed by interpolating from a table produced by the US National Oceanic and Atmospheric Administration (NOAA). A portion of the table is shown in Table 10.1.

Table 10.1. Heat index I_H as a function of temperature (T) and relative humidity (R)

Heat index in °F

T \ R	80	82	84	86	88	90	92	94
40%	80	81	83	85	88	91	94	97
45%	80	82	84	87	89	93	96	100
50%	81	83	85	88	91	95	99	103
55%	81	84	86	89	93	97	101	106
60%	82	84	88	91	95	100	105	110
65%	82	85	89	93	98	103	108	114
70%	83	86	90	95	100	105	112	119
75%	84	88	92	97	103	109	116	124
80%	84	89	94	100	106	113	121	129
85%	85	90	96	102	110	117	126	135

---

[1] Air conditioners both lower the temperature and remove humidity from the air. The hotter it is outside, the more your air conditioner runs to bring the indoor temperature down to the setting on your thermostat, the dryer it is indoors, and the cooler you feel!

10.1. Partial Derivatives

For instance, when it is 90° out and the relative humidity is 50%, the heat index is 95°. At that point, raising or lowering the temperature by 2° raises or lowers the heat index by 4°, while raising or lowering the relative humidity by 5% raises or lowers the heat index by 2°.

To study how the heat index I_H depends on humidity, we hold the temperature steady at a single value (say, 90°) and consider the function

(10.3) $g(R) = I_H(R, 90) =$ Heat index at temperature 90° and relative humidity R.

We can then approximate $g'(R)$ at a particular value (say, 50% humidity) by a forward difference quotient

(10.4) $$\frac{g(55) - g(50)}{5} = \frac{97 - 95}{5} = 0.4,$$

or by a backward difference quotient

(10.5) $$\frac{g(50) - g(45)}{5} = \frac{95 - 93}{5} = 0.4,$$

or by a centered difference quotient

(10.6) $$\frac{g(55) - g(45)}{10} = \frac{97 - 93}{10} = 0.4.$$

Conveniently, these all give the same answer, and we estimate

(10.7) $$g'(50) \approx 0.4° \text{ Fahrenheit/percent humidity}.$$

Similarly, to study the effect of temperature on heat index, we could define a function

(10.8) $h(T) = I_H(50, T) =$ heat index at temperature T and relative humidity 50%

and compute

(10.9) $$h'(90) \approx 2.$$

(The units of h' are degrees/degree, so our answer is dimensionless.)

The derivatives g' and h' give us partial information about how $I_H(R, T)$ depends on R and T, since each one tells us what happens when only one variable is changed. For that reason, g' and h' are called **partial derivatives** of the original function $I_H(R, T)$ and are denoted $\partial I_H/\partial R$ and $\partial I_H/\partial T$.[2] That is,

(10.10) $$\frac{\partial I_H}{\partial R}(50, 90) = g'(50) = \lim_{h \to 0} \frac{I_H(50 + h, 90) - I_H(50, 90)}{h},$$
$$\frac{\partial I_H}{\partial T}(50, 90) = h'(90) = \lim_{k \to 0} \frac{I_H(50, 90 + k) - I_H(50, 90)}{k}.$$

In the second equation, we used the letter k rather than h to distinguish between a change in R (called h) and a change in T (called k). We could also write $h = \Delta R$, $k = \Delta T$.

---

[2] The symbol ∂ for partial derivatives, usually called "partial", dates to 1770 and is intended to be a stylized "d".

> A **partial derivative** of a function of two or more variables is the ordinary derivative of the function that you get by holding all but one variable fixed.

We just computed $\partial I_H/\partial R$ and $\partial I_H/\partial T$ at $(R, T) = (50, 90)$, but the same ideas apply everywhere. At $(45, 82)$ we have $\partial I_H/\partial R \approx 0.2°/\%$ and $\partial I_H/\partial T \approx 1$. At low humidity and (relatively) low temperature, the humidity doesn't have much of an impact and each degree of temperature translates into just one degree of heat index. However, at $(80, 92)$ we have $\partial I_H/\partial R \approx 1°/\%$ and $\partial I_H/\partial T \approx 4$. At high temperatures and very high humidity, small increases in the temperature and humidity can be dangerous, and small decreases can feel wonderful.

> **Reality Check:** Some of the numbers in the heat index table are downright scary. In a city like Houston, Texas, the relative humidity will often hit 80%, and the temperature will usually get into the 90's or 100's in the summer. A temperature of just 94 with 80% humidity would mean a deadly heat index of 129°.
>
> Fortunately, high temperatures and high relative humidities don't usually happen at the same time. The *relative* humidity is the amount of water vapor in the air divided by the maximum amount that the air can hold. Air at 94° can hold about 30% more water vapor than air at 86°. A morning in Houston might well have a temperature of 86° and a relative humidity of 80%, for a heat index of 100°. If the absolute humidity stayed the same in the afternoon when the temperature hit 94°, then the *relative* humidity would drop to about 62%, for a heat index of about 111°. That's plenty hot, but it's nothing like 129°.
>
> On the flip side, falling temperatures make the relative humidity rise. Evening thunderstorms can appear in the summer, seemingly from nowhere, as the relative humidity approaches 100% and the water vapor already in the air coalesces, first into clouds and then into rain.

Partial Derivatives in General. Now that we've endured the heat index, we can return to our air-conditioned classroom and study an arbitrary function $f(x, y)$ of two variables.

> **Definition of partial derivatives:**
>
> $$\frac{\partial f}{\partial x}(a, b) = \lim_{h \to 0} \frac{f(a+h, b) - f(a, b)}{h},$$
>
> (10.11) $$\frac{\partial f}{\partial y}(a, b) = \lim_{k \to 0} \frac{f(a, b+k) - f(a, b)}{k}.$$
>
> We can also think of $\frac{\partial f}{\partial x}$ and $\frac{\partial f}{\partial y}$ as functions of x and y:
>
> $$\frac{\partial f}{\partial x}(x, y) = \lim_{h \to 0} \frac{f(x+h, y) - f(x, y)}{h},$$
>
> (10.12) $$\frac{\partial f}{\partial y}(x, y) = \lim_{k \to 0} \frac{f(x, y+k) - f(x, y)}{k}.$$

10.1. Partial Derivatives

If we define $g(x) = f(x, b)$ and $h(y) = f(a, y)$, then $\frac{\partial f}{\partial x}(a, b) = g'(a)$ and $\frac{\partial f}{\partial y}(a, b) = h'(b)$. As with ordinary derivatives, the procedure we use for finding the derivative at a point (a *number*) can be applied everywhere at once to get the derivative as a *function* of x and y. To save space, we often denote partial derivatives with subscripts. f_x means $\partial f/\partial x$ and f_y means $\partial f/\partial y$.

Calculating partial derivatives is easy. To compute $\partial f/\partial x$, we *just take the derivative of $f(x, y)$ with respect to x while treating y as a constant*. This is because $\partial f/\partial x = g'(x)$, and $y = b$ is a constant in the function $g(x) = f(x, b)$. Likewise, to compute $\partial f/\partial y$, we treat x as a constant. Since the partial derivatives of $f(x, y)$ are *ordinary* derivatives of $g(x)$ and $h(y)$, all the formulas and rules that we developed for ordinary derivatives can be copied over verbatim for computing partial derivatives.

For instance, suppose that $f(x, y) = e^{xy^2 - 9}$. For any fixed $y = b$, we have

(10.13)
$$\frac{\partial e^{xy^2 - 9}}{\partial x} = \frac{d}{dx} e^{b^2 x - 9} = b^2 e^{b^2 x - 9} = y^2 e^{xy^2 - 9}.$$

Likewise, for any fixed $x = a$, we have

(10.14)
$$\frac{\partial e^{xy^2 - 9}}{\partial y} = \frac{d}{dy} e^{ay^2 - 9} = 2ay e^{ay^2 - 9} = 2xy e^{xy^2 - 9}.$$

We don't usually write the calculation out in such detail, replacing x and y in turn with constants a and b, taking an ordinary derivative with respect to the other variable and then replacing a and b with x and y. When taking a partial derivative with respect to x, we just *remember* that y is a constant. When taking a partial derivative with respect to y, we just *remember* that x is a constant. Using the chain rule (specifically, $(e^u)' = e^u u'$), we then write

$$\frac{\partial e^{xy^2 - 9}}{\partial x} = e^{xy^2 - 9} \frac{\partial (xy^2 - 9)}{\partial x} = y^2 e^{xy^2 - 9},$$

(10.15)
$$\frac{\partial e^{xy^2 - 9}}{\partial y} = e^{xy^2 - 9} \frac{\partial (xy^2 - 9)}{\partial y} = 2xy e^{xy^2 - 9}.$$

Higher Derivatives: Mixed Partials are Equal! The expressions f_x and f_y are functions in their own right, so we can take their derivatives to get second partial derivatives of $f(x, y)$. There are four of these, each of which can be written in several different ways:

$$f_{xx} = (f_x)_x = \frac{\partial^2 f}{\partial x^2} = \frac{\partial}{\partial x}\left(\frac{\partial f}{\partial x}\right),$$

$$f_{yy} = (f_y)_y = \frac{\partial^2 f}{\partial y^2} = \frac{\partial}{\partial y}\left(\frac{\partial f}{\partial y}\right),$$

$$f_{xy} = (f_x)_y = \frac{\partial^2 f}{\partial y \partial x} = \frac{\partial}{\partial y}\left(\frac{\partial f}{\partial x}\right),$$

(10.16)
$$f_{yx} = (f_y)_x = \frac{\partial^2 f}{\partial x \partial y} = \frac{\partial}{\partial x}\left(\frac{\partial f}{\partial y}\right).$$

The order of the variables is different in subscript and ∂ notation. f_{xy} means $\frac{\partial^2 f}{\partial y \partial x}$, not $\frac{\partial^2 f}{\partial x \partial y}$. Fortunately, it doesn't really matter, as we'll soon see.

To get a feel for these second derivatives, suppose that $f(x, y)$ is a quadratic function

(10.17) $$f(x, y) = Ax + By + Cx^2 + Dxy + Ey^2,$$

where A, B, C, D, and E are constants. Then

$$\begin{aligned} f_x(x, y) &= A + 2Cx + Dy, \\ f_y(x, y) &= B + Dx + 2Ey, \\ f_{xx} &= 2C, \\ f_{yy} &= 2E, \\ f_{xy} &= D, \\ f_{yx} &= D. \end{aligned}$$

(10.18)

Let's take a closer look at what these quantities mean

- f_x is just the ordinary first derivative of $g(x)$. It tells us how fast $f(x, y)$ is changing when we change x but don't change y. Likewise, f_y is just the ordinary first derivative of $h(y)$. It tells us how fast $f(x, y)$ is changing when we change y but don't change x.

- f_{xx} and f_{yy} are the ordinary second derivatives of $g(x)$ and $h(y)$. If we graphed $g(x)$ and $h(y)$, they would indicate the curvature of the graphs. They also indicate how far $f(x, y)$ is from the linear approximation $Ax + By$ when $x = 0$ or $y = 0$.

- The **mixed partial derivatives** f_{xy} and f_{yx} are more mysterious. To get f_{xy}, we first compute f_x by treating y as a constant, but then we take a derivative with respect to y! To get f_{yx}, we first compute f_y by treating x as a constant, but then we take a derivative with respect to x. In this example, both are equal to D and give a sense of the synergy between x and y. Both f_{xy} and f_{yx} capture what happens when we vary both x and y that we don't see by varying x and y separately.

Meanwhile, in our running example $f(x, y) = e^{xy^2 - 9}$,

$$\begin{aligned} f_{xx} &= \left(y^2 e^{xy^2-9}\right)_x = y^4 e^{xy^2-9}, \\ f_{yy} &= \left(2xy e^{xy^2-9}\right)_y = (2x + 4x^2 y^2) e^{xy^2-9}, \\ f_{xy} &= \left(y^2 e^{xy^2-9}\right)_y = (2y + 2xy^3) e^{xy^2-9}, \\ f_{yx} &= \left(2xy e^{xy^2-9}\right)_x = (2y + 2xy^3) e^{xy^2-9}. \end{aligned}$$

(10.19)

The last three calculations all used the product rule, taking the derivative of y^2 or $2xy$ as well as the derivative of e^{xy^2-9}.

In both of our examples, f_{xy} and f_{yx} happened to be equal. That wasn't a coincidence! The following theorem goes by several names. Some call it Clairaut's theorem, some call it Schwarz's theorem, and some call it Young's theorem, but most people just say "mixed partials are equal".

10.1. Partial Derivatives

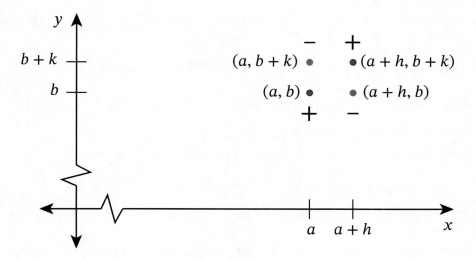

Figure 10.1. $Q(h, k)$ involves taking differences in both the x and y directions.

Theorem 10.1. *If f_{xy} and f_{yx} are continuous in a region around (a, b), then $f_{xy}(a, b) = f_{yx}(a, b)$.*

To see why this is true, we cook up a two-dimensional analogue $Q(h, k)$ of a difference quotient. We'll see that the limit of $Q(h, k)$, as h and k go to 0, is equal to both $f_{xy}(a, b)$ and $f_{yx}(a, b)$, so $f_{xy}(a, b)$ and $f_{yx}(a, b)$ must be equal. See Figure 10.1. We add the values of the function at the blue points marked $+$, subtract the values at the red points marked $-$, and divide by hk to get

(10.20) $$Q(h, k) = \frac{f(a+h, b+k) - f(a, b+k) - f(a+h, b) + f(a, b)}{hk}.$$

If $f(x, y)$ doesn't depend on y (that is, if $f(x, y)$ is really just a function of x), then $Q(h, k)$ is 0, since then $f(a+h, b+k) = f(a+h, b)$ and $f(a, b+k) = f(a, b)$. $Q(h, k)$ is also 0 for functions that don't depend on x, because in that case $f(a+h, b+k) = f(a, b+k)$ and $f(a+h, b) = f(a, b)$. This also means that $Q(h, k) = 0$ when $f(x, y)$ is the sum of a function of x and a function of y.

Before talking about functions in general, let's see how this works for the quadratic function (10.17). The only term that depends on both x and y is Dxy, so to get $Q(h, k)$ for the function (10.17), we can ignore all the other terms and compute

$$Q(h, k) = \frac{D(a+h)(b+k) - Da(b+k) - D(a+h)b + Dab}{hk}$$

$$= D\frac{ab + ak + hb + hk - ab - ak - ab - hb + ab}{hk}$$

(10.21) $$= D\frac{hk}{hk} = D,$$

which is both f_{xy} and f_{yx}.

Returning to a general function $f(x, y)$, we take the limit of $Q(h, k)$ as h and k go to 0. If we first take the limit as $h \to 0$, we get

$$\lim_{h \to 0} Q(h, k) = \frac{1}{k} \lim_{h \to 0} \left(\frac{f(a+h, b+k) - f(a, b+k)}{h} - \frac{f(a+h, b) - f(a, b)}{h} \right)$$
(10.22)
$$= \frac{1}{k} \left(f_x(a, b+k) - f_x(a, b) \right).$$

The limit of *that* as $k \to 0$ is, by definition, $(f_x)_y(a, b)$.

On the other hand, if we first take the limit of $Q(h, k)$ as $k \to 0$, we get

$$\lim_{k \to 0} Q(h, k) = \frac{1}{h} \lim_{k \to 0} \left(\frac{f(a+h, b+k) - f(a+h, b)}{k} - \frac{f(a, b+k) - f(a, b)}{k} \right)$$
(10.23)
$$= \frac{1}{h} \left(f_y(a+h, b) - f_y(a, b) \right).$$

Taking a limit of that as $h \to 0$ gives $(f_y)_x(a, b)$. Since both $f_{xy}(a, b)$ and $f_{yx}(a, b)$ are the limit of $Q(h, k)$ as h and k go to 0, we must have $f_{xy}(a, b) = f_{yx}(a, b)$. Mixed partials are equal!

10.2. Linear Approximations

The Two-Dimensional Microscope Equation. Returning to our running example $f(x, y) = e^{xy^2 - 9}$, we want to use information about the function and its derivatives at $(1, 3)$ to approximate $f(0.99, 3.03)$. We define $g(x) = f(x, 3)$ and $h(y) = f(1, y)$. We have already computed f_x and f_y:

$$f_x(x, y) = y^2 e^{xy^2 - 9},$$
(10.24)
$$f_y(x, y) = 2xy e^{xy^2 - 9}.$$

Plugging in $(a, b) = (1, 3)$ gives

$$f(1, 3) = 1,$$
$$f_x(1, 3) = 9,$$
(10.25)
$$f_y(1, 3) = 6.$$

The microscope equation for $g(x)$ tells us what happens when we decrease x from 1 to 0.99:

(10.26) $\quad f(0.99, 3) = g(0.99) \approx g(1) - 0.01 g'(1) = f(1, 3) - 0.01 f_x(1, 3) = 0.91.$

That is, $f(x, y)$ increases by $f_x(a, b) \Delta x$ (which is negative) when x changes from a to $a + \Delta x$. Likewise, we know what happens when we increase y from 3 to 3.03 without changing x:

(10.27) $\quad f(1, 3.03) = h(3.03) \approx h(3) + 0.03 h'(1) = f(1, 3) + 0.03 f_y(1, 3) = 1.18.$

In this case, $f(x, y)$ increases by $f_y(a, b) \Delta y$.

10.2. Linear Approximations

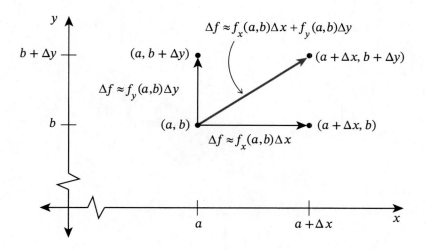

Figure 10.2. The change in $f(x, y)$ as we go along the diagonal arrow is approximately the sum of the changes along the horizontal and vertical arrows.

So what happens when we change both x and y, as in Figure 10.2? We just add up the pieces:

> **The two-dimensional microscope equation:**
> $$\Delta f \approx f_x \Delta x + f_y \Delta y,$$
> (10.28) $$f(x, y) \approx f(a, b) + f_x(a, b)(x - a) + f_y(a, b)(y - b).$$

In our particular example, with $x \approx 1$ and $y \approx 3$, we have

(10.29) $$f(x, y) \approx 1 + 9(x - 1) + 6(y - 3).$$

This is the best linear approximation to our function of two variables. Instead of giving the equation of the *line* tangent to the graph $y = f(x)$, it gives the equation of the *plane* tangent to the three-dimensional graph $z = f(x, y)$ at $(1, 3)$, as shown in Figure 10.3.

Of course, equation (10.28) is just an approximation. In our example,

$$f(0.99, 1.03) \approx 1 + 9(-0.01) + 6(0.03) = 1.09.$$

In fact, $f(0.99, 1.03) = 1.09318$ to five decimal places. There are three sources of error, and these are associated with the three second partial derivatives of $f(x, y)$, namely f_{xx}, f_{yy}, and f_{xy}. (We don't consider f_{yx} separately because it equals f_{xy}.)

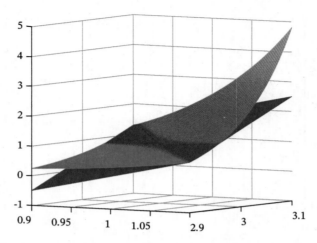

Figure 10.3. The graph of $f(x,y) = e^{xy^2-9}$ in red and the linear approximation (tangent plane) at $(1,3)$, $z = 1 + 9(x-1) + 6(y-3)$ in blue.

(1) The microscope equation for $g(x)$ isn't exact; $g''(1) = f_{xx}(1,3)$ gives a measure of how fast $g(x)$ is curving, and how far off our linear approximation for $g(0.99)$ is.

(2) The microscope equation for $h(y)$ isn't exact, and $f_{yy}(1,3)$ gives a measure of how far off our linear approximation for $h(1.03)$ is.

(3) $f(a+h, b+k) - f(a,b)$ isn't exactly the sum of $f(a+h, b) - f(a,b)$ and $f(a, b+k) - f(a,b)$. The difference between them is

$$\begin{aligned} f(a+h, b+k) - f(a+h, b) - f(a, b+k) + f(a,b) &= hkQ(h,k) \\ &\approx hk f_{xy}(a,b). \end{aligned}$$
(10.30)

If the linear approximation isn't accurate enough for us, we can take these errors into account and obtain a quadratic approximate that is more accurate. If (x,y) is close to (a,b), then the best quadratic approximation is

$$\begin{aligned} f(x,y) &\approx f(a,b) + f_x(a,b)\Delta x + f_y(a,b)\Delta y \\ &+ \frac{1}{2} f_{xx}(a,b) \Delta x^2 + \frac{1}{2} f_{yy}(a,b) \Delta y^2 + f_{xy}(a,b) \Delta x \Delta y, \end{aligned}$$
(10.31)

where $\Delta x = x - a$ and $\Delta y = y - b$. Incidentally, the best quadratic approximation to a function of one variable is

$$f(x) \approx f(a) + f'(a)\Delta x + \frac{f''(a)}{2} \Delta x^2.$$
(10.32)

We will cover quadratic, cubic, and higher-order approximations to functions of one variable in Chapter 11.

The Two-Dimensional Chain Rule and the Product Rule Revisited. Suppose that x and y are themselves functions of another variable, such as t. This makes $f(x,y)$ a function of t. What is df/dt? If we know the temperature and humidity, and we know how fast the temperature and humidity are changing, then how fast is the heat index changing?

To figure this out, we can divide the first line of (10.28) by Δt,

$$(10.33) \qquad \frac{\Delta f(x,y)}{\Delta t} \approx f_x(x,y)\frac{\Delta x}{\Delta t} + f_y(x,y)\frac{\Delta y}{\Delta t},$$

and take a limit as $t \to 0$ to get:

The two-dimensional chain rule:
$$(10.34) \qquad \frac{df(x,y)}{dt} = f_x(x,y)\frac{dx}{dt} + f_y(x,y)\frac{dy}{dt}.$$

The two-dimensional chain rule also tells us things about ordinary derivatives. If we apply the chain rule to the function $f(x,y) = xy$, then we have

$$(10.35) \qquad \frac{d(xy)}{dt} = y\frac{dx}{dt} + x\frac{dy}{dt},$$

since $f_x(x,y) = y$ and $f_y(x,y) = x$. This is our old friend the product rule! The notation is a little different from what we are used to, since usually our functions are $f(t)$ and $g(t)$ (or $f(x)$ and $g(x)$) instead of x and y, but the result is exactly the same. The derivative of a product is the second factor times the derivative of the first, plus the first factor times the derivative of the second.

10.3. Double Integrals and Iterated Integrals

Amber Waves of Grain. Kansas produces a lot of wheat, more than any other state in the nation, earning Kansas the nickname of "the wheat state". Based on its size, can we figure out how much wheat it produces?

The area of Kansas is about 82,300 square miles. The average wheat-growing farm in the United States produces about 50 bushels of wheat[3] per acre per year, or 32,000 bushels per square mile per year. So you might expect Kansas to produce around $32,000 \times 82,300 = 2.63$ billion bushels of wheat per year.

In fact, average annual wheat production in the 2010s was only about 345 million bushels/year. The reason is that not every acre in Kansas produces 50 bushels of wheat per year. The western half of the state is much drier than the eastern half, making it difficult to grow wheat there. (Agriculture in western Kansas is mostly ranching, not farming.) Even in the eastern half, not all farms grow wheat. Some grow corn or soybeans. Some parts of Kansas are forest or parkland or urban and don't grow any crops at all. Where wheat is grown, not all farms produce 50 bushels per acre per year. (The average in Kansas is about 40.) To understand wheat production in Kansas, we need to take all that regional variation into account.

Kansas is approximately a rectangle, 400 miles from east to west and 210 miles from north to south.[4] We can represent it as the rectangle in the x-y plane with vertices at $(0,0)$, $(400,0)$, $(0,210)$, and $(400,210)$, where x and y are measured in miles. Let $f(x,y)$ be the average wheat productivity, in bushels per square mile per year, *as a function of position*.

---

[3] 50 bushels weighs 3,000 pounds and provides about 4.5 million calories.
[4] That makes 84,000 square miles, but about 1700 square miles in the northeast corner of the rectangle are on the other side of the Missouri River and belong to Missouri.

Given the function $f(x, y)$, we could figure out the average annual wheat production as follows:

(1) Break Kansas into many little pieces, as in Figure 10.4, which we'll call boxes, each so small that $f(x, y)$ is nearly constant in that box.
(2) Pick a representative point (x_i^*, y_i^*) in each box.
(3) Estimate the wheat production of the ith box by multiplying $f(x_i^*, y_i^*)$ by the area of the ith box.
(4) Add up the contributions of all the boxes.
(5) For greater accuracy, use smaller boxes. For an exact answer, take a limit as the number of boxes goes to ∞ and the size of each box goes to 0.

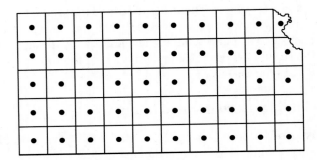

Figure 10.4. Breaking Kansas into smaller pieces

In other words, figuring out a bulk quantity like wheat production in a two-dimensional region follows *exactly* the same slice-and-dice strategy as in one dimension. The only differences are that we multiply our function by area, not by width, and that breaking a two-dimensional region into boxes is more complicated than breaking an interval into pieces.

As in one dimension, the sum is called a **Riemann sum**, the limit of the sum is called an **integral**, and we write

(10.36) $$\iint_K f(x, y) dA = \lim \sum f(x_i^*, y_i^*) \Delta A_i.$$

Here K (for Kansas) is the region we are integrating over, ΔA_i is the area of the ith box, and we use the symbol $\iint$ instead of $\int$ to emphasize that we are working in two dimensions instead of one. If we were integrating a function of three variables over a three-dimensional region R, we might write $\iiint_R f(x, y, z) dV$, where V stands for volume. We also sometimes call $\iint_K f(x, y) dA$ a **double integral** and $\iiint_R f(x, y, z) dV$ a **triple integral**. In general, we call a function of two (or more) variables **integrable** over a region K if the limit of the Riemann sum exists and doesn't depend on the choices we make. As in one dimension, all continuous functions are integrable, and many functions that aren't continuous are also integrable.

10.3. Double Integrals and Iterated Integrals

It's important to remember that double and triple integrals are *integrals*, meaning limits of sums, not *anti-derivatives*. The Fundamental Theorem of Calculus was just a tool for computing one-dimensional integrals, and it doesn't directly apply in two or more dimensions.

Evaluating a Double Integral Numerically. The actual function describing wheat production is complicated, so we'll switch our focus to a simpler (but more contrived) example. Let's estimate the integral

(10.37) $$\iint_R xe^{-xy} dA,$$

where $R = [0,2] \times [0,1]$ is the rectangle with vertices at the origin, $(0,1)$, $(2,0)$, and $(2,1)$. We divide R into twenty 0.4×0.25 boxes, as in Figure 10.5. We could number the boxes from 1 to 20, but it's more practical to label them with two indices i and j. The index i runs from 1 to 5 and says which column a box is in, while j runs from 1 to 4 and says which row it is in. As with our one-dimensional integrals, we pick points $x_0 = 0$, $x_1 = 0.4$, $x_2 = 0.8$, $x_3 = 1.2$, $x_4 = 1.6$, and $x_5 = 2$ that divide the interval $[0,2]$ into equal sized pieces. We similarly pick points $y_0 = 0$, $y_1 = 0.25$, $y_2 = 0.5$, $y_3 = 0.75$, and $y_4 = 1$. The ijth box is then a rectangle with vertices at (x_{i-1}, y_{j-1}), (x_{i-1}, y_j), (x_i, y_{j-1}), and (x_i, y_y). Each box has width $\Delta x = 0.4$, height $\Delta y = 0.25$, and area $\Delta x \Delta y = 0.1$.

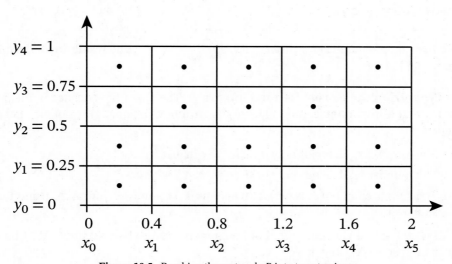

Figure 10.5. Breaking the rectangle R into twenty pieces

For each box, we need to pick a representative point (x_{ij}^*, y_{ij}^*). The simplest way to do this is to make x_{ij}^* depend only on i and y_{ij}^* depend only on j. That is, we pick just one value of x (call it x_i^*) that applies to every box in the ith column, and just one value of y (call it y_j^*) that applies to every box in the jth row. We will use the midpoint rule $x_i^* = (x_{i-1} + x_i)/2$, $y_j^* = (y_{j-1} + y_j)/2$, but we could also use left or right endpoints for

x_i^* or use top or bottom endpoints for y_j^*. Our two-dimensional Riemann sum is then

(10.38) $$\sum_{i=1}^{5}\sum_{j=1}^{4} x_i^* e^{-x_i^* y_j^*}\Delta x \Delta y.$$

To evaluate this sum with the midpoint rule, we use a simple MATLAB program. It is written in a way that is easily modified for other functions or other rectangles or other numbers of horizontal and vertical divisions.

```
program RIEMANN2D                      % No input parameters.
sum=0;                                  % Initialize our sum at 0
a=0; b=2;                               % x ranges from 0 to 2
c=0; d=1;                               % y ranges from 0 to 1
N=5; M=4;                               % 5 columns and 4 rows
deltax=(b-a)/N; deltay=(d-c)/M;         % Compute Δx and Δy
for k=1:N
    xstar=a+(k-1/2)*deltax;             % x_k* = (x_{k-1}+x_k)/2.
    for j=1:M
        ystar=c+(j-1/2)*deltay;         % y_j* = (y_{j-1}+y_j)/2.
        sum=sum+xstar*exp(-xstar*ystar); % Add f(x_{kj}*, y_{kj}*) to sum
end; end;                               % Finish the loops
sum=deltax*deltay*sum                   % Multiply sum by ΔxΔy.
```

Note that the program uses k instead of i, since $i = \sqrt{-1}$. The result, to six decimal places, is 1.135906.

Organizing the Boxes: Iterated Integrals. The algorithm encoded in the program has three important features that are good to follow in (almost) all double integral calculations:

(1) The number x_{ij}^* depends only on i, so we can just write x_i^*.
(2) The number y_{ij}^* depends only on j, so we can just write y_j^*.
(3) The sum over all boxes is nested, where we first sum over j for fixed i, and then sum over i.

If we want to divide our region of integration N times horizontally and M times vertically, instead of 5 and 4, we have

(10.39) $$\text{Riemann sum} = \sum_{i=1}^{N}\sum_{j=1}^{M} f(x_i^*, y_j^*)\Delta y \Delta x.$$

For each value of i, the sum over j gives the contribution of a column of boxes, as in Figure 10.6. All the boxes in a column have the same representative x value, and the contribution of the column is proportional to the width of the column, with the proportionality constant being a function of x. Summing over i means adding up all the columns to get the contribution of the entire region of integration.

10.3. Double Integrals and Iterated Integrals

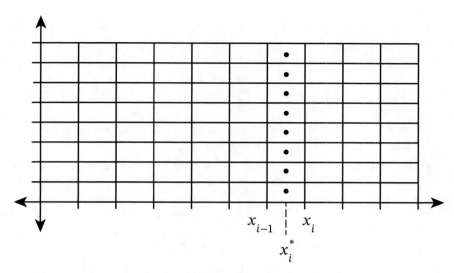

Figure 10.6. Adding up the boxes one column at a time

As with partial derivatives, our slogan is *one variable at a time!* We can rewrite (10.39) to make the iterated sum more explicit:

$$(10.40) \qquad \text{Riemann sum} = \sum_{i=1}^{N} \left(\sum_{j=1}^{M} f(x_i^*, y_j^*) \Delta y \right) \Delta x.$$

The sum in parentheses is a sum over y for a fixed value of x, namely x_i^*. As $M \to \infty$ it becomes a one-dimensional *integral* $\int_c^d f(x_i^*, y) dy$. The variable of integration is y, and y takes on all values from c to d. We want to do this for many different values of x_i^*, so we define the function

$$(10.41) \qquad g(x) = \int_c^d f(x, y) dy,$$

where in the integral we *treat x as a constant*.

For integrating $\iint_R xe^{-xy} \, dA$, our one-dimensional integral becomes

$$(10.42) \qquad g(x) = \int_0^1 xe^{-xy} dy.$$

Since the derivative of $-e^{-xy}$ with respect to y (treating x as a constant) is xe^{-xy}, the second Fundamental Theorem of Calculus tells us that

$$(10.43) \qquad g(x) = \int_0^1 xe^{-xy} dy = -e^{-xy} \Big|_{y=0}^{1} = 1 - e^{-x}.$$

Holding N fixed, the limit of our Riemann sum as $M \to \infty$ is then

$$(10.44) \qquad \lim_{M \to \infty} \text{Riemann sum} = \sum_{i=1}^{N} (1 - e^{-x_i^*}) \Delta x,$$

where we have simply replaced $\sum_{j=1}^{M} x_i^* e^{-x_i^* y_j^*} \Delta y$ with $g(x_i^*) = \int_0^1 x_i^* e^{-x_i^* y} dy = 1 - e^{-x_i^*}$. As $N \to \infty$, the remaining sum also becomes a one-dimensional integral, so

$$(10.45) \quad \iint_R xe^{-xy} dA = \int_0^2 (1 - e^{-x}) dx = x + e^{-x}\Big|_0^2 = 1 + e^{-2} \approx 1.135335.$$

This answer is exact. The approximate answer of 1.135906 that we got from using 20 boxes and midpoints was accurate to three decimal places.

The upshot of this procedure is that we can replace a **double** integral over a two-dimensional region with an **iterated integral**, where we first integrate $f(x, y) dy$ for each value of x to get a function $g(x)$, and then integrate $g(x) dx$ to get a number. To evaluate those one-dimensional integrals, we can use everything that we previously learned, and in particular we can use the Fundamental Theorem of Calculus. Everything that we did with our example can be summarized as

$$\iint_R xe^{-xy} dA = \int_{x=0}^2 \left(\int_{y=0}^1 xe^{-xy} dy \right) dx$$

$$= \int_{x=0}^2 \left(-e^{-xy} \Big|_{y=0}^1 \right) dx$$

$$= \int_{x=0}^2 (1 - e^{-x}) dx$$

$$= x + e^{-x}\Big|_{x=0}^2$$

$$(10.46) \qquad = 1 + e^{-2}.$$

You may (should!) wonder why we summed the boxes in each column and then added up the columns. Why didn't we sum the boxes in each row and then add up the rows? In fact, we could have. In general, either procedure is legal. This is expressed formally as the following.

Theorem 10.2 (Fubini's theorem). *If $R = [a, b] \times [c, d]$ is a rectangle and $f(x, y)$ is an integrable function on R, then the double integral $\iint_R f(x, y) dA$ can be expressed as an iterated integral in two ways, and both give the same answer:*

$$(10.47) \quad \iint_R f(x, y) dA = \int_{x=a}^b \left(\int_{y=c}^d f(x, y) dy \right) dx = \int_{y=c}^d \left(\int_{x=a}^b f(x, y) dx \right) dy,$$

where in the integrals inside the parentheses the other variable is treated as a constant.

Sometimes it's easier to integrate first over y and then over x. Other times it's easier to integrate first over x and then over y. Still other times both iterated integrals are equally easy or equally hard. In our example, we could have written

$$(10.48) \quad \iint_R f(x, y) dA = \int_{y=0}^1 \left(\int_{x=0}^2 xe^{-xy} dx \right) dy.$$

10.3. Double Integrals and Iterated Integrals

This is 100% correct, but is almost useless. The inner integral requires integration by parts and works out to be $\frac{1}{y^2} - \left(\frac{1}{y^2} - \frac{2}{y}\right)e^{-2y}$. Integrating *that* from 0 to 1 requires even more advanced methods. Integrating first over y and then over x is much, much easier.

To recap: Fubini's theorem says that we always *can* write a double integral as an iterated integral, integrating first over y and then over x, or integrating first over x and then over y. However, this doesn't mean that we always *should* do that. Most of the time we should, but not always. Sometimes it makes more sense to organize the boxes differently, or to use boxes that aren't even rectangles.

For instance, suppose that our region of integration is a dartboard, as in Figure 10.7, and that the function being integrated is the value of a throw. In that case, we should pick our boxes to be exactly the regions defined by the rules of the game! A similar division applies to any region of integration that has rotational symmetry. Instead of having columns and rows, we have pie wedges and rings. Instead of getting iterated integrals where you integrate first over y and then over x (or vice versa), this scheme gives us iterated integrals where we first integrate over the radius r and then over the angle θ, or vice versa. We will explore integration over such round regions in Exercises 10.30–10.32.

Figure 10.7. A region of integration that shouldn't be divided into rectangles. (Photo Xygo_bg]/ iStock/Getty Images Plus, via Getty Images.)

The details depend on the problem, but the general philosophy is always the same. To do double (and triple) integrals, organize your boxes so that you can do your Riemann sums, and ultimately your integrals, *one variable at a time*.

10.4. Chapter Summary

The Main Ideas. If $f(x, y)$ is a function of two variables, then:

- Almost all problems involving $f(x, y)$ are best handled *one variable at a time*.
- The partial derivative $f_x = \partial f / \partial x$ is the derivative of f with respect to x treating y as a constant. It is also the derivative of $g(x) = f(x, b)$, where y literally is a constant. Similarly, $f_y = \partial f / \partial y$ is the derivative of $h(y) = f(a, y)$.
- Computing partial derivatives is easy. Just use all the rules you already know, treating the other variable as a constant.
- f_x and f_y give us our best linear approximation to $f(x, y)$:

$$(10.49) \qquad f(x, y) \approx f(a, b) + f_x(a, b)(x - a) + f_y(a, b)(y - b).$$

- If $z = f(x, y)$ and both x and y are functions of some other variable t, then

$$(10.50) \qquad \frac{dz}{dt} = \frac{\partial f}{\partial x} \frac{dx}{dt} + \frac{\partial f}{\partial y} \frac{dy}{dt}.$$

This is the two-dimensional chain rule.

- f_x and f_y are functions in their own rights, whose derivatives f_{xx}, f_{xy}, f_{yx}, and f_{yy} are called the second partial derivatives of f.
- Mixed partials are equal. If f_{xy} and f_{yx} are continuous, then $f_{xy} = f_{yx}$.
- Bulk quantities in two dimensions come from double integrals. $\iint_R f(x, y) \, dA$ is what we get when we break the region R into boxes, add up $f(x, y)$ times the area of each box, and take a limit as the boxes get smaller and smaller.
- By definition, double integrals can be approximated by Riemann sums. The methods are very similar to those for one-dimensional integrals.
- By organizing the boxes intelligently, we can express our double integral as an iterated integral $\int_a^b \left(\int_c^d f(x, y) \, dy \right) dx$ or $\int_c^d \left(\int_a^b f(x, y) \, dx \right) dy$.
- The Fundamental Theorem of Calculus does not apply to double integrals. However, it can be applied to each stage of an iterated integral.

Expectations. You should be able to:

- Explain what partial derivatives, double integrals, and iterated integrals are.
- Explain the difference between double integrals and iterated integrals.
- Compute partial derivatives both numerically and from formulas.
- Use partial derivatives to approximate functions near known points.
- Express bulk quantities as double integrals.
- Approximate double integrals with Riemann sums.
- Program a computer to do the grungy calculations that this involves.
- Express a double integral as an iterated integral in two different ways.
- Use the Fundamental Theorem of Calculus to evaluate iterated integrals.

10.5. Exercises

10.1. Find the partial derivatives f_x and f_y of each function.
  (a) $f(x, y) = xy^3$
  (b) $f(x, y) = \sqrt{2x + 5y}$
  (c) $f(x, y) = \cos(xy)$
  (d) $f(x, y) = e^{xy}$
  (e) $f(x, y) = 5\dfrac{y^2}{x^3} - x^3 \sin(y) + \pi$
  (f) $f(x, y) = y \tan(x)$
  (g) $f(x, y) = \dfrac{xy + y^2}{x}$
  (h) $f(x, y) = \ln(x^2 + xe^y)$

10.2. In parts (a)–(d), find the second partial derivatives f_{xx}, f_{yy}, f_{xy}, and f_{yx} of each function.
  (a) $f(x, y) = y \tan(x)$
  (b) $f(x, y) = 5\dfrac{y^2}{x^3} - x^3 \sin(y) + \pi$
  (c) $f(x, y) = e^{xy}$
  (d) $f(x, y) = xy^3$
  (e) What did you notice about f_{xy} and f_{yx}?

10.3. Consider the function $f(x, y) = \ln\left(\dfrac{1 + xy^2}{2}\right)$.
  (a) Compute $f(1, 1)$, $f_x(1, 1)$, and $f_y(1, 1)$.
  (b) Use the two-dimensional microscope equation to estimate $f(1.03, 0.99)$.

10.4. Consider the function $f(x, y) = \dfrac{\sin(x)}{1 + y^2}$.
  (a) Compute $f(\pi/4, 1)$, $f_x(\pi/4, 1)$, and $f_y(\pi/4, 1)$.
  (b) Use the two-dimensional microscope equation to estimate $f(\pi/3, 1.1)$.

10.5. Ideal Gas Law. The volume V of a given quantity of gas is a function of the temperature T (in **kelvins**) and the pressure p. (Yes, chemists and physicists always use capital letters for temperature and volume but a lowercase p for pressure.) In a so-called ideal gas the relationship between volume and pressure is given by the ideal gas law:

$$pV = nRT, \quad \text{or equivalently} \quad V = nR\dfrac{T}{p},$$

where where n is the amount of gas (measured in **moles**) and R is a universal constant. In parts (a)–(c) of this exercise, we are thinking of V as a function of T and p, with n and R constant.
  (a) Find formulas for the partial derivatives $V_T(T, p)$ and $V_p(T, p)$.
  (b) The constant R equals 8.3×10^3 newton-meters per kelvin. (A **newton** is a unit of force in the MKS unit system.) Check that the units in the ideal gas law are consistent if V is measured in cubic meters, T in kelvins, and p in newtons per square meter, or **pascals**.
  (c) Suppose that one mole of gas at 280 kelvins is under a pressure of 30 newtons per square meter. What is its volume?

(d) The ideal gas law can also be used to express pressure as a function of temperature and volume. If the temperature of the gas in part (c) increases by 8 kelvins and the volume increases by 0.5 cubic meter, will the pressure increase or decrease? By about how much?

10.6. Write the formula for a linear function $f(x, y)$ with the following properties:

$$f_x(x, y) = -1.5 \quad \text{for all } x \text{ and } y,$$
$$f_y(x, y) = 1.8 \quad \text{for all } x \text{ and } y,$$
$$f(2, 1) = 6.0.$$

10.7. Suppose that $z = f(x, y)$ and

$$f(100, 100) = 30, \quad f_x(100, 100) = -4, \quad f_y(100, 100) = 5.$$

(a) Estimate $f(99, 100)$, $f(100, 101)$, $f(99, 101)$, and $f(98, 103)$.
(b) When $x = 100$ and $y = 100$, how much change in z does a 2% increase in x cause? How much does a 2% increase in y cause? Which has a larger effect: a 2% increase in x or a 2% increase in y? Why?

10.8. A forester who wants to know the height of a tree walks 200 feet from its base, sights to the top of the tree, and finds the resulting angle to be 50 degrees.
(a) Draw a diagram illustrating the given information.
(b) Use trigonometry and the forester's measurements to calculate the height of the tree.
(c) Suppose the 200-foot measurement has an error of up to 2 feet and the angle measurement has an error of up to 4 degrees. Use a linear approximation to estimate the biggest possible error in the calculated height of the tree. (*Hint*: Don't forget to convert your angle measures to radians!)
(d) Which would improve the accuracy of the forester's measurements more: a line that gives the exact distance to the tree, or a sight that gives the exact angle? Explain.

Capital and Labor Costs. The costs of running a business are often divided between capital costs (buildings, equipment, raw materials, etc.) and labor costs (salaries and benefits for workers). The revenue, and thus the profit, may be a complicated function of capital and labor, so financial managers often have to decide whether to buy more equipment, hire more workers, both, or neither.

10.9. Let $P = f(x, y)$ represent the functional relationship between the monthly profit P, the monthly labor costs x, and the monthly capital costs y (all measured in thousands of dollars) for a small manufacturing company. Current spending levels for labor and capital are $x = 25$ and $y = 30$. Company financial experts have determined that

$$\frac{\partial P}{\partial x}(25, 30) = -0.15 \quad \text{and} \quad \frac{\partial P}{\partial y}(25, 30) = -0.20.$$

(a) Estimate how the monthly profit changes if monthly capital expenses increase to $32,000.
(b) The company is considering adding labor to increase production. Each additional hire adds $3,000 to the monthly labor expense. Estimate how the

monthly profit changes if the company adds one person to the workforce. What is the rate of change of profit, in thousands of dollars per person, for adding labor? Is the rate positive or negative?

(c) The company's product is becoming very popular. To meet the increase in demand, management adds two workers. Approximately how much does this change the monthly profit?

(d) If the company wants to maintain the same profit level, they could try to cut capital expenses. What change in y is needed to keep profit unchanged when two workers are added? (This is called a **trade-off**.)

The Cobb-Douglas Production Equation. In manufacturing, managers must decide how to allocate funds between capital and labor to optimize production. Charles Cobb and Paul Douglas modeled the relationship between production (Q), labor (L), and capital (K) by the equation of the form

$$Q = f(L, K) = AL^\alpha K^\beta.$$

The parameter A represents the overall efficiency of production, and the parameters α and β (both between 0 and 1) determine the responsiveness of productivity to changes in labor and capital, respectively. Exercises 10.10–10.12 explore properties of this **Cobb-Douglas equation**.

10.10. Depending on the values of α and β, the Cobb-Douglas equation $Q = f(L, K) = AL^\alpha K^\beta$ can exhibit different scaling behavior. Let $Q_0 = AL_0^\alpha K_0^\beta$ be the initial production rate with initial labor input L_0 and initial capital input K_0.

(a) Suppose that $\alpha + \beta = 1$. Show that when the inputs K and L are increased by the same factor λ, to λK_0 and λL_0, the resulting production level Q_1 is equal to λQ_0. What does this imply about production if both inputs are doubled? (A production function with this property has "constant returns to scale".)

(b) Next suppose that $\alpha + \beta > 1$. Show that when both inputs K and L are increased by the same factor λ, the resulting production level Q_2 is greater than λQ_0. What does this imply about production if both inputs are doubled? (A production function with this property has "increasing returns to scale".)

(c) Finally, suppose that $\alpha + \beta < 1$. Show that when K and L are increased by the same factor λ, the resulting production level Q_3 is less than λQ_0. What does this imply about production if both inputs are doubled? (A production function with this property has "decreasing returns to scale".)

10.11. For a production function of the form $Q = f(L, K)$, we define the **marginal product of labor** (MPL) to be $\frac{\partial Q}{\partial L}$, and the **marginal product of capital** (MPC) to be $\frac{\partial Q}{\partial K}$. For the Cobb-Douglas equation $Q = f(L, K) = AL^\alpha K^\beta$:

(a) Show that the MPL is equal to $\alpha \frac{Q}{L}$. Explain why the MPL is always greater than 0.

(b) Show that the MPC is equal to $\beta \frac{Q}{K}$. Explain why the MPC is always greater than 0.

(c) Show that $\frac{\partial^2 Q}{\partial L^2}$ is equal to $-\alpha(1-\alpha)\frac{Y}{L^2}$. Why is this quantity negative? (A similar argument can be made for capital expenditures and β.)

10.12. A company manufactures medical supplies. The company's productivity depends on both labor (x) capital (y). The function

$$f(x, y) = 60\sqrt[3]{xy}$$

provides a good estimate of how this company's productivity $f(x, y)$ depends on both labor and capital.

(a) Explain why $f(x, y)$ is a Cobb-Douglas equation. What are the values of A, α, and β?
(b) Find $f_x(x, y)$ and $f_y(x, y)$.

For parts (c)–(f), suppose the company is currently using 324 units of labor and 144 units of capital.

(c) Find the marginal product of labor (MPL) and the marginal product of capital (MPC), as defined in Exercise 10.11.
(d) Suppose that labor is increased to 330 units and capital is increased to 150 units. Use the two-dimensional microscope equation to estimate the new production level. Compare that to the exact value of $f(330, 150)$.
(e) Suppose that labor is increased to 330 units and capital is increased to 150 units. Use the quadratic approximation to estimate the new production level. Compare that to the exact value of $f(330, 150)$.
(f) To get the greatest increase in production for the least cost, should management increase labor or capital? Explain.

Utility Functions. The **utility** of a commodity is the amount of benefit, or satisfaction, that it gives a consumer. This varies from consumer to consumer.

For example, suppose that you like both white chocolate and dark chocolate, but you like dark chocolate three times as much as white chocolate. You might have a utility function of the form $U(x_1, x_2) = x_1 + 3x_2$, where x_1 and x_2 measure ounces of white and dark chocolate. If you have 20 ounces of white chocolate and 10 ounces of dark chocolate, then your utility is $U(20, 10) = 50$. If you have 10 ounces of white chocolate and 20 ounces of dark chocolate, your utility is $U(10, 20) = 70$, which implies that you would be 20 units of utility ("utils") happier than you are with 20 ounces of white and 10 ounces of dark.

If you have 5 ounces of white chocolate and 15 ounces of dark chocolate, then your utility is $U(5, 15) = 50$, which is the same as having 20 ounces of white chocolate and 10 ounces of dark chocolate. You might say that you are **indifferent** to these two scenarios.

An **indifference curve** shows all the amounts of commodities that give the consumer the same level of satisfaction. The indifference curve is generally written as an implicit function of the quantities of each commodity.

10.13. Bill likes to eat apples and plums together and buys bundles of fruit with x_A apples and x_P plums, which we denote (x_A, x_P). Suppose that Bill's utility function happens to be $U(x_A, x_P) = x_A x_P$.

10.5. Exercises

(a) Bill has 30 apples and 8 plums. What is Bill's utility? That is, what is $U(30, 8)$?

(b) The indifference curve through $(30, 8)$ includes all bundles (x_A, x_P) such that $x_A x_P$ is equal to what value?

(c) Write the equation of the indifference curve through $(30, 8)$.

(d) Graph the indifference curve showing all bundles that Bill likes exactly as well as the bundle $(30, 8)$.

Marginal Rate of Substitution. Returning to our white and dark chocolate example, suppose that someone wants half an ounce of your dark chocolate. Giving up that dark chocolate makes you a little less happy, so you demand some white chocolate to make up for it. The **marginal rate of substitution** (MRS) of dark chocolate for white chocolate is the amount of white chocolate that you would insist on per amount of dark chocolate that you gave up.

More generally, when dealing with two quantities A and B, the MRS of A for B is the amount of B that would compensate a consumer for each lost unit of A, in the limit of small exchanges. That is, starting with quantities (x_1, x_2) of A and B and changing to $(x_1 - \Delta x_1, x_2 + \Delta x_2)$ in a way that keeps the utility the same, the MRS is approximately $\Delta x_2 / \Delta x_1$. The exact MRS of A for B is the *limit* of $\Delta x_2 / \Delta x_1$ as both of them get small.

10.14. Explain why the MRS at (x_1, x_2) is *minus* the slope of the indifference curve at (x_1, x_2).

10.15. For the utility function defined in Exercise 10.13, find the MRS at $(30, 8)$ and again at $(24, 10)$. Explain why the MRS is different at these points, despite their being on the same indifference curve.

10.16. What happens to the MRS as you move along the indifference curve found in Exercise 10.13? What you are observing is called the **diminishing rate of marginal substitution**. Explain why this name makes sense.

Marginal utility is the rate of change in the utility function $U(x, y)$ per small change in x_1 or x_2. For the white chocolate/dark chocolate example, if someone gives you a small additional amount Δx_1 of white chocolate, causing an increase ΔU in utility, then the marginal utility of white chocolate is approximately $\Delta U / \Delta x_1$. Unlike the MRS, marginal utility can be defined separately for each commodity. When working with two commodities, these are often denoted MU_1 and MU_2. In the language of calculus,

$$MU_1 = \frac{\partial U}{\partial x_1} \quad \text{and} \quad MU_2 = \frac{\partial U}{\partial x_2}.$$

In our chocolate example we had $U(x_1, x_2) = x_1 + 3x_2$, so $MU_1 = 1$ and $MU_2 = 3$.

10.17. Suppose that you like eating turkey jerky together with beef jerky and that your utility function is $U(x_1, x_2) = x_1^2 \sqrt{x_2}$, where x_1 and x_2 are the number of ounces of turkey and beef jerky, respectively.

(a) Find $MU_1(4, 16)$ and $MU_2(4, 16)$ and interpret the results.

(b) Suppose you have 4 ounces of turkey jerky and 16 ounces of beef jerky. Your friend Kris wants to trade you some of her turkey jerky for 1 ounce of your beef jerky. Since you and Kris are friends, you want to do a fair trade that neither improves nor decreases your utility. Use the microscope equation to determine how much turkey jerky Kris should give you in exchange for your giving her 1 ounce of your beef jerky. (*Hint*: Recall that for a function $f(x, y)$ the microscope equation at the point (x_0, y_0) is $\Delta f \approx f_x(x_0, y_0)\Delta x + f_y(x_0, y_0)\Delta y$.)

(c) Compute the MRS for beef jerky to turkey jerky as in Exercise 10.15. Use this to determine how much turkey jerky Kris should give you in exchange for one ounce of beef jerky.

(d) How do your results in parts (b) and (c) compare? In general, how is the MRS for two quantities related to the marginal utilities of those quantities?

Competitive and Complementary Products. Two products are said to be **competitive** if an increased demand for one of the products is associated with an decreased demand for the other product. For instance, Coke and Pepsi are competitive products, since the more people drink Coke, the fewer will drink Pepsi.

Two products are said to be **complementary** if an increased demand for one product is associated with an increased demand for the other product. Cellphones and protective phone covers are complementary, as are notepads and pens. Stores often stock complementary products next to each other, so buying one will prompt you to buy the other.

Exercises 10.18–10.20 explore competition and complementarity using partial derivatives. We denote the prices of the two products as p_1 and p_2 and the demands as D_1 and D_2. Note that D_1 and D_2 depend on both p_1 and p_2. If a cellphone is too expensive to buy, you won't need a new protective case.

10.18. (a) In business terms, what do $\frac{\partial D_1}{\partial p_1}$ and $\frac{\partial D_2}{\partial p_2}$ mean? Why would we expect both of these partial derivatives to be negative?

(b) If two products are competitive, and the price of the first product increases while the price of the second stays the same, what will happen to the demand for the two products?

(c) If two products are complementary, and the price of the first product increases while the price of the second stays the same, what will happen to the demand for the two products?

(d) If $\frac{\partial D_1}{\partial p_2}$ and $\frac{\partial D_2}{\partial p_1}$ are both negative, are the products competitive, complementary, or neither?

(e) If $\frac{\partial D_1}{\partial p_2}$ and $\frac{\partial D_2}{\partial p_1}$ are both positive, are the products competitive, complementary, or neither?

(f) Explain why products with $\frac{\partial D_1}{\partial p_2} > 0$ and $\frac{\partial D_2}{\partial p_1} < 0$ (or vice versa) are neither competitive nor complementary.

10.19. List three pairs of products that are competitive, three that are complementary, and three that are neither.

10.20. For each pair of demand functions, indicate whether the products are competitive, complementary, or neither. Explain your reasoning.

(a) $D_1 = \dfrac{10 - 2p_1}{1 + p_2^2}$, $D_2 = \dfrac{3p_1}{2 + \sqrt{p_2}}$, where $p_1 < 5$.

10.5. Exercises

(b) $D_1 = 100 - 4p_1 + 3p_2,\qquad D_2 = 100 + 3p_1 - 3p_2.$
(c) $d_1 = 60 - 0.2p_1 - \sqrt{p_2},\qquad D_2 = 8000 - 0.5p_1 - 5p_2^2.$

Iterated Integrals. The remaining exercises all involve evaluating double integrals by expressing them as iterated integrals. Always remember to do your calculations *one variable at a time*.

10.21. Evaluate each iterated integral.
   (a) $\int_{-2}^{1} \int_{0}^{3} 12x^3 y^2 \, dy \, dx$
   (b) $\int_{0}^{\ln 2} \int_{0}^{2} x^3 e^y \, dx \, dy$

10.22. Use the MATLAB program Riemann2D to estimate the iterated integral
$$\int_{0}^{3} \int_{1}^{2} y e^{x^2} \, dx \, dy,$$
using $\Delta x = \Delta y = 0.1$.

10.23. (a) Write down and compute the Riemann sum that approximates
$$\iint_R x^3 + 2xy^2 \, dA$$
over the rectangle $R = [0, 3] \times [1, 3]$ by partitioning R into six squares, each with an area of one square unit, and selecting (x_i^*, y_i^*) to be the lower left corner of R_{ij}.
   (b) Convert the double integral from part (a) into an iterated integral.
   (c) Evaluate the iterated integral from part (b).

10.24. Let R be the region inside the unit circle in the first quadrant. That is, R is defined by $x \geq 0$, $y \geq 0$, and $x^2 + y^2 \leq 1$.

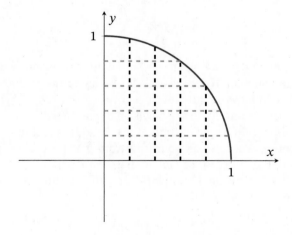

(a) Express $\iint_R xy \, dA$ as an iterated integral, where you integrate first over y and then over x. The subtlety is that our region of integration is not a rectangle. We are still arranging our boxes in columns, summing all the boxes in each column, and then adding up the columns, but the columns don't all have the same height. As a result, the limits of integration for your integral over y will depend on x.

(b) Evaluate the iterated integral from part (a).

10.25. The downtown region of a city consists of a 64-square-block grid that is laid out around two main streets, which are modeled by the x and y axis, with the center of the city at the origin. By using such a grid, each point in this downtown area can be described by coordinates (x, y) with $-4 \leq x \leq 4$ and $-4 \leq y \leq 4$. Property values (measured in millions of dollars per square block) are a function of location and are given by

$$V(x, y) = (10 + x + y)2^{-x}.$$

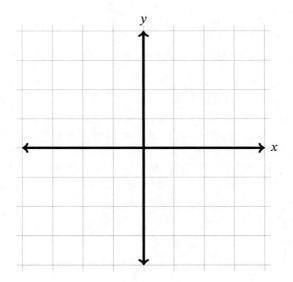

(a) To estimate the total value of the downtown region, we will partition it into 16 equal-sized squares. Make a copy of the grid and outline the 16 squares on it. Make sure to use plenty of space to draw your grid.

(b) What are Δx and Δy in the partition of part (a)?

(c) To estimate the value of each of the 16 square regions, we will use their midpoints. Indicate on your grid from part (a) the location and coordinates of the midpoint of each of the 16 squares.

(d) The two-dimensional Riemann sum which estimates total property value is given by

$$\sum_{j=1}^{m}\sum_{i=1}^{n} V\left(x_i^*, y_j^*\right) \Delta x_i \Delta y_j.$$

For your squares as described in part (c):

(i) What are the values of m and n?

(ii) List the coordinates $\left(x_j^*, y_i^*\right)$ for each of the squares.

(e) Write down a Riemann sum of the form $\sum_{i=1}^{m}\sum_{j=1}^{n} V(x_j^*, y_i^*)\Delta x_j \Delta y_i$ that estimates the total property value of the downtown region.

Average Values. In Chapter 9 we defined the average value of a function $f(x)$ on an interval $[a,b]$ to be $\frac{1}{b-a}\int_a^b f(x)\,dx$. The same idea works in two dimensions, only replacing the length of the region of integration with its area. If $f(x,y)$ is any function, then the average value of f over a region R, denoted $\bar{f}_R$, is given by

(10.51) $$\bar{f}_R = \frac{1}{\text{area of } R}\iint_R f(x,y)\,dA.$$

Exercises 10.26–10.29 involve computing average values.

10.26. Suppose an industry invests L million dollars in labor per month with $2 \leq L \leq 3$ and K million dollars of capital per month with $0.5 \leq K \leq 3$ in the production of an electric bicycle. The production of Q bicycles is given by the Cobb-Douglas equation
$$Q(L,K) = 1000 L^{0.7} K^{0.3}.$$
Find the average monthly output for the factory. Why are the units for your answer bicycles?

10.27. The state of Wyoming has a rectangular shape, which we model as a rectangle with vertices as (1,1), (4,1), (1,3), and (4,3). Most of Wyoming is desert, but the northwest corner gets a fair amount of rain. We model the average annual rainfall at position (x,y), in inches per year, by the (not very accurate!) function
$$W(x,y) = 5(5-x)^{0.3} y^{0.7}.$$
Find the average rainfall across the state.

10.28. In dry regions, animals tend to live near watering holes. Suppose that a watering hole is located at the origin and that the density of animals (number per square kilometer) is approximately
$$n(x,y) = 2000 - 30(x^2 + y^2),$$
where x and y are measured in kilometers.
 (a) What is the average density of animals in an 8 km × 8 km square region centered at the watering hole?
 (b) What is the total number of animals in that square?

10.29. The chance of a person with a contagious disease infecting somebody else in a social setting is generally assumed to be a function $f(r)$ of the distance r between them.
 (a) Explain why $f(r)$ should be a decreasing function of r.
 (b) Suppose you are located at the origin $(0,0)$ of the x,y plane. Explain why $r = \sqrt{x^2 + y^2}$.
 (c) For Covid, it was initially believed (without any good empirical data) that the virus could only be transmitted within a distance of $R = 6$ feet. (The parameter R is called the **radius of infectivity**.) Explain why it would be

reasonable to use $f(r) = c\left(1 - \frac{r^2}{R^2}\right)$, where c is a constant, for the risk of being a distance r away from an infected person.

(d) For simplicity, take $c = 1$ and $R = 6$ feet. Find the average value of $f(r)$ in the rectangle B_1 defined by $-3 \le x \le 3$ and $-1 \le y \le 2$.

(e) Find the average value of f in the rectangle B_2 defined by $-2 \le x \le 3$ and $1 \le y \le 5$.

(f) Which is riskier to your health, talking to an infected person in rectangle B_1 or an infected person in rectangle B_2?

Integration in Polar Coordinates. Exercises 10.30–10.32 involve integrating over circular regions. In Section 10.4, we saw that when integrating over some regions (like a dartboard), it's a bad idea to integrate first over x and then over y, or vice versa. Instead, we need different kinds of boxes.

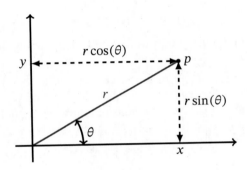

For a circular region centered at the origin, such as the unit disk, it's often useful to describe points by their distance from the origin and their angle above the x axis, instead of by their x and y coordinates. That is, we use **polar coordinates** r and θ, where

(10.52) $$x = r\cos(\theta), \qquad y = r\sin(\theta).$$

We can also express r and θ in terms of x and y:

(10.53) $$r = \sqrt{x^2 + y^2}, \qquad \theta = \tan^{-1}(x/y).$$

The second equation of (10.53) only determines θ up to a multiple of π. For instance, x/y is the same for $x = y = 1$ as for $x = y = -1$, but the first has $r = 1$ and $\theta = \pi/4$, while the second has $r = 1$ and $\theta = 5\pi/4$. Equations (10.52) make sense for any values of r and θ, but we usually restrict our attention to $r \ge 0$ and $0 \le \theta \le 2\pi$.

10.5. Exercises

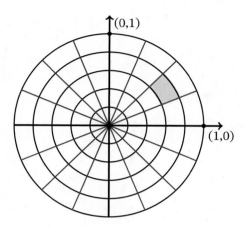

When working with rectangular regions, our boxes are usually smaller rectangles, where each box has a limited range of x-values, of size Δx, and a limited range of y-values, of size Δy. When working with polar coordinates, we divide things into so-called **polar rectangles**. These boxes have a limited range of r-values (running from r_{i-1} to r_i in the ith ring) and a limited range of θ-values (of size $\Delta\theta$). Instead of assembling into columns and rows, they assemble into rings and sectors. When Δr and $\Delta\theta$ are small, each box is roughly rectangular in shape, with the straight edges having length Δr and the (slightly) curved edges having length $r_{i-1}\Delta\theta$ and $r_i\Delta\theta$. This means that the area of the box is between $r_{i-1}\Delta r\Delta\theta$ and $r_i\Delta r\Delta\theta$, and so is approximately $r_i^*\Delta r\Delta\theta$, where r_i^* is our favorite number between r_{i-1} and r_i.

Note that the area of a box is *not* $\Delta r\Delta\theta$! If we are measuring length in meters, then $r_i^*\Delta r\Delta\theta$ has units of square meters, which makes sense for an area. On the other hand, $\Delta r\Delta\theta$ has units of meters, which doesn't make sense. Compared to the formula for integrating $dx\,dy$, there is an extra factor of r in the formula for integration using polar coordinates.

To recap: If we have a function $f(r,\theta)$ that we want to integrate over a polar region, the contribution of each box is approximately

(10.54) $$f(r_i^*,\theta_j^*)r_i^*\,\Delta r\,\Delta\theta.$$

Adding up the boxes in each ring and then adding up the rings gives a nested sum

(10.55) $$\sum_{i=1}^{N}\sum_{j=1}^{M} f(r_i^*,\theta_j^*)r_i^*\,\Delta\theta\,\Delta r.$$

Taking a limit as $M\to\infty$ turns this into

(10.56) $$\sum_{i=1}^{N} r_i^*\left(\int_0^{2\pi} f(r_i^*,\theta)\,d\theta\right)\Delta r.$$

Taking a limit as $N\to\infty$, this becomes

(10.57) $$\int_0^1\int_0^{2\pi} f(r,\theta)r\,d\theta\,dr.$$

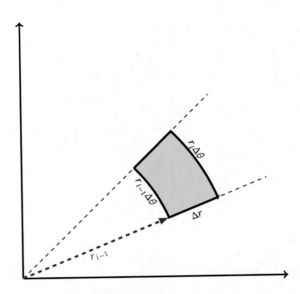

If we had nested our sum the other way, adding up all the boxes in each sector and then adding up the sectors, we would have gotten

(10.58) $$\int_0^{2\theta} \int_0^1 f(r,\theta) r\, dr\, d\theta.$$

That is, if D is the unit disk, then

$$\iint_D f(r,\theta)\, dA = \int_0^1 \int_0^{2\pi} f(r,\theta) r\, d\theta\, dr$$

(10.59) $$= \int_0^{2\theta} \int_0^1 f(r,\theta) r\, dr\, d\theta.$$

10.30. In Exercise 10.29 we used the function $f(r) = 1 - \frac{r^2}{R^2}$, with $R = 6$, to model the risk of infection at distance r from an infected person. We integrated that function over a rectangle to compute an average risk. Now we will compute the risk of infection in two circular regions. Let D_3 be the disk of radius 3 around the origin, and let D_6 be the disk of radius 6.

(a) Compute $\iint_{D_3} f(r)\, dA$. Don't forget the factor of r in $r\, dr\, d\theta$! Then find the average value of $f(r)$ in this disk.

(b) Compute $\iint_{D_6} f(r)\, dA$. Then find the average value of $f(r)$ in this disk.

(c) Based on this model, how much riskier is it to talk to somebody within a distance of three feet than it is to talk to somebody within a distance of six feet?

(d) In general, what fraction of your risk of infection comes from people getting within three feet of you?

10.31. Consider the function $f = e^{-(x^2+y^2)} = e^{-r^2}$. We are going to integrate $f\, dA$ over the entire plane $\mathbb{R}^2$ by integrating over a large disk and taking a limit as the radius of that disk goes to infinity.

(a) Compute $\iint_{D_a} e^{-r^2}\, dA$, where D_a is a disk of radius a centered at the origin.

(b) Taking a limit as $a \to \infty$, what is $\iint_{\mathbb{R}^2} e^{-r^2}\, dA$?

(c) Since e^{-r^2} can be written in Cartesian coordinates as $e^{-(x^2+y^2)} = e^{-x^2}e^{-y^2}$, Fubini's theorem tells us that $\iint_{\mathbb{R}^2} e^{-r^2}\, dA = \int_{-\infty}^{\infty} \int_{-\infty}^{\infty} e^{-x^2} e^{-y^2}\, dx\, dy$. Comparing this to your answer to part (b), what is $\int_{-\infty}^{\infty} e^{-x^2}\, dx$?

10.32. In polar coordinates, the region R of Exercise 10.24 is given by $0 \le r \le 1$, $0 \le \theta \le \pi/2$.

(a) Express the function $f(x, y) = xy$ in polar coordinates.

(b) Using polar coordinates, compute $\iint_R xy\, dA$. Compare your result to your answer to Exercise 10.24.

Chapter 11

Taylor Series

Wouldn't it be nice if all functions were polynomials? Polynomials are easy to evaluate, easy to differentiate, and easy to integrate. The bad news, of course, is that there are plenty of functions that aren't polynomials, including exponentials, logs, trig functions, inverse trig functions, and so on. The good news is that most of those other functions can be *approximated* by polynomials, making calculations much easier.

In this chapter we'll see how to do that. We'll start by looking at ways to represent numbers as infinite sums, or **series**. In Section 11.1 we'll see that adding up an infinite list of numbers requires taking a limit: saying that a number *equals* an infinite sum means that it is *approximated* by a finite sum. Some series have limits and are said to **converge**. Others don't and are said to **diverge**. In particular, we study convergence and divergence of geometric series.

In Section 11.2 we'll look at power series, where each term is a coefficient times a power of x. These are similar to polynomials, except with infinitely many terms, and they give us a practical way to represent many common functions. We'll use what we learned about geometric series to express a number of common functions as power series. In other words, we'll learn how to approximate those functions by polynomials.

In Section 11.3 we'll go beyond those special examples. We'll learn how to approximate arbitrary smooth functions by polynomials, called **Taylor polynomials**, and see how accurate those approximations are. By taking a limit we'll see how to express functions as power series called **Taylor series**. In Section 11.4 we'll turn to the most common and most important Taylor series, for e^x, $\sin(x)$ and $\cos(x)$, and make sense of Euler's strange formula $e^{ix} = \cos(x) + i\sin(x)$.

Finally, we consider questions of convergence. An infinite series that doesn't converge doesn't really make sense. In Section 11.5 we'll develop a lot of tools to determine whether a series converges or not, and in Section 11.6 we'll apply these tools to figure out for which values of x a Taylor series converges.

11.1. What Does $\pi = 3.14159265\cdots$ Mean?

Examples. You have probably seen expressions like

(11.1) $$0.333333333333333\cdots$$

and

(11.2) $$3.14159265358979323846\cdots.$$

These are examples of **infinite sums**. A terminating decimal like 0.333 means $3/10 + 3/100 + 3/1000$, so an infinite decimal like $0.333333333\cdots$ means

(11.3) $$\frac{3}{10} + \frac{3}{100} + \frac{3}{1000} + \frac{3}{10,000} + \cdots,$$

where "$+\cdots$" means that we keep adding more and more terms forever. Similarly, $3.14159\cdots$ means

(11.4) $$3 + \frac{1}{10} + \frac{4}{100} + \frac{1}{1000} + \frac{5}{10,000} + \frac{9}{100,000} + \cdots.$$

Likewise, we can talk about infinite sums that aren't decimals, like

(11.5) $$1 + \frac{1}{2} + \frac{1}{4} + \frac{1}{8} + \frac{1}{16} + \cdots,$$

(11.6) $$1 + 2 + 4 + 8 + 16 + 32 + \cdots,$$

and

(11.7) $$1 - 2 + 4 - 8 + 16 - 32 + \cdots.$$

It takes some work to make sense of any of this, because addition is a *binary* operation. Addition takes two inputs, like 3 and 5, and gives an output of $3 + 5 = 8$. When we add three things we're really adding two things twice. $2 + 3 + 5$ means $(2 + 3) + 5$. (This happens to equal $2 + (3 + 5)$, so we usually don't worry about what order we do the addition.) Likewise, when we add $2 + 3 + 5 + 7 + 11 + 13$, we say something like

2 plus 3 makes 5, 5 plus 5 makes 10, 10 plus 7 makes 17, 17 plus 11 makes 28, and 28 plus 13 makes 41, so our answer is 41.

In general, whenever we are faced with a finite sum

(11.8) $$a_1 + a_2 + \cdots + a_n,$$

we generate a sequence of **partial sums**

$$\begin{aligned} s_1 &= a_1, \\ s_2 &= s_1 + a_2, \\ s_3 &= s_2 + a_3, \\ &\vdots \end{aligned}$$

(11.9) $$s_n = s_{n-1} + a_n,$$

and say that the number at the end of the sequence is the sum of all the a_i's:

(11.10) $$s_n = \sum_{i=1}^{n} a_i.$$

But what if the a_i's never end? What do we do then?

11.1. What Does $\pi = 3.14159265 \cdots$ Mean?

Convergence. The answer comes from the 1st Pillar of Calculus: *Close is good enough!* If a sequence of numbers $s_1, s_2, \ldots$ gets closer and closer to some number S, then we say that the sequence **converges** to S, and we write

(11.11) $$S = \lim_{n \to \infty} s_n.$$

If those numbers s_n are partial sums as in (11.9), coming from trying to add up an infinite list of numbers, then we say that the infinite sum $a_1 + a_2 + \cdots$ **converges** to S, and we write

(11.12) $$S = \sum_{i=1}^{\infty} a_i.$$

That is,

(11.13) $$\sum_{i=1}^{\infty} a_i \quad \text{means} \quad \lim_{n \to \infty} \sum_{i=1}^{n} a_i.$$

An infinite sum is also called an **infinite series**. If it doesn't converge to any number, we say that the series **diverges**.

In our examples, (11.1) converges to $1/3$, since by taking enough digits we can get as close to $1/3$ as we like. Similarly, (11.2) converges to π and (11.5) converges to 2. The partial sums in (11.5) never actually *reach* 2, but they home in on 2, and that's good enough for us. However, (11.6) just gets bigger and bigger, and doesn't home in on any number at all. The partial sums in the series (11.7) not only get bigger and bigger in size, but they alternate between positive and negative!

It's tempting to say that (11.6) "goes to infinity" and to write something like

(11.14) $$\sum_{j=0}^{\infty} 2^j = +\infty.$$

That isn't wrong, but please be careful, because ∞ is not a number like 17 or -35! The word "infinity" comes from the Latin words *in*, meaning "not", and *finis*, meaning "finish" or "end", so *infinity* just means "continue without end". When something keeps growing without end, eventually getting bigger than any number you can name, we say it "goes to infinity". When we write

(11.15) $$\int_a^b f(x)\,dx = \lim_{N \to \infty} \sum_{i=1}^{N} f(x_i^*) \Delta x,$$

we imagine a process in which N gets bigger and bigger and bigger without ever stopping. Likewise, when something gets more and more negative, eventually getting more negative than any number we can name, we say that it "goes to $-\infty$", as in

(11.16) $$\lim_{x \to 0} \frac{-1}{x^2} = -\infty.$$

We will use the symbols ∞ and $-\infty$, but we must always remember that

> *Infinity is a process, not a number!*

In particular, when we write (11.14), we do *not* mean that the series (11.6) converges to infinity. The partial sums don't home in on a number called "infinity". They don't home in on any number at all! The series *diverges*, and "infinity" is just our way of describing *how* it diverges. The series (11.7) also diverges, but in a way that we can't describe as easily.

In summary,

- An **infinite sequence** is an infinite list of numbers $a_1, a_2, \ldots$. We say that the sequence **converges** to a number L if the numbers a_i get closer and closer to L, in which case we write $\lim_{i \to \infty} a_i = L$. If the sequence doesn't converge to a number, then it **diverges**. If the sequence diverges by getting more and more positive or more and more negative, we might say that it "goes to ∞" or "goes to $-\infty$", but it still diverges.

- An **infinite series** is a sum

$$a_1 + a_2 + a_3 + \cdots \tag{11.17}$$

involving infinitely many terms. The only way to make sense of an infinite sum is to think of the sequence of partial sums s_n from (11.9). If the sequence of partial sums converges to a number S, we say that the series converges to S and write $\sum_{i=1}^{\infty} a_i = S$. Otherwise the series diverges.

- Representing something as an infinite sum is *exactly* the same as approximating it, better and better, by finite sums.

- Every infinite decimal, and lots of other useful expressions in mathematics, are actually convergent infinite series. The rest of this chapter is about exploring how they work and what they are good for.

Geometric Series. A **geometric series** is a series where the ratio between successive terms is fixed. We have already seen four examples of this, namely (11.1), (11.5), (11.6) and (11.7). We usually call the first term a and the common ratio r, giving us

$$a + ar + ar^2 + ar^3 + \cdots . \tag{11.18}$$

We also usually label the first term as a_0 rather than a_1, so that we can write

$$a_i = ar^i. \tag{11.19}$$

As long as $r \neq 1$, there is a simple trick for evaluating the partial sums s_n:

$$\begin{aligned} s_n &= a + ar + ar^2 + \cdots + ar^n, \\ rs_n &= \phantom{a + {}} ar + ar^2 + \cdots + ar^n + ar^{n+1}, \\ (1-r)s_n &= a \phantom{{}+ ar + ar^2 + \cdots + ar^n} - ar^{n+1}, \\ s_n &= \frac{a - ar^{n+1}}{1-r} = a\frac{1 - r^{n+1}}{1-r}. \end{aligned} \tag{11.20}$$

For (11.1) we have $a = 0.3$ and $r = 0.1$, so $1 - r = 0.9$ and $s_n = \frac{1}{3}(1 - 10^{-(n+1)})$. Since this gets closer and closer to $1/3$ as $n \to \infty$, the series converges to $1/3$. Similarly, for

11.2. Power Series

(11.5) we have $a = 1$ and $r = 1/2$, so $1 - r = 1/2$ and $s_n = 2(1 - 2^{-(n+1)}) = 2 - 2^{-n}$, which approaches 2. The series[1] (11.6) and (11.7) have $r = 2$ and $r = -2$, respectively, so $1 - r^{n+1}$ grows (in magnitude) without bound. Since s_n doesn't approach a definite number, these geometric series diverge.

In general, whenever $|r| < 1$, $1 - r^{n+1}$ approaches 1 as $n \to \infty$, and we have

$$(11.21) \qquad \sum_{i=0}^{\infty} ar^i = \lim_{n \to \infty} a \frac{1 - r^{n+1}}{1 - r} = \frac{a}{1 - r}.$$

When $|r| > 1$, s_n grows without bound in magnitude, and the infinite series $\sum_{i=1}^{\infty} ar^i$ diverges. If $r > 1$, then all terms have the same sign and the sum goes to $\pm\infty$, depending on the sign of a, while if $r < -1$ the partial sums s_n grow in magnitude and alternate in sign, as in (11.7).

When $r = 1$, we have the series

$$(11.22) \qquad a + a + a + a + \cdots,$$

which diverges to $\pm\infty$, depending on the sign of a. Finally, when $r = -1$, we have

$$(11.23) \qquad a - a + a - a + a - a + \cdots.$$

The partial sums alternate between $s_n = a$ (when n is even) and $s_n = 0$ (when n is odd). This doesn't approach a *single* fixed number, so the series diverges. We summarize these results as a theorem.

Theorem 11.1 (Geometric series). *Suppose that $a \neq 0$. The geometric series $\sum_{i=0}^{\infty} ar^i$ converges if and only if $|r| < 1$, in which case it converges to $\frac{a}{1-r}$.*

11.2. Power Series

In a geometric series, the parameter r can itself be a variable, or a function of a variable. For instance, as long as $|x| < 1$,

$$(11.24) \qquad 1 + x + x^2 + x^3 + \cdots = \frac{1}{1 - x}$$

and

$$(11.25) \qquad 1 - x^2 + x^4 - x^6 + \cdots = \frac{1}{1 - (-x^2)} = \frac{1}{1 + x^2},$$

since these are just geometric series with $r = x$ or $r = -x^2$. Series like these, where each term is a multiple of a power of x, or of a power of $x - a$ for some fixed constant a, are called **power series**.

Read left-to-right, equations (11.24) and (11.25) say that certain power series add up to certain simple functions. Read right-to-left, they say that certain functions can be represented as power series. (In Sections 11.3 and 11.4, we'll see how to find power series for other functions.)

In particular, equation (11.24) doesn't just tell us that we can add up $1 + x + x^2 + \cdots$ to get $\frac{1}{1-x}$ (as long as $|x| < 1$), it tells us the polynomials $T_n(x)$ that we get by keeping

---

[1] The plural of *series* is *series*.

the first n powers of x in the series are *approximations* to the function $f(x) = \frac{1}{1-x}$. The approximation

(11.26) $$\frac{1}{1-x} \approx 1 + x$$

is our old friend the microscope equation $f(x) \approx f(a) + f'(a)(x-a)$ with $a = 0$, since $f(0) = 1$ and $f'(0) = 1$. This is the best *linear* approximation to our function near $x = 0$. The best *quadratic* approximation is

(11.27) $$\frac{1}{1-x} \approx 1 + x + x^2,$$

the best *cubic* approximation is

(11.28) $$\frac{1}{1-x} \approx 1 + x + x^2 + x^3,$$

and so on. We get a sequence of polynomials

(11.29) $$T_n(x) = 1 + x + x^2 + \cdots + x^n$$

that approximate $f(x)$ better and better on the interval $(-1, 1)$. These are called **Taylor polynomials** for the function $f(x) = \frac{1}{1-x}$. The graphs of the first few of these polynomials are shown in Figure 11.1.

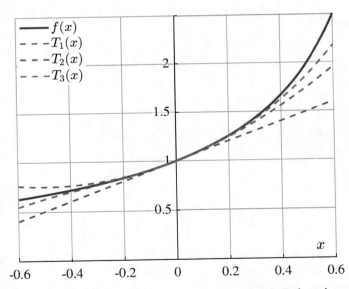

Figure 11.1. The graphs of $f(x) = 1/(1-x)$ and approximating Taylor polynomials $T_1(x) = 1 + x$, $T_2(x) = 1 + x + x^2$, and $T_3(x) = 1 + x + x^2 + x^3$

Since $T_n(x)$ is a good approximation to $f(x)$, $\int T_n(x)\,dx$ is a good approximation to $\int f(x)\,dx$. Similarly, $T_n'(x)$ is a good approximation to $f'(x)$.[2] This means that $\int f(x)\,dx$ and $f'(x)$ are the limits as $n \to \infty$ of $\int T_n(x)\,dx$ and $T_n'(x)$. But those are

---

[2] This second fact takes some extra work to prove, but it's still true.

exactly the series that we get by integrating or differentiating every term in the series for $f(x)$. For instance, by differentiating every term in the series (11.24), we get

$$\frac{1}{(1-x)^2} = \frac{d}{dx}\frac{1}{1-x}$$

$$= \frac{d}{dx}\left(1 + x + x^2 + x^3 + x^4 + \cdots\right)$$

$$= 0 + 1 + 2x + 3x^2 + 4x^3 \cdots$$

(11.30)
$$= \sum_{n=0}^{\infty}(n+1)x^n.$$

By integrating each term we get

$$\ln(1-x) = \int_0^x \frac{-ds}{1-s}$$

$$= -\int_0^x \left(1 + s + s^2 + \cdots\right) ds$$

$$= -\left(x + \frac{x^2}{2} + \frac{x^3}{3} + \cdots\right)$$

(11.31)
$$= -\sum_{n=1}^{\infty}\frac{x^n}{n}.$$

Similarly, by integrating each term in the series (11.25), we get

(11.32) $$\tan^{-1}(x) = \int_0^x \frac{ds}{1+s^2} = x - \frac{x^3}{3} + \frac{x^5}{5} - \frac{x^7}{7} + \cdots = \sum_{n=0}^{\infty}\frac{(-1)^n x^{2n+1}}{2n+1}.$$

The formulas for $\ln(1-x)$ and $\tan^{-1}(x)$, which apply whenever $|x| < 1$, give us an effective way to estimate logs and arctangents. For instance, letting $x = \pm 0.1$, we get

$$\ln(0.9) \approx -\left(0.1 + \frac{0.01}{2} + \frac{0.001}{3} + \frac{0.0001}{4}\right) \approx -0.10535833,$$

$$\ln(1.1) \approx 0.1 - \frac{0.01}{2} + \frac{0.001}{3} - \frac{0.0001}{4} \approx .09530833,$$

$$\tan^{-1}(0.1) \approx 0.1 - \frac{0.001}{3} + \frac{0.00001}{5} \approx 0.09966867, \quad \text{and}$$

(11.33) $$\tan^{-1}(-0.1) \approx -0.1 + \frac{0.001}{3} - \frac{0.00001}{5} \approx -0.09966867.$$

11.3. Taylor Polynomials and Taylor Series

So far we have expressed a number of functions, including $\ln(1-x)$ and $\tan^{-1}(x)$, as power series. However, our methods were completely ad hoc and didn't tell us anything about how to find the series expansions of e^x or $\sin(x)$ or $\cos(x)$ or any other function. In this section we're going to see how to find polynomials that approximate an arbitrary function that has derivatives to all orders. We call such functions **smooth**. We'll also see how to approximate a function as a polynomial in $(x-1)$ or $(x-2)$ or $(x-a)$, which is more useful if we're trying to understand the function near $x = 1$ or $x = 2$ or $x = a$.

Finding the Coefficients. The key idea comes from the Fundamental Theorem of Calculus, specifically FTC2. Suppose that $f(x)$ is a smooth function. Then

$$(11.34) \qquad f(x) - f(a) = \int_a^x f'(s)\,ds \quad \text{or} \quad f(x) = f(a) + \int_a^x f'(s)\,ds.$$

If we can find a good approximation for the function $f'(x)$, we can integrate that approximation to get an approximation for $f(x)$. If our estimate for $f'(s)$ is accurate to within a quantity ϵ for all s between a and x, then our estimate for $f(x)$ will be accurate to within $\epsilon|x - a|$. When x is close to a, this will give us an estimate for $f(x)$ that is more accurate than our estimate for $f'(x)$.

The roughest estimate for $f'(s)$ is simply that $f'(x) \approx f'(a)$ whenever $x \approx a$. That is, we pretend that $f'(x)$ is a constant function. We than have

$$(11.35) \qquad f(x) \approx f(a) + \int_a^x f'(a)\,ds = f(a) + f'(a)(x - a).$$

This is our old friend the microscope equation. We get a better approximation for $f'(x)$ by applying a linear approximation to f':

$$(11.36) \qquad f'(x) \approx f'(a) + f''(a)(x - a), \quad \text{so} \quad f'(s) \approx f'(a) + f''(a)(s - a).$$

Integrating this linear approximation for f' gives us a quadratic approximation for f:

$$(11.37) \quad f(x) \approx f(a) + \int_a^x (f'(a) + f''(a)(x-a))\,ds = f(a) + f'(a)(x-a) + \frac{f''(a)}{2!}(x-a)^2.$$

Likewise, we have a quadratic approximation for $f'(s)$:

$$(11.38) \qquad f'(s) \approx f'(a) + f''(a)(s - a) + \frac{f'''(a)}{2!}(s - a)^2.$$

Plugging this into (11.34) then gives a cubic approximation for $f(x)$:

$$f(x) \approx f(a) + \int_a^x \left(f'(a) + f''(a)(x - a) + \frac{f'''(a)}{2!}(s - a)^2 \right) ds$$

$$(11.39) \qquad = f(a) + f'(a)(x - a) + \frac{f''(a)}{2!}(x - a)^2 + \frac{f'''(a)}{3!}(x - a)^3.$$

We can continue this bootstrap argument forever. Once we have an nth order approximation for $f(x)$, we can apply the same procedure to $f'(x)$ to get an nth order approximation for $f'(x)$, and likewise for $f'(s)$. We then integrate our nth order approximation for $f'(s)$ to get an $(n+1)$-st order approximation to $f(x)$. In general, the nth order polynomial that best approximates $f(x)$ is

$$(11.40) \qquad T_n(x) = \sum_{k=0}^n \frac{f^{(k)}(a)}{k!}(x - a)^k = f(a) + f'(a)(x - a) + \cdots + f^{(n)}(a)\frac{(x - a)^n}{n!},$$

11.3. Taylor Polynomials and Taylor Series

where the notation $f^{(k)}(a)$ means the kth derivative of $f(x)$, evaluated at $x = a$. In particular, the right-hand sides of (11.35), (11.37), and (11.39), which give the best linear, quadratic, and cubic approximations to $f(x)$ near $x = a$, are $T_1(x)$, $T_2(x)$, and $T_3(x)$. Taking a limit as $n \to \infty$, we express $f(x)$ as an infinite power series:

$$(11.41) \qquad \sum_{k=0}^{\infty} \frac{f^{(k)}(a)}{k!}(x-a)^k = f(a) + f'(a)(x-a) + \frac{f''(a)}{2!}(x-a)^2 + \cdots.$$

The polynomials $T_n(x)$ are called **Taylor polynomials**, after the 18th century English mathematician Brook Taylor, while the infinite series (11.41) is called a **Taylor series**. When $a = 0$, as with our approximations of $\ln(1-x)$ and $\tan^{-1}(x)$, they are often called MacLaurin polynomials and MacLaurin series, after the 18th century Scottish mathematician Colin MacLaurin. However, many examples of MacLaurin series were already known to Isaac Newton in the 17th century, and some were even known to Indian mathematicians in the 14th century!

An Alternate Derivation. Another way to understand Taylor polynomials is by matching derivatives. Suppose we want an nth order polynomial

$$(11.42) \qquad \begin{aligned} T(x) &= c_0 + c_1(x-a) + c_2(x-a)^2 + \cdots + c_n(x-a)^n \\ &= \sum_{k=0}^{n} c_k(x-a)^k \end{aligned}$$

that does a good job of approximating $f(x)$ near $x = a$. One measure of this is how well the derivatives of $T(x)$ at $x = a$ match the derivatives of $f(x)$ at $x = a$. We start by computing derivatives of $T(x)$:

$$\begin{aligned} T(x) &= c_0 + c_1(x-a) + c_2(x-a)^2 + \cdots, \\ T'(x) &= c_1 + 2c_2(x-a) + 3c_3(x-a)^2 + \cdots, \\ T''(x) &= 2c_2 + 6c_3(x-a) + \cdots, \\ T'''(x) &= 6c_3 + 24c_4(x-a) + \cdots, \\ &\vdots \\ T^{(k)}(x) &= k!\, c_k + \text{terms with at least one power of } (x-a), \\ &\vdots \end{aligned}$$

$$(11.43) \qquad T^{(n)}(x) = n!\, c_n.$$

Since $T^{(n)}(x)$ is a constant, all subsequent derivatives are 0. Plugging in $x = a$ gives

$$(11.44) \qquad T^{(k)}(a) = \begin{cases} k!\, c_k & k \le n \\ 0 & k > n. \end{cases}$$

There's no hope of matching any derivatives higher than nth order, but we can adjust the coefficients c_k to match the value and first n derivatives of $f(x)$ at $x = a$. Since we want

$$(11.45) \qquad k!\, c_k = f^{(k)}(a),$$

we must have

(11.46) $$c_k = \frac{f^{(k)}(a)}{k!}.$$

Plugging this back into the second line of equation (11.42) then gives

(11.47) $$T(x) = \sum_{k=0}^{n} \frac{f^{(k)}(a)}{k!}(x-a)^k.$$

In other words, $T(x)$ is *exactly* what we previously called $T_n(x)$.

A Test Case. As an example of Taylor polynomials, let's approximate $\sqrt{102}$ by hand. (Maybe you left your calculator and phone at home, or maybe you're trying to figure out how an engineer would program a calculator in the first place.) We pick $a = 100$, since that's a place where we understand $f(x) = \sqrt{x}$. We compute the first few derivatives of $f(x)$ at $x = 100$:

k	$f^{(k)}(x)$	$f^{(k)}(100)$	$f^{(k)}(100)/k!$
0	$x^{1/2}$	10	10
1	$\frac{1}{2}x^{-1/2}$	1/20	1/20
2	$-\frac{1}{4}x^{-3/2}$	$-1/4000$	$-1/8000$
3	$\frac{3}{8}x^{-5/2}$	$3/800{,}000$	$1/1{,}600{,}000$

Plugging these into equation (11.47) gives us our first few Taylor polynomials:

$$T_1(x) = 10 + \frac{x-100}{20},$$

$$T_2(x) = 10 + \frac{x-100}{20} - \frac{(x-100)^2}{8000},$$

(11.48) $$T_3(x) = 10 + \frac{x-100}{20} - \frac{(x-100)^2}{8000} + \frac{(x-100)^3}{1{,}600{,}000}.$$

Setting $x = 102$ then gives

$$T_1(102) = 10 + \tfrac{2}{20} = 10.1,$$

$$T_2(102) = 10 + \tfrac{2}{20} - \tfrac{4}{800} = 10.0995,$$

(11.49) $$T_3(102) = 10 + \tfrac{2}{20} - \tfrac{4}{800} + \tfrac{8}{1{,}600{,}000} = 10.099505.$$

In fact, $\sqrt{102} = 10.0995049384$ to ten decimal places. The linear approximation $T_1(102)$ is accurate to within 5×10^{-4}, the quadratic approximation $T_2(102)$ is accurate to within 5×10^{-6}, and the cubic approximation $T_3(102)$ is accurate to within 7×10^{-8}. Each additional term gives us roughly two more decimal places of accuracy.

The bottom line is that the computations needed to use Taylor polynomials to approximate a function near a known data point are very much like using the microscope equation (a.k.a. a linear approximation), except that we compute several derivatives of $f(x)$ at $x = a$ instead of just $f'(a)$, and we get rewarded for our effort with greater accuracy. The linear approximation $T_1(x)$ is good, but $T_2(x)$ is better, $T_3(x)$ is better still, and so on.

Accuracy and Taylor's Theorem. But how much better? Just how accurate *are* Taylor polynomials? The answer is given below.

Theorem 11.2 (Taylor's theorem). *Suppose that $f(x)$ is a function that is differentiable at least $n + 1$ times on an interval I around a. Then for every $x \in I$ there is a number c, somewhere between a and x, such that*

(11.50) $$f(x) = T_n(x) + \frac{f^{(n+1)}(c)}{(n+1)!}(x-a)^{n+1}.$$

In other words, when we approximate $f(x)$ by $T_n(x)$, our error looks just like the last term in $T_{n+1}(x)$, except that the derivative is evaluated at a mystery point, somewhere between a and x. Most of the time, $f^{(n+1)}(c)$ is fairly close to $f^{(n+1)}(a)$, and our error is roughly the same size as the next term in our series. In any case, the error is bounded by $\frac{M(x-a)^{n+1}}{(n+1)!}$, where M is the largest value of $|f^{(n+1)}(x)|$ on I.

Note that the error scales like $(x-a)^{n+1}$. When we used $T_3(x)$ to approximate $\sqrt{102}$, we found that our answer was accurate to within 7×10^{-8}. If we had used $T_3(x)$ to approximate $\sqrt{101}$, it would have been $2^4 = 16$ times more accurate, or good to within 5×10^{-9}. If we had used $T_3(x)$ to approximate $\sqrt{106}$, it would have been $3^4 = 81$ times less accurate, or only good to within 6×10^{-6}.

The proof of Taylor's theorem is beyond the scope of this book, involving repeated application of the mean value theorem, which we also aren't covering.

11.4. Sines, Cosines, Exponentials, and Logs

Now that we have a formula (11.41) for the Taylor series of a function $f(x)$, let's find the Taylor series of some common functions.

Compound Interest and the Series for e^x. If $f(x) = e^x$, then $f'(x) = e^x$, $f''(x) = e^x$, and the kth derivative of $f(x)$, evaluated at $x = 0$, is

(11.51) $$f^{(k)}(0) = e^0 = 1.$$

This makes the nth order Taylor polynomial

(11.52) $$T_n(x) = 1 + x + \frac{x^2}{2!} + \cdots + \frac{x^n}{n!},$$

and makes the Taylor series

(11.53) $$e^x = \sum_{n=0}^{\infty} \frac{x^n}{n!} = 1 + x + \frac{x^2}{2!} + \frac{x^3}{3!} + \cdots.$$

More generally, if A is any expression, then

(11.54) $$e^A = \sum_{n=0}^{\infty} \frac{A^n}{n!},$$

and in particular

(11.55) $$e^{rt} = 1 + rt + \frac{(rt)^2}{2!} + \frac{(rt)^3}{3!} + \cdots = \sum_{n=0}^{\infty} \frac{(rt)^n}{n!}.$$

We can understand each term in this series via compound interest. Suppose that we invest an amount of money P at an interest rate r. How much money will we have at time t? We have already solved this problem with differential equations. If $Y(t)$ represents our bank balance at time t, then $Y(t)$ solves the IVP

(11.56) $$\frac{dY}{dt} = rY, \qquad Y(0) = P.$$

The solution to this IVP is $Y(t) = Pe^{rt}$, as can be easily checked.

However, there's another way of looking at the problem. Let $Y_0(t)$ be our initial investment, also known as the **principal**. Let $Y_1(t)$ be the interest earned on the principal up to time t. Let $Y_2(t)$ be the interest earned on the interest on the principal, in other words the interest earned by Y_1. Let $Y_3(t)$ be the interest earned by Y_2, and so on. We then have

(11.57) $$Y(t) = Y_0(t) + Y_1(t) + Y_2(t) + \cdots.$$

The function $Y_0(t) = P$ is a constant. Since this generates interest at rate $rY_0(t) = rP$, we have

(11.58) $$Y_1(t) = \int_0^t rY_0(s)\,ds = \int_0^t rP\,ds = rtP.$$

That is, $Y_1(t)$ is the accumulation function for $rY_0(t)$. Similarly,

$$Y_2(t) = \int_0^t rY_1(s)\,ds = \int_0^t r^2 sP\,ds = \frac{r^2 t^2}{2}P,$$

$$Y_3(t) = \int_0^t rY_2(s)\,ds = \int_0^t \frac{r^3 s^2}{2}P\,ds = \frac{r^3 t^3}{3!}P,$$

$$Y_4(t) = \int_0^t rY_3(s)\,ds = \int_0^t \frac{r^4 s^3}{3!}P\,ds = \frac{r^4 t^4}{4!}P,$$

$$\vdots$$

$$Y_n(t) = \int_0^t rY_{n-1}(s)\,ds = \int_0^t \frac{r^n s^{n-1}}{(n-1)!}P\,ds = \frac{r^n t^n}{n!}P,$$

(11.59) $$Y(t) = P\sum_{n=0}^{\infty} \frac{(rt)^n}{n!},$$

where the $0!$ in the $n = 0$ term means 1. $Y_n(t)$ is exactly the nth term in the Taylor series of Pe^{rt}.

11.4. Sines, Cosines, Exponentials, and Logs

For instance, if we invest $1000 at an interest rate of $r = 6\% = 0.06$, then after one year we will have $Y_0(1) = P = \$1000$. The interest on this $1,000 is $Y_1(1) = 0.06P = \$60$. The interest on the interest is $Y_2(1) = (0.06)^2 P/2! = \$1.80$. The interest on the interest on the interest is $Y_3(1) = (0.06)^3 P/3! = \$0.036$, or less than 4 cents. This, and all subsequent terms, are small enough to be ignored. For all practical purposes, the total interest we earn in one year is

$$\text{(11.60)} \qquad \left(0.06 + \frac{(0.06)^2}{2}\right) P = 0.0618 P.$$

The multiplier 0.0618 is called the **yield**. That is, an account with interest *rate r* = 0.06 generates a *yield* of approximately $r + \frac{r^2}{2} = 0.0618$. (The exact yield is $e^r - 1$.)

Another way to understand the series (11.53) is through the IVP $y' = y$, $y(0) = 1$. If we define $f(x)$ to be the right-hand side of equation (11.53), then

$$f'(x) = \frac{d}{dx}\left(1 + x + \frac{x^2}{2!} + \frac{x^3}{3!} + \frac{x^4}{4!} + \cdots\right)$$
$$= 0 + 1 + x + \frac{x^2}{2!} + \frac{x^3}{3!} + \cdots$$
$$\text{(11.61)} \qquad = f(x),$$

since the derivative of each term of (11.53) is exactly the previous term. Of course $f(0) = 1 + 0 + 0^2/2! + \cdots = 1$, so $f(x)$ satisfies our IVP. Since e^x is the unique function that satisfies this IVP, $f(x)$ must be e^x.

Figure 11.2 shows the graph of e^x and its first few approximating polynomials on two length scales. We have already seen that the linear approximation does well when x is close to 0, and the quadratic approximation does even better. However, they start to fail when we get farther away. To maintain accuracy, we need to keep more and more terms as we get farther and farther away.

Let's see what Taylor's theorem has to say about the approximation

$$\text{(11.62)} \qquad e \approx 1 + 1 + \frac{1}{2} + \frac{1}{3!} + \frac{1}{4!} + \cdots + \frac{1}{n!},$$

which we get by evaluating the Taylor polynomial $T_n(x)$ at $x = 1$. Since the $(n+1)$-st derivative of e^x is e^x and the biggest value of e^x between 0 and 1 is $e^1 = e$, we know that our error is bounded by $e(1-0)^{n+1}/(n+1)! = e/(n+1)!$. That is, our error as a fraction of e is at most $1/(n+1)!$. The approximation

$$\text{(11.63)} \qquad e \approx 2 + \frac{1}{2} + \frac{1}{6} + \frac{1}{24} = 2.708333$$

is good to one part in $5! = 120$, and the approximation

$$\text{(11.64)} \qquad e \approx 2 + \frac{1}{2} + \frac{1}{6} + \frac{1}{24} + \frac{1}{120} + \frac{1}{720} = 2.718056$$

is good to one part in $7! = 5040$.

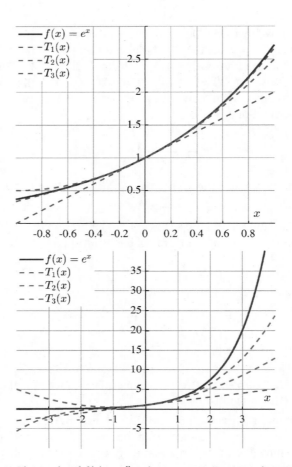

Figure 11.2. The graphs of $f(x) = e^x$ and some approximating polynomials. $T_3(x)$ gives a very good approximation on the interval $[-1,1]$, but to approximate e^x on $[-4,4]$, we need to go at least to $T_7(x)$.

An "Impossible" Integral. Can anybody evaluate the indefinite integral $\int e^{-x^2}\,dx$? Integration by substitution doesn't work. Neither does integration by parts, or any other known trick. In fact, the answer cannot be expressed as *any* finite formula involving elementary functions. However, it *can* be expressed as a power series.

Applying (11.54) with $A = -x^2$, we get

$$(11.65) \qquad e^{-x^2} = \sum_{n=0}^{\infty} \frac{(-1)^n x^{2n}}{n!} = 1 - x^2 + \frac{x^4}{2!} - \frac{x^6}{3!} + \frac{x^8}{4!} - \cdots.$$

Integrating term-by-term, an anti-derivative of e^{-x^2} is

$$(11.66) \qquad A(x) = \int_0^x e^{-s^2}\,ds = \sum_{n=0}^{\infty} \frac{(-1)^n x^{2n+1}}{(2n+1)n!} = x - \frac{x^3}{3} + \frac{x^5}{10} - \frac{x^7}{42} + \frac{x^9}{216} - \cdots.$$

11.4. Sines, Cosines, Exponentials, and Logs

By evaluating the first few terms, we can accurately approximate quantities such as $\int_{-0.3}^{0.2} e^{-x^2} dx$. Approximating integrals on a wider scale, such as $\int_{-3}^{2} e^{-x^2} dx$, is similar, except that more terms are needed to be accurate.

Sines and Cosines. We now turn to figuring out the Taylor series for $\sin(x)$ and $\cos(x)$ (around $x = 0$). The first few derivatives of $f(x) = \sin(x)$ are

k	$f^{(k)}(x)$	$f^{(k)}(0)$
0	$\sin(x)$	0
1	$\cos(x)$	1
2	$-\sin(x)$	0
3	$-\cos(x)$	-1
4	$\sin(x)$	0

after which the pattern repeats. Since $f''''(x) = f(x)$, $f^{(4n+k)}(x) = f^{(k)}(x)$, and

$$(11.67) \qquad f^{(4n+k)}(0) = \begin{cases} 0 & k = 0 \text{ or } 2, \\ 1 & k = 1, \\ -1 & k = 3. \end{cases}$$

The Taylor series for $\sin(x)$ with $a = 0$ is then

$$(11.68) \qquad 0 + x + 0\frac{x^2}{2!} - \frac{x^3}{3!} + 0\frac{x^4}{4!} + \frac{x^5}{5!} + 0\frac{x^6}{6!} + \cdots.$$

After eliminating the terms that are 0, this simplifies to

$$(11.69) \qquad \sin(x) = \sum_{n=0}^{\infty} \frac{(-1)^n}{(2n+1)!} x^{2n+1} = x - \frac{x^3}{3!} + \frac{x^5}{5!} + \cdots.$$

Note that $T_1(x) = T_2(x) = x$. This is the *small angle approximation* that we saw at the end of Chapter 5. The approximation $\sin(x) \approx x$ is accurate for x up to half a radian or so (28.65 degrees). When $x = 1/2$, the error is roughly the size of the next term, namely $(1/2)^3/6 \approx 0.021$, and $\sin(1/2) = 0.479$.

To compute the series for $\cos(x)$ we could compute successive derivatives of $\cos(x)$ and apply equation (11.46). However, since $\cos(x) = (\sin(x))'$, we can get the series for $\cos(x)$ by taking the derivative of each term in the series for $\sin(x)$. The derivative of $(-1)^n x^{2n+1}/(2n+1)!$ is $(-1)^n x^{2n}/(2n)!$, so

$$(11.70) \qquad \cos(x) = \sum_{n=0}^{\infty} \frac{(-1)^n}{(2n)!} x^{2n} = 1 - \frac{x^2}{2!} + \frac{x^4}{4!} + \cdots.$$

Here $T_2(x) = T_3(x) = 1 - \frac{x^2}{2}$, which agrees with the small angle approximation $\cos(x) \approx 1 - \frac{x^2}{2}$. Note that the derivative of the series for $\cos(x)$ is minus the series for $\sin(x)$, as of course it must be.

Complex Exponentials and Euler's Formula. The Swiss mathematician Leonhard Euler, after whom Euler's method and the constant e are named, proposed an audacious formula using complex numbers that combines exponentials and trigonometric functions:

(11.71) $$e^{ix} = \cos(x) + i\sin(x),$$

where $i = \sqrt{-1}$. This is illustrated in Figure 11.3.

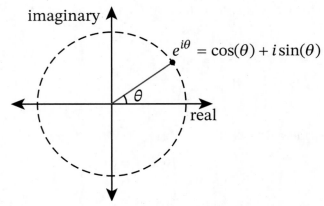

Figure 11.3. In the complex plane, the point $e^{i\theta}$ sits on the unit circle with angle θ.

This looks like nonsense. How can you possibly multiply e by itself an imaginary number of times?! Then again, we had the same issues with $e^{\sqrt{2}}$ and $e^{1/2}$ and e^{-1}, none of which are e multiplied by itself a certain number of times. We managed to define all of those numbers using reciprocals, roots, and limits, but a better method is to simply define e^x to be the unique solution to the IVP $\frac{dy}{dx} = y$, $y(0) = 1$. Likewise, for any constant k we can define e^{kx} to be the unique solution to the IVP $\frac{dy}{dx} = ky$, $y(0) = 1$.

Let's do that for $k = i$. The derivative of $\cos(x) + i\sin(x)$ is

$$\frac{d}{dx}\Big(\cos(x) + i\sin(x)\Big) = \frac{d(\cos(x))}{dx} + i\frac{d(\sin(x))}{dx}$$
$$= -\sin(x) + i\cos(x)$$
(11.72) $$= i\Big(\cos(x) + i\sin(x)\Big)$$

and

(11.73) $$\cos(0) + i\sin(0) = 1 + i0 = 1,$$

as required. Since $\cos(x) + i\sin(x)$ solves the IVP, we must have $e^{ix} = \cos(x) + i\sin(x)$.

11.4. Sines, Cosines, Exponentials, and Logs

We can also understand Euler's formula (11.71) with Taylor series. Plugging $A = ix$ into the formula (11.54) for e^A, we get

$$e^{ix} = \sum_{n=0}^{\infty} \frac{(ix)^n}{n!}$$

$$= 1 + ix - \frac{x^2}{2} - i\frac{x^3}{3!} + \frac{x^4}{4!} + i\frac{x^5}{5!} + \cdots$$

$$= \left(1 - \frac{x^2}{2} + \frac{x^4}{4!} - \frac{x^6}{6!} + \cdots\right)$$

$$+ i\left(x - \frac{x^3}{3!} + \frac{x^5}{5!} - \frac{x^7}{7!} + \cdots\right)$$

(11.74)
$$= \cos(x) + i\sin(x).$$

Euler's formula allows us to view exponential growth, exponential decay, oscillations like $\sin(x)$ and $\cos(x)$, and oscillations that grow or shrink (such as $e^{\pm x}\cos(x)$) on the same footing. It also gives us a simple way to remember (or derive) the addition of angle formulas for sine and cosine. If A and B are any two angles, then $e^{i(A+B)} = \cos(A+B) + i\sin(A+B)$. However, $e^{i(A+B)}$ also equals e^{iA} times e^{iB}, so

$$\cos(A+B) + i\sin(A+B) = (\cos(A) + i\sin(A))(\cos(B) + i\sin(B))$$
$$= \cos(A)\cos(B) - \sin(A)\sin(B)$$
(11.75)
$$+ i(\cos(A)\sin(B) + \sin(A)\cos(B)).$$

This means that

$$\cos(A+B) = \cos(A)\cos(B) - \sin(A)\sin(B) \quad \text{and}$$
(11.76)
$$\sin(A+B) = \cos(A)\sin(B) + \sin(A)\cos(B),$$

since if two complex numbers are equal, then their real and imaginary parts must also be equal.

Logarithms. Finally, we compute the Taylor series for $\ln(x)$ around $x = 1$ in two ways. First, we compute derivatives of $\ln(x)$ at $x = 1$ and apply (11.46) to find the coefficients c_k. If $f(x) = \ln(x)$, then we get the following values.

k	$f^{(k)}(x)$	$f^{(k)}(1)$	c_k
0	$\ln(x)$	0	0
1	x^{-1}	1	1
2	$-x^{-2}$	-1	$-1/2$
3	$2x^{-3}$	$2!$	$1/3$
4	$-3!\,x^{-4}$	$-3!$	$-1/4$
k	$(-1)^{k-1}(k-1)!\,x^{-k}$	$(-1)^{k-1}(k-1)!$	$(-1)^{k-1}/k$

This gives us

$$\ln(x) = \sum_{k=1}^{\infty} \frac{(-1)^{k-1}}{k}(x-1)^k = (x-1) - \frac{(x-1)^2}{2} + \frac{(x-1)^3}{3} + \cdots. \quad (11.77)$$

The second derivation uses the series we already developed for $\ln(1-x)$ in powers of x, which becomes a series for $\ln(1-A)$ in powers of A for any quantity A. Setting $A = 1 - x$, we then have

$$\begin{aligned}
\ln(x) &= \ln(1-(1-x)) \\
&= -\sum_{n=1}^{\infty} (1-x)^n/n \\
&= \sum_{n=1}^{\infty} \frac{(-1)^{n-1}}{n}(x-1)^n \\
&= (x-1) - \frac{(x-1)^2}{2} + \frac{(x-1)^3}{3} + \cdots. \quad (11.78)
\end{aligned}$$

We summarize the expansions we have derived for common functions in Table 11.1.

Table 11.1. Taylor series for common functions

Function	$\sum$ form	First 3 nonzero terms	Where?
e^x	$\sum_{n=0}^{\infty} x^n/n!$	$1 + x + x^2/2$	All x
$\sin(x)$	$\sum_{n=0}^{\infty} \frac{(-1)^n}{(2n+1)!} x^{2n+1}$	$x - \frac{x^3}{3!} + \frac{x^5}{5!}$	All x
$\cos(x)$	$\sum_{n=0}^{\infty} \frac{(-1)^n}{(2n)!} x^{2n}$	$1 - \frac{x^2}{2!} + \frac{x^4}{4!}$	All x
$\ln(1-x)$	$-\sum_{n=1}^{\infty} x^n/n$	$-x - \frac{x^2}{2} - \frac{x^3}{3}$	$-1 \leq x < 1$
$\tan^{-1}(x)$	$\sum_{n=0}^{\infty} \frac{(-1)^n}{(2n+1)} x^{2n+1}$	$x - \frac{x^3}{3} + \frac{x^5}{5}$	$-1 \leq x \leq 1$
$\ln(x)$	$\sum_{n=1}^{\infty} \frac{(-1)^{n-1}}{n}(x-1)^n$	$(x-1) - \frac{(x-1)^2}{2} + \frac{(x-1)^3}{3}$	$0 < x \leq 2$

11.5. Tests for Convergence

When faced with an infinite series $\sum a_n$, the most basic question is *does it converge?* When faced with a power series $\sum a_n x^n$ or $\sum c_n(x-a)^n$, the question becomes *for what values of x does it converge?* In this section, we start to develop some tests to answer the first question. In the next section we address the second.

The Divergence Test. If a series[3] $\sum_{n=1}^{\infty} a_n$ converges to a value L, then the partial sums s_n all get close to L, and for n large we have

$$a_n = s_n - s_{n-1} \approx L - L = 0. \quad (11.79)$$

---

[3]Sometimes we label our terms a_i, and sometimes we label them a_n. The name we give to the index doesn't matter.

11.5. Tests for Convergence

That is, in a convergent series the individual terms a_n must approach 0. Conversely, if the individual terms a_n do *not* approach 0, then the series must diverge. For instance, if a geometric series has $|r| \geq 1$, then the series must diverge.

Note that this does *not* say that a series with $\lim_{n \to \infty} a_n = 0$ automatically converges. It just says that a series whose individual terms don't approach 0 automatically *diverges*. For example, the **harmonic series**

$$\sum_{n=1}^{\infty} \frac{1}{n} = 1 + \frac{1}{2} + \frac{1}{3} + \cdots \tag{11.80}$$

has terms that approach 0, but the sum does not converge. To see this, note that

$$\begin{aligned} s_{2^n} &= 1 + \frac{1}{2} + \left(\frac{1}{3} + \frac{1}{4}\right) \\ &\quad + \left(\frac{1}{5} + \frac{1}{6} + \frac{1}{7} + \frac{1}{8}\right) + \cdots \\ &\quad + \left(\frac{1}{2^{n-1}+1} + \cdots + \frac{1}{2^n}\right). \end{aligned} \tag{11.81}$$

The first term in parentheses contains two numbers, each at least $1/4$, so it adds up to something at least as big as $1/2$. The second term in parentheses contains four numbers, each at least $1/8$, so it also adds up to something at least as big as $1/2$. The same reasoning applies to each term in parentheses. Since s_{2^n} is $\frac{3}{2}$ plus the sum of $n-1$ such terms, we have $s_{2^n} \geq 1 + \frac{n}{2}$, which is eventually bigger than any number you can name.

> **Theorem 11.3** (Divergence test). *If $\lim_{i \to \infty} a_i \neq 0$ or $\lim_{i \to \infty} a_i$ does not exist, then $\sum_{i=1}^{\infty} a_i$ diverges.*

The Comparison and Limit Comparison Tests. If the terms a_i of a series are all nonnegative, then the partial sums s_n can only go up. Either they stay bounded and approach a limit, in which case the series converges, or they grow without bound, in which case the series diverges to ∞.

A **positive series** is a series where each term is nonnegative. (Despite the plain meaning of the word "positive", some terms are allowed to be 0.) Not having to worry about oscillations in s_n makes positive series simpler to deal with than arbitrary series.

Suppose that $\sum_{i=1}^{\infty} a_i$ and $\sum_{i=1}^{\infty} b_i$ are positive series, and that for each value of i we have $a_i \leq b_i$. Then the partial sum

$$s_n = \sum_{i=1}^{n} a_i \tag{11.82}$$

is less than or equal to the partial sum

$$t_n = \sum_{i=1}^{n} b_i. \tag{11.83}$$

If the s_n's grow without bound, then the t_n's also grow without bound. If the t_n's are bounded, then so are the s_n's. In other words, if $\sum_{i=1}^{\infty} a_i$ diverges, then so does $\sum_{i=1}^{\infty} b_i$, and if $\sum_{i=1}^{\infty} b_i$ converges, then so does $\sum_{i=1}^{\infty} a_i$.

For instance, the series

$$\sum_{i=i}^{\infty} \frac{1}{2^i + 1} = \frac{1}{3} + \frac{1}{5} + \frac{1}{9} + \cdots \tag{11.84}$$

converges, by comparison to the convergent geometric series

$$\sum_{i=1}^{\infty} \frac{1}{2^i} = \frac{1}{2} + \frac{1}{4} + \frac{1}{8} + \cdots. \tag{11.85}$$

Each term in the first series is less than or equal to the corresponding term in the second series, and the second series converges, so the first series converges.

The convergence or divergence of a series doesn't depend on what happens in the first ten terms, or the first thousand terms, or even the first million terms. It depends on what happens in the long run, and we can wait as long as we want for that to happen. We say that a property involving a_i and b_i is **eventually true** if there is a number N such that it is true for all $i > N$. When applying the comparison test, we don't need all the a_i's and b_i's to be nonnegative, and we don't need to have $a_i \leq b_i$ for every i. We just need the series $\sum a_i$ and $\sum b_i$ to be *eventually* positive, and for $a_i \leq b_i$ to be *eventually* true.

> **Theorem 11.4** (Comparison test). *Suppose that $\sum_{i=1}^{\infty} a_i$ and $\sum_{i=1}^{\infty} b_i$ are eventually positive series, and that we eventually have $a_i \leq b_i$. Then*
>
> (1) *If $\sum_{i=1}^{\infty} b_i$ converges, then so does $\sum_{i=1}^{\infty} a_i$.*
>
> (2) *If $\sum_{i=1}^{\infty} a_i$ diverges, then so does $\sum_{i=1}^{\infty} b_i$.*
>
> (3) *If $\sum_{i=1}^{\infty} a_i$ converges, or if $\sum_{i=1}^{\infty} b_i$ diverges, that does not tell us anything about the other series.*

If there is a positive constant c such that a_i is eventually less than cb_i, then we can apply the comparison test to $\sum a_i$ and $\sum cb_i$. If $\sum b_i$ converges, then $\sum cb_i$ converges, so $\sum a_i$ converges. If $\sum a_i$ diverges, then $\sum cb_i$ diverges, so $\sum b_i$ diverges. The extra factor of c doesn't make one bit of difference.

For instance, consider the series

$$\sum_{i=1}^{\infty} \frac{1}{2^i - 1} = \frac{1}{1} + \frac{1}{3} + \frac{1}{7} + \cdots. \tag{11.86}$$

At first glance, comparing (11.86) to the convergent series (11.85) doesn't seem to work because each term in (11.86) is bigger than the corresponding term in (11.85). In the comparison test, being bigger than a convergent series doesn't tell us anything! However, each term in (11.86) is less than or equal to *twice* the corresponding term in (11.85), and that's enough to show that (11.86) converges.

Next suppose that we have two eventually positive series, $\sum a_i$ and $\sum b_i$, such that the ratio b_i/a_i approaches a positive number L. (For instance, in the last example the ratio of $\frac{1}{2^n-1}$ and $\frac{1}{2^n}$ approached 1.) In that case, b_i/a_i is eventually close to L and, in particular, is eventually somewhere between $L/2$ and $2L$. Since b_i is eventually less than $2La_i$ and is eventually greater than $La_i/2$, we can do comparisons between $\sum La_i/2$,

11.5. Tests for Convergence

$\sum b_i$, and $\sum 2La_i$. If $\sum a_i$ converges, then $\sum 2La_i$ converges, so $\sum b_i$ converges. On the other hand, if $\sum a_i$ diverges, then $\sum La_i/2$ diverges, so $\sum b_i$ diverges. This observation is summarized below.

> **Theorem 11.5** (Limit comparison test). *Suppose that $\sum_{i=1}^{\infty} a_i$ and $\sum_{i=1}^{\infty} b_i$ are eventually positive series, and that $\lim_{i \to \infty} b_i/a_i$ is a positive number L. Then $\sum_{i=1}^{\infty} a_i$ and $\sum_{i=1}^{\infty} b_i$ either both converge or both diverge.*

Absolute Convergence. So far we have been talking about (eventually) positive series. Next we turn to series whose terms are sometimes negative. Of course, no matter what the sign of a_i is, the series $\sum_{i=1}^{\infty} |a_i|$ is a positive series. We say that $\sum_{i=1}^{\infty} a_i$ is **absolutely convergent** if $\sum_{i=1}^{\infty} |a_i|$ converges. In most cases of interest, we can figure out whether $\sum |a_i|$ converges using our other tests, and then apply the following theorem.

> **Theorem 11.6** (Absolute convergence). *If $\sum_{i=1}^{\infty} |a_i|$ converges, then so does $\sum_{i=1}^{\infty} a_i$.*

That is, absolute convergence implies convergence. However, convergence does not imply absolute convergence. When a series $\sum a_i$ converges but $\sum |a_i|$ does not, we say that $\sum a_i$ is **conditionally convergent**.

The proof of Theorem 11.6 uses the comparison test. The quantity $a_i + |a_i|$ is either 0 (if $a_i \leq 0$) or is $2|a_i|$ (if $a_i \geq 0$). This makes $\sum(a_i + |a_i|)$ a positive series that we can compare to $\sum 2|a_i|$. If $\sum |a_i|$ converges, $\sum 2|a_i|$ converges, so $\sum a_i + |a_i|$ converges. Then, since $\sum(a_i + |a_i|)$ and $\sum |a_i|$ both converge, their difference $\sum a_i = \sum(a_i + |a_i| - |a_i|)$ also converges.

Absolute convergence helps us to understand series like

$$\text{(11.87)} \qquad \sum_{n=1}^{\infty} \frac{\sin(n\pi/6)}{2^n}.$$

The numerator oscillates, but that doesn't faze us, since the numerator never gets bigger than 1 or less than -1. We always have $|a_n| \leq 2^{-n}$. Since $\sum 2^{-n}$ converges, our series converges absolutely.

The Root and Ratio Tests. The two most powerful and commonly used tests for convergence, called the **root test** and the **ratio test**, come from comparison to a geometric series.

In the ratio test, we look at the limit

$$\text{(11.88)} \qquad L = \lim_{n \to \infty} \left| \frac{a_{n+1}}{a_n} \right|.$$

> **Theorem 11.7** (Ratio test). *Let $\sum_{n=1}^{\infty} a_n$ be an infinite series, and let L be defined as in (11.88).*
>
> (1) *If $L < 1$, then $\sum_{n=1}^{\infty} a_n$ converges absolutely.*
>
> (2) *If $L > 1$ or if the ratio $|a_{n+1}/a_n|$ goes to $+\infty$ as $n \to \infty$, then $\sum_{n=1}^{\infty} a_n$ diverges.*
>
> (3) *If the limit fails to exist in some other way or if $L = 1$, then the test is inconclusive.*

The root test is almost identical, except that we look at the limit

$$\rho = \lim_{n \to \infty} |a_n|^{1/n} = \lim_{n \to \infty} \sqrt[n]{|a_n|}. \tag{11.89}$$

> **Theorem 11.8** (Root test). *Let $\sum_{n=1}^{\infty} a_n$ be an infinite series, and let ρ be defined as in (11.89).*
>
> (1) *If $\rho < 1$, then $\sum_{n=1}^{\infty} a_n$ converges absolutely.*
>
> (2) *If $\rho > 1$ or if $|a_n|^{1/n}$ diverges to $+\infty$, then $\sum_{n=1}^{\infty} a_n$ diverges.*
>
> (3) *If the limit fails to exist in some other way or if $\rho = 1$, then the test is inconclusive.*

The ratio test is usually easier to apply, but the root test is more general. If L exists, then ρ also exists and $\rho = L$. However, we will soon see an example where ρ exists but L doesn't. Before proving these theorems, let's apply them to some examples.

- In the series

$$\sum_{n=1}^{\infty} \frac{n^4}{2^n}, \tag{11.90}$$

we have $a_n = n^4/2^n$, so

$$L = \lim_{n \to \infty} \frac{(n+1)^4/2^{n+1}}{n^4/2^n} = \lim_{n \to \infty} \frac{1}{2}\left(\frac{n+1}{n}\right)^4 = \frac{1}{2}. \tag{11.91}$$

Since $L < 1$, the series converges. After a while, each term is roughly half the previous term, so the series behaves much like the (convergent!) geometric series with $r = 1/2$.

We could also apply the root test to this series, getting

$$\rho = \lim_{n \to \infty} \left(\frac{n^4}{2^n}\right)^{1/n} = \lim_{n \to \infty} \frac{1}{2}(n^{1/n})^4. \tag{11.92}$$

Evaluating this limit requires knowing that $\lim_{n \to \infty} n^{1/n} = 1$, which is far from obvious (but easy to check numerically). In this example, both the ratio and root tests work, but the ratio test is easier.

- If we were faced with the series $\sum(5n^4 + 37n^3 - 21n^2 + 452n - 1832)2^{-n}$, we could try applying the ratio or root tests directly. However, it is much simpler to apply the limit comparison test, comparing this series to the series $\sum n^4 2^{-n}$ that we just studied. Since $\sum n^4 2^{-n}$ converges, and since the ratio

(11.93)
$$\frac{(5n^4 + 37n^3 - 21n^2 + 452n - 1832)2^{-n}}{n^4 2^{-n}}$$
$$= 5 + 37n^{-1} - 21n^{-2} + 452n^{-3} - 1832n^{-4}$$

approaches 5, the series $\sum(5n^4 + 37n^3 - 21n^2 + 452n - 1832)2^{-n}$ converges.

- Next consider the series

(11.94)
$$\sum_{n=1}^{\infty} \frac{n!}{50^n}.$$

We compute

(11.95)
$$\frac{a_{n+1}}{a_n} = \frac{(n+1)!/50^{n+1}}{n!/50^n} = \frac{n+1}{50},$$

which goes to $+\infty$ as $n \to \infty$. This series therefore diverges. If we tried the root test we would have

(11.96)
$$a_n^{1/n} = (n!)^{1/n}/50.$$

This also goes to $+\infty$, but the fact that $\lim_{n\to\infty}(n!)^{1/n} = \infty$ is *not* obvious. (See Exercise 11.37.)

- On the other hand, the series

(11.97)
$$\sum_{n=1}^{\infty} \frac{50^n}{n!}$$

has $L = \lim_{n\to\infty} 50/(n+1) = 0$ and $\rho = \lim_{n\to\infty} 50/(n!)^{1/n} = 0$, so the series converges absolutely, despite growing rapidly for the first 50 terms.

- Finally, consider the series

(11.98)
$$\sum_{n=1}^{\infty} \frac{3 + 2(-1)^n}{2^n}.$$

That is, a_n equals 2^{-n} when n is odd and equals 5×2^{-n} when n is even. The ratio a_{n+1}/a_n alternates between 5/2 and 1/10 and has no overall limit, so the ratio test doesn't tell us anything. However, $1^{1/n}$ and $5^{1/n}$ are both close to (or equal to) 1 when n is large, so

(11.99)
$$\rho = \lim_{n\to\infty} \left(\frac{1 \text{ or } 5}{2^n}\right)^{1/n} = \lim_{n\to\infty} \frac{1}{2}(1 \text{ or } 5)^{1/n} = \frac{1}{2}.$$

The ratio test fails, but the root test still works, and it tells us that the series converges.

Proof of the Ratio and Root Tests. We now prove Theorems 11.7 and 11.8, the ratio and root tests, starting with Theorem 11.7.

Suppose that $L < 1$. Let M be a number (say, $M = (1+L)/2$) that is bigger than L but less than 1. Since $|a_{n+1}/a_n|$ gets closer and closer to L, it must eventually be less than M. But that means that $|a_n|$ must eventually be less than a constant times M^n. The series $\sum |a_n|$ then converges by comparison to the geometric series $\sum M^n$, so $\sum a_n$ converges absolutely.

If $L > 1$ or if $|a_{n+1}/a_n|$ goes to $+\infty$, then the terms a_n are eventually growing in size, and so can't be approaching 0. By the divergence test, the series $\sum a_n$ diverges.

To see that the test is inconclusive if $L = 1$, we just need examples that converge absolutely, converge conditionally, and diverge. The series $\sum n^2$ has $L = 1$ and diverges. The series $\sum n^{-2}$ has $L = 1$ and converges. The series $\sum (-1)^n/n$ has $L = 1$ and converges conditionally.

The proof of Theorem 11.8 is very similar. If $\rho < 1$, then we can find a number μ such that $\rho < \mu < 1$. Since $|a_n|^{1/n}$ approaches ρ, it is eventually less than μ, so $|a_n|$ is eventually less than μ^n. Our series then converges absolutely by comparison to $\sum \mu^n$. If $\rho > 1$ or if $|a_n|^{1/n}$ goes to infinity, then $|a_n|^{1/n}$ is eventually bigger than 1, so $|a_n|$ is eventually bigger than 1, and in particular does not approach 0, so the series diverges by the divergence test. The series $\sum n^2$, $\sum n^{-2}$, and $\sum (-1)^n/n$ all have $\rho = 1$, confirming the third statement.

11.6. Intervals of Convergence

When looking at a series $\sum a_n$, the natural question is *does it converge?* When looking at a power series $\sum c_n x^n$ or $\sum c_n (x-a)^n$, the natural question is **where** *does it converge?* The set of all x's such that the series converges is called the **interval of convergence**. First we will explore the intervals of convergence of series of the form $\sum c_n x^n$. Once we understand them, the more general case of $\sum c_n (x-a)^n$ will be easy.

Suppose that the series $\sum c_n x^n$ converges when $x = x_0$. This implies that the numbers $c_n x_0^n$ eventually approach 0, and in particular are bounded. That is, there is a number M such that $|c_n x_0^n| < M$ for every natural number n. Then, for every x with $|x| < |x_0|$, we have

$$(11.100) \qquad |c_n x^n| = |c_n x_0^n| \left|\frac{x}{x_0}\right|^n < M \left|\frac{x}{x_0}\right|^n.$$

Since $|x| < |x_0|$, the series $\sum c_n x^n$ converges absolutely by comparison to the geometric series $\sum M|x/x_0|^n$.

In other words, if the interval of convergence contains x_0, then it also contains the entire open interval $(-|x_0|, |x_0|)$. Since the series always converges at $x = 0$ (where it is $c_0 + 0 + 0 + 0 + \cdots$),[4] this leaves only three possibilities for what the interval of convergence is.

---
[4]The notation $\sum_{n=0}^{\infty} c_n x^n$ is a little misleading when $x = 0$, since 0^0 doesn't make sense. A better notation would be $c_0 + \sum_{n=1}^{\infty} c_n x^n$.

11.6. Intervals of Convergence

Theorem 11.9. *If $\sum c_n x^n$ is a power series, then either*

(1) *the series converges only when $x = 0$;*

(2) *the series converges for every real number x; or*

(3) *there is a positive number R such that the interval of convergence is either $(-R, R)$, $(-R, R]$, $[-R, R)$, or $[-R, R]$.*

The number R in case (3) is called the **radius of convergence** of the series. See Figure 11.4. If $|x| < R$, then the series converges, and if $|x| > R$, then it diverges. The situations where $x = \pm R$ are tricky. There are series that converge at both of these endpoints, series that converge at only one, and series that diverge at both endpoints. For instance, $\sum n x^n$, $\sum x^n/n$, and $\sum x^n/n^2$ all have $R = 1$, but the first diverges at $x = \pm 1$, the second converges at $x = -1$ and diverges at $x = 1$, and the third converges at $x = \pm 1$. Depending on what happens at the endpoints of the interval of convergence, R is either the largest value of $|x|$ for which the series converges at x, or the smallest value of $|x|$ where it diverges, or both. We also use the language of radius of convergence for the first two cases. In case (1) we say that $R = 0$, and in case (2) we say that R is infinite.

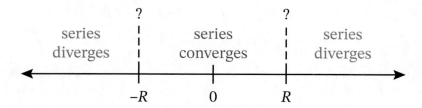

Figure 11.4. A power series $\sum c_n x^n$ with radius of convergence R converges when $|x| < R$ and diverges when $|x| > R$. What happens at $x = \pm R$ is more complicated.

Computing R. So how do we actually compute a radius of convergence? There are many techniques, but the two most common, and the most powerful, are the root and ratio test.

As before, we begin with the ratio test. Theorem 11.7 says that the criterion for $\sum c_n x^n$ converging is for the limit

(11.101) $$\lim_{n \to \infty} \left| \frac{c_{n+1} x^{n+1}}{c_n x^n} \right|$$

to be less than 1. However, that ratio is just the limit of $\left|\frac{c_{n+1}}{c_n} x\right|$, which is less than 1 precisely when $|x|$ is less than the limit of $\left|\frac{c_n}{c_{n+1}}\right|$. If that last ratio approaches a finite number, then that is our radius of convergence. If it diverges to ∞, then the series converges for all x. If it goes to 0, then the series diverges for all $x \neq 0$. In other words,

> **Theorem 11.10** (Ratio test for power series). *If the sequence $|c_n/c_{n+1}|$ has a finite limit, then the radius of convergence of the power series $\sum c_n x^n$ is*
>
> (11.102) $$R = \lim_{n\to\infty} \left|\frac{c_n}{c_{n+1}}\right|.$$
>
> *If the sequence of ratios goes to ∞, then the radius of convergence is also infinite.*

Note: Many people look at the limit of $|c_{n+1}/c_n|$ rather than the limit of $|c_n/c_{n+1}|$. When that limit is a positive number, the radius of convergence is equal to the *reciprocal* of the limit of $|c_{n+1}/c_n|$, and is equal to the limit of $|c_n/c_{n+1}|$.

- Let's apply the ratio test to the series for e^x.

(11.103) $$\lim_{n\to\infty} \left|\frac{c_n}{c_{n+1}}\right| = \lim_{n\to\infty} \frac{1/n!}{1/(n+1)!} = \lim_{n\to\infty} n+1.$$

  Since this goes to ∞, the series for e^x converges for every x. That is, the radius of convergence is infinite.

- The terms in the series for $\sin(x)$ and $\cos(x)$ are either 0 or are plus or minus the corresponding terms in the series for e^x. For every x, the series for $\sin(x)$ and $\cos(x)$ converge (absolutely) by comparison to the series for e^x.

- The ratio test also gives us the radius of convergence for the series for $\ln(1-x)$:

(11.104) $$R = \lim_{n\to\infty}\left|\frac{c_n}{c_{n+1}}\right| = \lim_{n\to\infty} \frac{-1/n}{-1/(n+1)} = \lim_{n\to\infty} \frac{n+1}{n} = 1.$$

- In the series for $\tan^{-1}(x)$, namely $x - x^3/3 + x^5/5 - x^7/7 + \cdots$, every other coefficient is 0, so the ratio of successive coefficients is either 0 or undefined. However, we can find the radius of convergence by looking at the ratio of successive *nonzero* coefficients, and the answer is 1.

- Finally, we look at the series for $\ln(x)$. The ratio of successive coefficients is 1, as before, so the radius of convergence is 1. However, this does not mean that the series converges when $|x| < 1$! It converges when $|x-1| < 1$, in other words when $0 < x < 2$. When doing a Taylor series in powers of a, the interval of convergence runs from $a - R$ to $a + R$, not from $-R$ to R. See Figure 11.5.

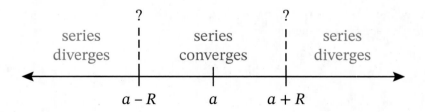

Figure 11.5. A power series $\sum c_n(x-a)^n$ with radius of convergence R converges when $|x-a|<R$ and diverges when $|x-a|>R$. What happens at $x=a\pm R$ is more complicated.

The root test is almost identical.

11.6. Intervals of Convergence

> **Theorem 11.11** (Root test for power series). *If the sequence $1/\sqrt[n]{|c_n|}$ has a finite limit, then the radius of convergence of the power series $\sum c_n x^n$ is*
>
> $$\text{(11.105)} \qquad R = \lim_{n \to \infty} \left| \frac{1}{\sqrt[n]{|c_n|}} \right|.$$
>
> *If $1/\sqrt[n]{|c_n|}$ goes to ∞, then the radius of convergence is also infinite.*

So far we have only looked at the *radius* of convergence. To fully understand the *interval* of convergence, we have to consider what happens at the boundary points $x = a \pm R$. This can be subtle, and it involves tests that are not covered in this text, such as the **alternating series test** and the **integral test**, and the fact that $\sum 1/n^p$ (the so-called *p*-series) converges for $p > 1$ and diverges for $p \le 1$.

In our last example, the series for $\ln(x)$ does not converge at $x = 0$. This should come as no surprise, because $\ln(0)$ does not exist. However, it turns out that the series converges at $x = 2$, and that it converges to

$$\text{(11.106)} \qquad \ln(2) = 1 - \frac{1}{2} + \frac{1}{3} - \frac{1}{4} + \frac{1}{5} - \frac{1}{6} + \cdots.$$

The convergence is extremely slow; it takes 1000 terms to get three digits of accuracy, 1,000,000 terms to get six digits of accuracy, and 10^k terms to get k digits of accuracy. Similarly, the series for $\ln(1-x)$ converges at $x = -1$ and diverges at $x = 1$.

The series for $\tan^{-1}(x)$ converges when $x = \pm 1$. Since $\tan^{-1}(1) = \pi/4$, this gives a curious formula for π:

$$\text{(11.107)} \qquad \pi = 4\left(1 - \frac{1}{3} + \frac{1}{5} - \frac{1}{7} + \cdots\right).$$

Like the series for $\ln(2)$, this series converges ridiculously slowly, with over 10^k terms needed to get k digits of accuracy. It is a nice mathematical curiosity, but it's useless for estimating π.

Analytic Functions. Many functions can be represented by convergent Taylor series, just like e^x, $\sin(x)$, $\cos(x)$, $\ln(x)$, and $\tan^{-1}(x)$. Such functions are called **analytic**. In fact, so many of the functions that we encounter in calculus, science, business, and engineering are analytic that it sometimes seems that all functions are analytic. However, that isn't true. There are three ways that a function can fail to be analytic.

(1) It may not have a Taylor series. In a Taylor series, $c_n = f^{(n)}(a)/n!$. That only makes sense if the function is n-times differentiable. For instance, the function $f(x) = |x|$ is not differentiable at $x = 0$. Near $x = 0$ there is no linear approximation to the function, much less quadratic, cubic, quartic, etc., approximations.

(2) It may have a Taylor series, but the series may not converge. If the derivatives $f^{(n)}(a)$ grow too quickly with n, then the radius of convergence of the Taylor series may be 0.

(3) It may have a convergent Taylor series, but the series may not converge to the right answer! Sometimes this is because the function changes behavior: you can compute a Taylor series for $f(x) = |x|$ around $a = 1$, getting $1 + (x - 1) + 0(x-1)^2 + 0(x-1)^3 + \cdots$, and this converges for every value of x, but it gives the

wrong answer for $f(-1)$. However, even smooth functions can have problems. The error in the nth Taylor polynomial is $f^{(n+1)}(c)(x-a)^{n+1}/(n+1)!$. If $f^{(n+1)}(c)$ grows too rapidly with n, this may not go to 0 as $n \to \infty$. If the error doesn't go to 0, the series isn't actually converging to $f(x)$.

11.7. Chapter Summary

The Main Ideas.

- An infinite sum $\sum_{k=1}^{\infty} a_k$, also called an infinite series, is the limit of partial sums $s_n = \sum_{k=1}^{n} a_k$. If this limit exists, we say that the series converges. Otherwise it diverges.
- Representing something as an infinite sum is the same as approximating it better and better by finite sums.
- The geometric series $\sum_{n=0}^{\infty} ar^n$ converges to $a/(1-r)$ if $|r| < 1$, and diverges if $|r| \geq 1$.
- A power series is a series where each term is a coefficient times a power of x or $x - a$. While ordinary series can converge to numbers, power series can converge to functions.
- Many common functions, including e^x, $\sin(x)$, $\cos(x)$, $\ln(x)$, and $\tan^{-1}(x)$, can be expressed as power series. These series are given in Table 11.1 on page 330.
- The Taylor **series** for a function $f(x)$ around $x = a$ is $\sum_{n=0}^{\infty} c_n(x-a)^n$, where $c_n = f^{(n)}(a)/n!$.
- The Taylor **polynomial** $T_n(x)$ is the first $n+1$ terms of a Taylor series, up through x^n or $(x-a)^n$. It is the nth order polynomial that best approximates the original function. $T_1(x)$ is the best linear approximation, which is the equation of the tangent line. $T_2(x)$, $T_3(x)$, etc., are improvements on $T_1(x)$. Since we know how to integrate polynomials easily, Taylor polynomials allow us to approximate the integral of a general function.
- The error in approximating $f(x)$ by $T_n(x)$ is

$$(11.108) \qquad f(x) - T_n(x) = \frac{f^{(n+1)}(c)}{(n+1)!}(x-a)^{n+1},$$

where c is a mystery number somewhere between a and x.

- There are many tests for convergence, including the divergence test, the comparison test, the limit comparison test, the absolute convergence test, the ratio test, and the root test. The ratio and root tests are based on doing comparisons to geometric series.
- A power series either converges only at $x = a$, or converges everywhere, or converges on an interval of convergence centered at $x = a$. The radius of this interval of convergence is called the radius of convergence.

- We can often compute the radius of convergence from the ratio and root tests. In the ratio test, we look at $\lim_{n\to\infty} |c_n/c_{n+1}|$. If this limit exists, it is the radius of convergence of $\sum c_n(x-a)^n$. The root test is similar, but looks at $\lim_{n\to\infty} 1/\sqrt[n]{|c_n|}$.
- Most common functions are analytic, meaning they can be expressed as convergent power series. However, not every function is analytic.

Expectations. You should be able to:

- Explain what it means for a series to converge, what it means to diverge, and what it means for it to "go to infinity".
- Find the sum of a finite geometric series or a convergent infinite geometric series.
- Recite from memory the series expansions for $1/(1-x)$, e^x, $\sin(x)$, and $\cos(x)$.
- Manipulate series to get expansions for additional functions, e.g., getting the series for e^{-2x^2} by taking the series for e^x and replacing x with $-2x^2$.
- Differentiate and integrate series term-by-term.
- Relate the kth coefficient c_k in a Taylor series to the kth derivative of the function being approximated.
- Use Taylor polynomials for approximating functions near points where the first few derivatives are known and for approximating integrals such as $\int_{-0.1}^{0.2} e^{-x^2}\,dx$.
- Use a variety of tests to determine whether a series converges or diverges.
- Use the ratio and root tests to determine the radius of convergence of a power series.

11.8. Exercises

Geometric Series

11.1. Find the sum of each infinite series, if it exists, or indicate that the series diverges.
(a) $16 + 12 + 9 + \cdots$
(b) $1 + \frac{3}{4} + \frac{9}{16} + \cdots$
(c) $\frac{3}{2} - \frac{3}{4} + \frac{3}{8} - \cdots$
(d) $7 + 1.6 + 1.06 + 1.006 + \cdots$
(e) $0.9 + 0.09 + 0.009 + 0.0009 + \cdots$

11.2. The sum of an infinite geometric series is 16 and the common ratio is $1/3$. Find the first three terms of the series.

11.3. The sum of an infinite geometric series is 24 and the common ratio is $-1/4$. Find the first three terms of the series.

11.4. Alex and Charlie have a pet bird, who we'll call Birdie. One day, Alex and Charlie rode bikes toward each other, starting 30 miles apart, with each one riding at 10 mph. At the beginning of the ride, Birdie was next to Alex. Birdie then flew at 20 mph toward Charlie. When Birdie reached Charlie, Birdie turned around and flew at 20 mph back to Alex, then turned around and flew to Charlie, and

so on until the three of them met in the middle, where they shared a picnic of sandwiches and birdseed.
   (a) How long did it take Birdie to reach Charlie the first time? Where were Alex, Birdie, and Charlie at that time, relative to Alex's starting point? How far did Birdie fly on that first leg?
   (b) How long did it take Birdie to fly back to Alex? Where was everybody when Birdie reached Alex? How far did Birdie fly on that second leg?
   (c) How far did Birdie fly on the nth leg?
   (d) Write down an infinite series for the total distance that Birdie flew. Then find the sum of the series.

11.5. Here's a completely different way to understand the total distance that Birdie flew in Exercise 11.4. How long did it take for Alex and Charlie to meet? How far did Birdie fly in that time?

11.6. Arquala and Nathan are driving a 10-foot fence post into the ground using a sledgehammer. The first strike of the hammer sinks the post 10 inches, but every subsequent strike only drives the post 60% as far as the previous strike. For example, the second strike drives it only 6 inches deeper. Arquala argues that the depth of the post will eventually converge, with approximately 8 feet of the post remaining above ground, which is plenty to attach a tall fence to. Nathan argues that with each strike the post is being driven deeper. Since the post has a finite length (10 feet), it will eventually be completely driven into the ground. Who (if either) is right, and why?

11.7. **Aquarium Cleaning.** Fish living in an aquarium produce waste. It isn't feasible to clean a large aquarium thoroughly each day. Instead, part of the water is removed at the end of each day and replaced by clean water. Let P be the fraction of the water (and fraction of the existing waste) that is removed. Let W be the daily waste output from the fish in the tank. Starting with a clean tank, the amount W_1 of waste in the tank after one day is
$$W_1 = W(1 - P).$$
Similarly, the waste remaining after two days is
$$W_2 = (W + W_1)(1 - P) = W(1 - P) + W(1 - P)^2.$$
The first term is today's waste, while the second term is what's left of yesterday's waste.
   (a) Write down an expression for the amount of waste in the tank after five days.
   (b) Write down an expression using Sigma notation ($\sum$) for the amount of waste in the tank after n days.
   (c) If a third of the water is removed each day, how much waste is in the tank after ten days?
   (d) For the fish to stay healthy, the waste in the tank after each cleaning must remain under $3W$. What proportion of the water must be replaced each day to achieve this?

11.8. Exercises

The Multiplier Effect. A country's gross domestic product (GDP) is the total value of all the goods and services sold in that country. An economic recession occurs when the GDP falls for two successive quarters. To try to avoid a recession, or to recover from one, governments often provide stimulus money to their citizens. They hope that each dollar in stimulus will increase economic activity by more than $1. The way this happens is called the **multiplier effect**.

The GDP of the small (fictional) country of Calculania was 100 million dollars. During a severe recession, the GDP fell to 90 million dollars. The government intervened by providing 5 million dollars in stimulus money to individuals. Of every dollar a person receives, an average of 30% goes to taxes. Of the remainder, 10% goes into savings, 20% is spent on imported goods, and the rest is spent on domestically produced goods and services. Exercises 11.8–11.11 explore the multiplier effect for this attempt to stimulate Calculania's economy.

11.8. (a) Explain why every dollar that the government of Calculania gives out in stimulus payments results directly in 49 cents of additional spending on domestically produced goods and services.
   (b) The additional spending on domestically produced goods and services is income to others in society. Those who receive that extra income pay 30% in taxes, save 10% of the after-tax income, spend 20% of the after-tax income on imported goods, and spend the rest on domestic goods and services. How much of the 49 cents (per original dollar of stimulus money) is spent again on domestic goods and services?
   (c) Write down the first five terms of a geometric series that describes the total impact of a dollar of stimulus on Calculania's economy. Find the partial sum s_5.
   (d) Write down a geometric series $\sum_{i=0}^{\infty} ar^i$ that gives the total additional income for people in Calculania for every dollar spent on the stimulus. What number S does it converge to? This number is called the **spending multiplier**.

11.9. How much extra tax revenue did the government of Calculania get to offset the cost of the stimulus program?

Income vs. GDP. A country's GDP isn't the same as the total income of all the people in that country; while every dollar spent in the country is income to somebody, not every dollar in income comes from sales or salary. In particular, the stimulus money itself is income but is not part of the GDP.

11.10. (a) Was the $5 million that the government spent on stimulus money enough to offset the $10 million loss of *income* due to the drop in the GDP? Explain.
   (b) Was the $5 million that the government spent on stimulus money enough to offset the $10 million drop in the GDP?

11.11. The government of Calculania could have spent $5 million directly on domestic goods and services, for instance by paying people to rebuild roads and bridges, instead of sending out stimulus money to individuals. How much would that stimulus have increased the GDP? Would it have been enough to offset the recession? How much would it have increased the total income for the country?

Food for Thought: The kind of stimulus described in Exercise 11.11, involving direct government spending rather than payouts or tax cuts, is preferred by Keynesian economists. However, there are big political obstacles to implementing this sort of stimulus. Since government spending programs are great for businesses that get the contracts, but not so great for the ones that don't, there's a lot of fighting involved in the decisions. People (especially the losers!) object to the government picking winners and losers and are quick to argue (sometimes correctly) that the winners got their contracts through corruption. Giving money to everybody may not be as effective economically, but it is often fairer and is usually a lot more popular.

11.12. In this exercise, you will develop a formula for the spending multiplier. The spending multiplier determined in Exercise 11.8 depended on taxes, savings, and import spending. Economists call these three factors **leakages**, as they determine how much money leaves, or "leaks out", of the economy each time it changes hands. Economists commonly use the following terminology and notation when discussing spending multipliers:

Marginal Rate of Taxation	MRT,
Marginal Propensity to Save (after taxes)	MPS,
Marginal Propensity to Import (after taxes)	MPM,
Marginal Propensity to Consume	MPC.

Leakages include the MRT, MPS, and MPM, while MPC is whatever is left.
  (a) Find the formula for MPC in terms of MRT, MPS, and MPM.
  (b) Find an expression for the spending multiplier in terms of MPC. Using your answer to part (a), rewrite that formula in terms of MRT, MPS, and MPM.
  (c) If MRT = 0.25, MPS = 0.05, and MPM = 0.1, compute the spending multiplier.
  (d) Discuss how changes in MRT, MPS, and MPM affect the multiplier. What are the partial derivatives of the multiplier with respect to MRT, MPS, and MPM?

11.13. Suppose that the GDP of another small country has fallen by 40 million dollars, prompting calls for a government stimulus program. The leakages for the country are MRT = 0.30, MPS = 0.08, and MPM = 0.05. The finance minister of the country thoroughly understands the multiplier effect and wants to spend as little money as possible to make up the 40 million dollar drop.
  (a) If the stimulus is done by sending money to individuals, as in Exercise 11.8, how much money must be spent?
  (b) If the stimulus is done via direct government spending, as in Exercise 11.11, how much money must be spent?

11.8. Exercises

Amortization is the process of paying off debt with a fixed repayment schedule, such as in installment plans for large purchases, especially houses and cars. Typical questions include:

- How big a monthly payment is needed to pay off the loan in a certain amount of time?
- How long will it take to pay off the loan on a certain payment schedule?
- How much will you still owe after a certain number of payments?

As with continuous income streams, these questions are answered using present value and future value.

In previous chapters, we considered present value and future value when interest is compounded continuously. If the interest rate is r, then $1 today is worth e^{rt} dollars t years from now, and $1 t years from now has a present value of e^{-rt} dollars. However, interest is often compounded monthly or yearly, not continuously. Exercises 11.14–11.20 walk you through the present and future values of monthly payments.

11.14. Suppose that the yearly interest rate is r and that interest is compounded monthly. How much will a dollar grow into m months from now?

11.15. Suppose the yearly interest rate is r and interest is compounded monthly. What is the present value of a dollar that is paid m months from now?

11.16. Suppose that you buy a $10,000 car on a 24-month payment plan at 3% annual interest, compounded monthly. That is, you make 24 monthly payments of $P, with the first payment on the day you buy the car. What is the present value of all of your payments put together? How big does P have to be to pay for the car?

11.17. In the setup of Exercise 11.16, what would the monthly payment be if you were on a 120-month plan? Or, thanks to a clerical error, a 1200-month plan?

11.18. Daniela has just graduated from college with $20,000 of student loan debt. Upon graduation, Daniela begins to make monthly payments of $230 on the loan at an annual interest rate of 6.8%, compounded monthly. That is, the monthly interest rate is 6.8%/12 or 0.567%.
   (a) What is the present value of Daniela's kth payment?
   (b) What is the total present value of Daniela's first m payments? (This involves summing a geometric series.)
   (c) After 24 payments, what is the present value of Daniela's remaining debt?
   (d) After 24 payments, how many dollars does Daniela still owe? (*Hint*: Convert your answer to part (c) into a future value.)
   (e) After 24 payments, how much has Daniela spent on interest and how much on reducing the principal of her loan?

11.19. On the payment schedule described in Exercise 11.18, how long will it take for Daniela to pay off her loan?

11.20. Continuing with the situation of Exercise 11.18, use MATLAB to make a graph showing the following quantities as functions of time:
   - the present value of Daniela's remaining debt,

- the number of dollars that Daniela still owes,
- how much she has has paid so far in interest, and
- the total (principal plus interest) she has paid so far.

Taylor Polynomials

11.21. Let $f(x) = \tan(x)$, and let $a = \pi/4$.
   (a) Compute $f(a)$, $f'(a)$, $f''(a)$, and $f'''(a)$.
   (b) Find the first three Taylor polynomials $T_1(x)$, $T_2(x)$, and $T_3(x)$ around $x = a$.
   (c) Use the polynomials from part (b) to estimate $\tan(50°)$. (Don't forget to convert 50° to radians!)
   (d) Now use the polynomials from part (b) to estimate $\tan(45.5°)$. How does the rate at which these estimates approach the true value compare to the rate for the estimates in part (c)?

11.22. Let $f(x) = e^{x^3}$.
   (a) Find the Taylor polynomial $T_9(x)$ for $f(x)$ around $x = 0$. (*Hint*: This is easier than it looks. You do not have to compute nine derivatives of $f(x)$!)
   (b) Estimate $\int_0^{0.5} e^{x^3}\, dx$ to four decimal places.

11.23. Let $f(x) = e^{-x^2}$. Find $f^{(50)}(0)$ and $f^{(51)}(0)$. No, you aren't expected to compute 51 derivatives of $f(x)$! Instead, look at the coefficients of x^{50} and x^{51} in the MacLaurin series for $f(x)$.

11.24. Consider the function $f(x) = \sqrt{x+4}$. We will compare MacLaurin polynomials (Taylor polynomials around $x = 0$) to Taylor polynomials around $x = 45$.
   (a) Compute the third-order MacLaurin polynomial $M_3(x) = a_0 + a_1 x + a_2 x^2 + a_3 x^3$ that approximates $f(x)$ near $x = 0$.
   (b) Compute the third-order Taylor polynomial $T_3(x) = b_0 + b_1(x - 45) + b_2(x - 45)^2 + b_3(x - 45)^3$ that approximates $f(x)$ near $x = 45$.
   (c) Make a graph showing $f(x)$, $M_3(x)$, and $T_3(x)$ for $0 \le x \le 60$. For which values of x is M_3 better than T_3? For which values is T_3 better than M_3?

11.25. For the supply curve $p = S(x) = \sqrt{x+4}$:
   (a) find the producer surplus (to the nearest dollar) at a price of $p = \$6$.
   (b) use the first three terms of the Maclaurin series for $S(x)$ to estimate the producer surplus at a price of $p = \$10$.
   (c) use the first three terms of the Taylor series centered at $x = 45$ for $S(x)$ to estimate the producer surplus at a price of $p = \$10$.
   (d) discuss the accuracy of parts (b) and (c) and explain why the estimate in part (c) is much more accurate.

11.26. Find the sixth-order MacLaurin polynomial $M_6(x)$ for each function.
   (a) $f(x) = x^2 \sin x$
   (b) $f(x) = e^{\pi x}$
   (c) $f(x) = \ln(1 + x)$
   (d) $f(x) = \frac{1}{1+x}$

Convergence and Divergence

11.27. Determine whether each series converges or diverges by using either the comparison test or the limit comparison test.

(a) $\sum_{k=1}^{\infty} \dfrac{5}{k - \frac{1}{2}}$

(b) $\sum_{k=1}^{\infty} \dfrac{2}{2^k + 2}$

(c) $\sum_{k=1}^{\infty} \dfrac{\ln k}{k}$

11.28. Determine whether each series converges or diverges by using either the root test or the ratio test.

(a) $\sum_{k=1}^{\infty} \dfrac{k!}{k^3}$

(b) $\sum_{k=1}^{\infty} \left(\dfrac{k}{50}\right)^k$

(c) $\sum_{k=1}^{\infty} k\left(\dfrac{1}{2}\right)^k$

11.29. Consider the series $\sum_{k=1}^{\infty} a_k$, where $a_k = (1 - e^{-k})^k$.
 (a) Apply the root test to this series. What does it tell you?
 (b) Show that, whenever A is small and positive, $\ln(1 - A)$ is approximately $-A$. In particular, show that $-2A \le \ln(1 - A) \le 0$.
 (c) Show that, whenever k is big enough, $-2ke^{-k} \le \ln(a_k) \le 0$.
 (d) We already know that $\lim_{k \to \infty} ke^{-k} = 0$. What is $\lim_{k \to \infty} a_k$?
 (e) Does $\sum_{k=1}^{\infty} a_k$ converge or diverge? Why?

11.30. Classify each series as absolutely convergent, conditionally convergent, or divergent.

(a) $\sum_{k=1}^{\infty} 5\sin\left(\dfrac{k\pi}{2}\right)$

(b) $\sum_{k=1}^{\infty} \dfrac{\cos(k\pi)}{2k}$

(c) $\sum_{k=1}^{\infty} \dfrac{(-1)^k}{k!}$

11.31. Find the radius and interval of convergence for each power series.

(a) $\sum_{k=1}^{\infty} 3^k x^k$

(b) $\sum_{k=1}^{\infty} \dfrac{(-1)^k x^k}{k!}$

(c) $\sum_{k=1}^{\infty} \left(\dfrac{3}{4}\right)^k (x+5)^k$

Singularities. In general, the radius of convergence is never greater than the distance to the nearest **singularity**, or the place where the function stops existing. For instance, you already know that the Taylor series (11.77) for $\ln(x)$ around $x = 1$ has a radius of convergence of 1. This is because the log function doesn't make sense at $x = 1 - 1 = 0$. We explore this further in Exercises 11.32 and 11.33.

11.32. Consider the function $f(x) = \sqrt{x+4}$.
   (a) Find a formula for $f^{(n)}(0)$. You may find the following notation useful. If k is an odd integer, then $k!!$ is the product of all the *odd* integers from 1 to k, so $3!! = 1 \times 3 = 3$, $5!! = 1 \times 3 \times 5 = 15$, and so on. (*Note*: $k!!$ is not the same as $(k!)!$. For instance, $3!! = 3$ but $(3!)! = 6! = 720$.)
   (b) Use the ratio test to find the radius of convergence of the MacLaurin series.
   (c) Does it make sense to you that the radius of convergence is less than 5? Explain. (*Hint*: What is $f(-5)$?)

11.33. Find the radius of convergence of the MacLaurin series for $f(x) = \frac{1}{1+x^2}$. How does that compare to the distance from the origin to the nearest singularity?

Food for Thought: The precise rule, which we won't prove here, is if the formula for a function can be extended to complex inputs, then the radius of convergence is the distance to the nearest singularity *in the complex plane*. For instance, the function $f(x) = \frac{1}{1+x^2}$ has a singularity at $x = i$, which is at distance 1 from the origin. However, dealing with complex inputs can be confusing and, well, complex. If we stick to real inputs and *analytic* functions, then the rule is

> The radius of convergence is *at most* the distance to the nearest singularity.

11.34. **Geodetic Surveys.** We saw how to find the height of a nearby tree in Exercise 10.8. When computing the height of more distant objects, such as a tree on the opposite shore of an ecologically sensitive lake that you're not allowed to cross, we also have to take into account the Earth's curvature, since an object that is barely visible on the horizon is already some distance h above the ground. Let r be the radius of the Earth, and let d be the distance, along the surface of the Earth, from you to the object whose height you are measuring.

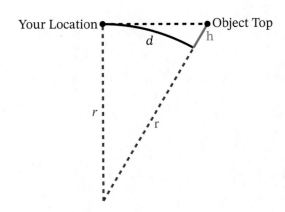

(a) Show that the height of the object is given by
$$h = r \sec\left(\frac{d}{r}\right) - r.$$
(b) Find the first two terms in the MacLaurin expansion for $\sec(x)$.
(c) Explain why $h \approx d^2/2r$.

Food for Thought: The h found in Exercise 11.34 is called the **curvature correction** in **geodetic surveying**. If an object at horizontal distance d appears at an angle θ above the horizon, then its true height is approximately $d \tan(\theta) + d^2/2r$. There are also corrections due to the refraction of light, since the index of refraction of air varies with elevation, and due to other factors. Accurate surveying is a complicated business!

11.35. Find the Taylor series around $x = 10$ for the function $f(x) = \ln(x)$.

11.36. Use the first four terms of an appropriate Taylor series for $f(x) = \sqrt[3]{x}$ to approximate $\sqrt[3]{28}$. Compare your answer to the value of $\sqrt[3]{28}$ given by a calculator.

11.37. In the text, we claimed that $\lim_{n \to \infty}(n!)^{1/n} = \infty$. In this exercise we're going to see why.
(a) When writing $n! = n \times (n-1) \times (n-2) \times \cdots \times 3 \times 2 \times 1$, how many of the factors are $n/2$ or bigger? (*Hint*: The answer depends slightly on whether n is even or odd.)
(b) Show that for all positive integers n, $n! \geq (n/2)^{n/2}$.
(c) Show that for all positive integers n, $(n!)^{1/n} \geq \sqrt{n/2}$.
(d) Finally, show that $\lim_{n \to \infty}(n!)^{1/n} = \infty$.

Solving Differential Equations. One use of Taylor (or MacLaurin) polynomials is to find approximate solutions to differential equations. For instance, consider the IVP

(11.109) $\qquad y' = \sin(y), \qquad y(0) = 0.3.$

We already know how to get a linear approximation by computing $y'(0) = \sin(0.3) \approx 0.29552$ and setting $y(t) \approx 0.3 + 0.29552t$. To do better than that, we need $y''(0), y'''(0)$, and so on.

We can get $y''(0)$ by taking the derivative of the differential equation itself:

(11.110) $\quad y'' = \cos(y)y' = \cos(y)\sin(y), \qquad y''(0) = \cos(0.3)\sin(0.3) \approx 0.28232.$

Taking another derivative gives us y''':

$$y''' = (\cos(y)\sin(y))' = (\cos^2(y) - \sin^2(y))y'$$
$$= \sin(y)(\cos^2(y) - \sin^2(y)),$$

(11.111) $\qquad y'''(0) = \sin(0.3)(\cos^2(0.3) - \sin^2(0.3)) \approx 0.24390.$

Now that we know $y(0), y'(0), y''(0)$, and $y'''(0)$, we can write down a third-order MacLaurin polynomial that approximates $y(t)$:

(11.112) $\qquad y(t) \approx 0.3 + 0.29552t + 0.14116t^2 + 0.04065t^3.$

We will apply this method in Exercises 11.38–11.40.

11.38. Find a third-order MacLaurin polynomial that approximates the solution to the IVP $y' = e^y, y(0) = 0$.

11.39. Recall the SIR model for estimating market penetration.
$$S' = -aSI,$$
$$I' = aSI - bI,$$
$$R' = bI,$$
with initial conditions $S(0) = S_0$, $I(0) = I_0$, and $R(0) = R_0$.
  (a) Find the third-order MacLaurin polynomial for $S(t)$.
  (b) Find the third-order MacLaurin polynomial for $I(t)$.
  (c) Find the third-order MacLaurin polynomial for $R(t)$.

11.40. In Exercise 11.39, you used the SIR model to find the third-order MacLaurin polynomial for $S(t)$. Using the parameters $a = -0.0001$, $b = 0.05$, $S_0 = 1000$, and $I_0 = 10$, complete the following steps.
  (a) Use MATLAB to plot the third-order MacLaurin approximation for $S(t)$ on the time interval $[0, 100]$.
  (b) Use the SIREulers program to graph an approximation for $S(t)$ on the time interval $[0, 100]$.
  (c) How do the two approximations in parts (a) and (b) compare? Which one do you think is more accurate? Why?

Index

acceleration, 138, 234
accumulation function, xvi, 205, 228, 240
    definition, 206, 223
    examples, 205, 207
    graph, 229, 230
addition of angle formulas, 88, 166, 329
amortization, 345
analysis, 8, 23
    max-min, 16
    numerical, 13, 46
anti-derivative, xvi, 225, 228, 233
    definition, 224
    lack of uniqueness, 224
    vs. indefinite integral, 230
approximation, 31
Australian rabbits, 157
average cost, 62, 63
average profit, 128

Babylonian method, xv, 50
ballistics, 233
beef jerky, 128, 303
bisection method, xv, 32, 49
budget deficit, 3, 61

Celsius scale, 55
circular motion, 137
Clairaut's theorem, 287
Cobb-Douglas, 301
competitive products, 304
complementary angles, 163
complementary products, 304
concavity, 109, 132
consumer surplus, 219

convergence tests, xvii, 330
    comparison test, 331, 332
    divergence test, 330, 331
    limit comparison test, 331, 333
    ratio test, xvii, 333, 334
    root test, xvii, 333, 334
conversion factor, 98
Covid-19, 5, 19, 176
critical point, 107, 112
curve sketching, 105, 131

demand curve, 219
demand function, 102
depreciation, 243
    double-declining balance, 244
    remaining years, 244
    straight-line, 244
derivative, xv, 2, 59, 67
    definition, 68, 76
    examples, 70
    notation, 75
    partial, xvi, 282, 284
        definition, 284
        higher-order, 285
derivative function, 72, 75
derivative rules
    basic functions, 86
        cosine, 88, 171
        exponentials, 90, 148, 149
        logarithms, 148, 149
        power law (Newton's hammer), 86
        sine, 88, 171
        sine and cosine, 88

chain, 97
    examples, 100
    hybrid form, 100
    in two dimensions, 290
    Leibniz form, 98
    Newton form, 100
    product, 92, 93, 95, 124
        common mistake, 93
        examples, 95
    quotient, 92, 95, 96
        examples, 96
difference quotient, xv, 76
    backward, xv, 60, 71
    centered, xv, 60, 71
    forward, xv, 60, 70
differentiability, 68
differential calculus, 68
differential equation, 10, 23, 136, 137
    first-order, 139
    general solution, 137, 143
    initial condition, 137, 139, 140
    initial value problem (IVP), 137, 139, 141
        examples, 141
    particular solution, 143
    second-order, 139
    solution, 137, 141
        existence and uniqueness, 142
        numerical, 137, 142
double angle formulas, 166

e, 91
epidemic, 18
equation of a line
    point-slope form, 14, 40, 52, 54
    slope-intercept form, 41, 54
Euler's formula, xvii, 328
Euler's method, xv, 15, 43, 46, 56, 142
Euler, Leonhard, 43, 328
exponential growth, 29
exponential growth and decay, 137, 145, 150, 157
    doubling time, 151
    half-life, 151
    law of 70, 151

Fahrenheit scale, 55
first derivative test, 108
fish, 203
Fubini's theorem, 296
functions, 31, 34
    algorithm, 35, 75
    analytic, 339

composition, 36
cosine, 122, 153, 163
    derivative, 171
    graph, 168
    symmetry, 169
exponential, 90, 137, 144
    derivative, 148, 149
    graph, 145, 146
    laws, 145
formula, 35, 75
graph, 36, 37, 75
increasing vs. decreasing, 108, 130
inverse sine (arcsine)
    definition, 171
    derivative, 172
    graph, 172
inverse tangent (arctangent)
    definition, 171
    derivative, 172
    graph, 172
linear, 31, 37, 38, 40
logarithm, 144, 146
    definition, 146
    derivative, 148, 149
    graph, 147
    laws, 147
logarithms, 90, 137
rise and run, 37
secant, 164
    derivative, 171
sine, 122, 153, 163
    derivative, 171
    graph, 168
    symmetry, 169
table, 35, 75
tangent, 164
    derivative, 171
    graph, 170
Fundamental Theorem of Calculus, 6, 221
    FTC1, xvi, 221, 228, 231, 240
    FTC2, xvi, 221, 225, 231
        and chain rule, 232
    net change theorem, 227
future value, 273

geodetic survey, 348
geometric series, xvi, 316
    convergence, 317
gross domestic product (GDP), 126, 343

heat index, 6, 281, 282
Hooke's law, 155

ideal gas, 299
indifference curve, 302
infinite series, xvi, 313
    convergence, 313, 315
        absolute, 333
        conditional, 333
    divergence, 313
    partial sums, 314
infinity, 315, 316
integrability, 201
integral
    definite, xvi, 201, 225
        definition, 201, 222
        notation, 201, 230
        properties, 202
    double, 282, 291–293
    improper, 270
    indefinite, xvi, 205, 233
        examples, 237
        notation, 230
        table, 252
    integral, xvi
    iterated, 282, 291, 294
integral calculus, 68, 185
integration, 185
    examples, 186, 213
        area under a curve, 188, 193, 195
        distance traveled, 187
        electrical energy, 215
        moment of inertia, 191
        probability, 218
        start-up losses, 186
        volume, 192
        work, 214
    methods, 249
        by parts, 249, 254
        corrected trapezoidal rule, 278
        geometry, 189
        midpoint rule, 249, 263
        Simpson's rule, xvi, 249, 263, 265
        slice-and-dice, 186, 194
        substitution, 249, 250
        trapezoidal rule, xvi, 190, 200, 249, 260
    Riemann sum, xvi, 195, 226, 295
        definition, 198
        in two dimensions, 292
interest
    compound, 43, 44, 150, 323
        annual, 45
        continuous, 45
        monthly, 45
    rate, 138
    simple, 44
interpretation, 8, 23
interval of convergence, 336, 339

leakages, 344
Leibniz, Gottfried, 1, 75
limit, xii, 2, 70, 77, 122, 123, 145, 195, 198, 201, 203, 209, 222, 229, 231, 238, 270, 282, 287, 291, 292, 313, 315, 321, 333, 334, 337, 340
    definition, 68
linear approximation, 13, 23, 31, 44, 64, 76
    extrapolation, 41, 42
    in two dimensions, 288
    interpolation, 41, 42
    microscope equation, xv, 41, 64, 76
        in two dimensions, 288
logistic growth, 137, 157, 158
Lotka-Volterra, xv, 137, 159

MacLaurin series, 321
marginal quantities, 62, 76, 102
    cost, xv, 62, 63
    price, 179
    profit, xv, 62, 64
    revenue, xv, 62, 64
    utility, 303
marginal rate of substitution, 303
market penetration, 7
MATLAB, xii, 24
    fplot, 77
    plot, 49, 77
    sample programs
        Babylonian method, 51
        Newton's method, 118
        Riemann, 200, 217
        Riemann2D, 294
        Simp, 276
        SIREulers, 48
    zoom, 77
max-min (optimization), 102
    critical point, 132
    local maximum, 73, 103, 108, 132
    local minimum, 73, 108, 112, 132
mixed partials are equal, 287
modeling, 8, 19, 23, 27, 135, 137, 141, 150, 153, 157
models, xiv
moment of inertia, 191
mortgage, 240

multiplier, 38, 41
multiplier effect, 343

national debt, 3
Newton's method, xv, 51, 114
   algorithm, 116
   and Babylonian method, 117
   basin of attraction, 118
Newton, Isaac, 1, 51, 75

optimization, 102
optimizing, 85

parameter, 11, 138
per capita, 126
point of inflection, 110, 112
polar coordinates, 308
population, 150
   Puerto Rico, 78
   Texas, 78
position and velocity, 39, 56, 126, 138, 234, 245
power series, xvi, 313, 317
present value, 210, 240, 269
price, 102, 128
   as control parameter, 103
probability
   cumulative probability function, 271
   density function, 270
   exponential distribution, 218, 270, 274
   mean, 218, 273
   normal distribution, 218, 274
   standard deviation, 218, 273
   uniform distribution, 274
   variance, 273
producer surplus, 246, 273
product bust, 27
Pythagorean identities, 165

radioactivity, 57, 150
radius of convergence, xvii, 336, 337
   and ratio test, 338
   and root test, 339
random walk, 66, 67
rate equation, 10, 23, 24, 136
rate of change, 2, 24, 38, 59
   average, 38
reality check, 14
recovery coefficient, 10
related rates, 99
relative humidity, 281, 282
right triangles, 163

second derivative, 86, 105
second derivative test, 110
SEIR model, 20
Sigma notation ($\sum$), 196
sign chart, 106, 107, 112, 113
simple harmonic motion, 156
singularity, 348
SIR model, xiv, 7, 9, 18, 19, 46, 136, 142, 150, 159
   active users, 9
   attrition, 10, 11
   attrition coefficient, 10
   herd immunity, 28
   infected, 18
   potential users, 9
   recovered, 18
   recovery coefficient, 18
   rejected users, 9
   removed, 18
   replication number, 19, 28
   susceptible, 18
   transmission, 11
   transmission coefficient, 11, 18
SIRS model, 21
Six Pillars of Calculus, xii, 2
   1st Pillar, xii, xv, 2, 31
   2nd Pillar, xii, xv, 2, 59
   3rd Pillar, xii, 2, 16, 73, 85
   4th Pillar, xii, xvi, 3, 185
   5th Pillar, xii, xvi, 5, 16, 221, 226
   6th Pillar, xii, xvi, 6, 281
slope, 2, 38
small angle approximation, 122
Social Security, 61
SOH-CAH-TOA, 164
Solow growth model, 57
spending multiplier, 343, 344
Stewart, James, xi
subcritical point, 110, 112
supply and demand, 102, 180, 247
supply curve, 179
supply function, 179

Taylor polynomial, xvi, 313, 318, 319
   accuracy, 323
   cubic approximation, 318, 320
   formula, 320, 322
   quadratic approximation, 318, 320
Taylor series, xvi, 313, 319
   arctangent, 319
   cosine, 323
   exponential, 323

    logarithm, 319
    sine, 323
Taylor's theorem, xvi, 323
Thomas, George, xi
threshold, 17
trigonometry, 163
trolls, 27

unit circle, 167
unit cost, 129
utility function, 302

vaccination, 18
variable, 8, 11

wheat, 291

zombies, 158

Selected Published Titles in This Series

61 **Bennett Chow,** Introduction to Proof Through Number Theory, 2023
59 **Oscar Gonzalez,** Topics in Applied Mathematics and Modeling, 2023
58 **Sebastian M. Cioabă and Werner Linde,** A Bridge to Advanced Mathematics, 2023
57 **Meighan I. Dillon,** Linear Algebra, 2023
56 **Lorenzo Sadun,** The Six Pillars of Calculus: Business Edition, 2023
55 **Joseph H. Silverman,** Abstract Algebra, 2022
54 **Rustum Choksi,** Partial Differential Equations, 2022
53 **Louis-Pierre Arguin,** A First Course in Stochastic Calculus, 2022
52 **Michael E. Taylor,** Introduction to Differential Equations, Second Edition, 2022
51 **James R. King,** Geometry Transformed, 2021
50 **James P. Keener,** Biology in Time and Space, 2021
49 **Carl G. Wagner,** A First Course in Enumerative Combinatorics, 2020
48 **Róbert Freud and Edit Gyarmati,** Number Theory, 2020
47 **Michael E. Taylor,** Introduction to Analysis in One Variable, 2020
46 **Michael E. Taylor,** Introduction to Analysis in Several Variables, 2020
45 **Michael E. Taylor,** Linear Algebra, 2020
44 **Alejandro Uribe A. and Daniel A. Visscher,** Explorations in Analysis, Topology, and Dynamics, 2020
43 **Allan Bickle,** Fundamentals of Graph Theory, 2020
42 **Steven H. Weintraub,** Linear Algebra for the Young Mathematician, 2019
41 **William J. Terrell,** A Passage to Modern Analysis, 2019
40 **Heiko Knospe,** A Course in Cryptography, 2019
39 **Andrew D. Hwang,** Sets, Groups, and Mappings, 2019
38 **Mark Bridger,** Real Analysis, 2019
37 **Mike Mesterton-Gibbons,** An Introduction to Game-Theoretic Modelling, Third Edition, 2019
36 **Cesar E. Silva,** Invitation to Real Analysis, 2019
35 **Álvaro Lozano-Robledo,** Number Theory and Geometry, 2019
34 **C. Herbert Clemens,** Two-Dimensional Geometries, 2019
33 **Brad G. Osgood,** Lectures on the Fourier Transform and Its Applications, 2019
32 **John M. Erdman,** A Problems Based Course in Advanced Calculus, 2018
31 **Benjamin Hutz,** An Experimental Introduction to Number Theory, 2018
30 **Steven J. Miller,** Mathematics of Optimization: How to do Things Faster, 2017
29 **Tom L. Lindstrøm,** Spaces, 2017
28 **Randall Pruim,** Foundations and Applications of Statistics: An Introduction Using R, Second Edition, 2018
27 **Shahriar Shahriari,** Algebra in Action, 2017
26 **Tamara J. Lakins,** The Tools of Mathematical Reasoning, 2016
25 **Hossein Hosseini Giv,** Mathematical Analysis and Its Inherent Nature, 2016
24 **Helene Shapiro,** Linear Algebra and Matrices, 2015
23 **Sergei Ovchinnikov,** Number Systems, 2015
22 **Hugh L. Montgomery,** Early Fourier Analysis, 2014
21 **John M. Lee,** Axiomatic Geometry, 2013
20 **Paul J. Sally, Jr.,** Fundamentals of Mathematical Analysis, 2013

For a complete list of titles in this series, visit the
AMS Bookstore at **www.ams.org/bookstore/amstextseries/**.